Phase-Transfer Catalysis

Phase-Transfer Catalysis

Mechanisms and Syntheses

Marc E. Halpern, EDITOR

PTC Technology

Developed from a symposium sponsored by the
International Chemical Congress of Pacific Basin Societies at the
1995 International Chemical Congress of Pacific Basin Societies

American Chemical Society, Washington, DC

Library of Congress Cataloging-in-Publication Data

Phase-transfer catalysis: mechanisms and syntheses/ Marc E. Halpern, editor.

p.   cm.—(ACS symposium series, ISSN 0097–6156; 659)

"Developed from a symposium sponsored by the International Chemical Congress of Pacific Basin Societies at the 1995 International Chemical Congress of Pacific Basin Societies, Honolulu, Hawaii, December 17–22, 1995."

Includes bibliographical references and indexes.

ISBN 0–8412–3491–4

1. Phase-transfer catalysis—Congresses.

I. Halpern, Marc, 1954–   . II. International Chemical Congress of Pacific Basin Societies (1995: Honolulu, Hawaii). III. Series

QD505.P488   1996
547'.1395—dc21
                                            96–49775
                                                CIP

This book is printed on acid-free, recycled paper.

Foreword

THE ACS SYMPOSIUM SERIES was first published in 1974 to provide a mechanism for publishing symposia quickly in book form. The purpose of this series is to publish comprehensive books developed from symposia, which are usually "snapshots in time" of the current research being done on a topic, plus some review material on the topic. For this reason, it is necessary that the papers be published as quickly as possible.

Before a symposium-based book is put under contract, the proposed table of contents is reviewed for appropriateness to the topic and for comprehensiveness of the collection. Some papers are excluded at this point, and others are added to round out the scope of the volume. In addition, a draft of each paper is peer-reviewed prior to final acceptance or rejection. This anonymous review process is supervised by the organizer(s) of the symposium, who become the editor(s) of the book. The authors then revise their papers according to the recommendations of both the reviewers and the editors, prepare camera-ready copy, and submit the final papers to the editors, who check that all necessary revisions have been made.

As a rule, only original research papers and original review papers are included in the volumes. Verbatim reproductions of previously published papers are not accepted.

ACS BOOKS DEPARTMENT

Contents

PHASE-TRANSFER CATALYSIS
IN ORGANIC AND POLYMER SYNTHESIS

Preface

CHARLES STARKS predicted 10 years ago that a vast number of new applications and more complex catalyst systems based on phase-transfer catalysis (PTC) await discovery and exploitation. Starks made this prediction 10 years ago in his overview in ACS Symposium Series 326, *Phase-Transfer Catalysis: New Chemistry, Catalysts, and Applications*. In his preface, Dr. Starks noted the exponential growth of PTC up to 1986. A decade has passed, and Dr. Starks's predictions have come true. The exponential growth of PTC continues in the laboratory synthesis and commercial manufacture of organic chemicals and polymers. This growth appears to sustain itself as a result of strong fundamental driving forces for high yield, short reaction time, replacement or elimination of solvent, and much more, all of which PTC provides. The driving forces of industry in the mid-1990s coincide with the capabilities of PTC to deliver the desired outcomes.

This book documents the research of many top contributors to the growth of PTC. Tracking down and recruiting these scientists was no easy task. Martin O'Donnell from Purdue University performed an outstanding service (investing truly exemplary planning and effort) to the academic and industrial PTC community by organizing the PTC symposium, upon which this book is based, at Pacifichem 95 in Honolulu, Hawaii, December 17–22, 1995. The speakers recruited by Professor O'Donnell and the papers presented addressed most of the major areas of PTC innovation, including mechanisms, research guidelines, chiral PTC, synthesis of organic chemicals, modification of polymers, triphase catalysis, and new catalysts. Several chapters in this book are landmarks in the field, including the chapters by Professor O'Donnell and coworkers. Other chapters provide excellent and unique historical and scientific perspective, particularly the mechanistic chapters in this book written by the founders of PTC.

Acknowledgments

The credit for this book goes to the authors, who invested great effort and elegant creativity; to Professor O'Donnell, who made it all happen;

and to the many reviewers, who executed the peer-review process with dedication and effort.

MARC E. HALPERN
PTC Technology
1040 North Kings Highway
Suite 627
Cherry Hill, NJ 08034

September 4, 1996

Chapter 1

Recent Trends in Industrial and Academic Phase-Transfer Catalysis

Marc E. Halpern[1]

PTC Technology, 1040 North Kings Highway, Suite 627,
Cherry Hill, NJ 08034

Phase-Transfer Catalysis ("PTC") continued to grow significantly during the past decade. The driving forces for industrial and academic PTC research are stronger than ever. Even in light of the impressive growth of PTC, much progress has yet to be made both in the fundamentals of PTC and its industrial application. Chemical companies which focus on cutting cost of manufacture and pollution prevention have much to gain by implementing PTC to reap great benefit. PTC trends, barriers and opportunities are discussed.

The last ACS Symposium Series book published on the subject of phase-transfer catalysis, "PTC," is about a decade old. During this decade, there have been changes in the chemical industry, changes in academic chemical research and changes in the state of the art of PTC. This paper represents a non-comprehensive attempt to characterize some of the progress of PTC over the last decade while stimulating thought about how PTC has or has not meshed with the needs of the changing environment in the chemical industry and academia.

Exactly 10 years ago, Dr. Charles Starks, who coined the term "Phase-Transfer Catalysis," wrote in the preface to ACS Symposium Series 326 on PTC that in the previous 15 years PTC "expanded greatly." Dr. Starks estimated that the "volume of phase-transfer catalysts totaled about 40,000 lb of catalysts per year in 1980 but grew to more than one million lbs of catalyst per year in 1985." In 1996, there are several commercial processes which use approximately one million lbs per year each of phase-transfer catalyst and I estimate that the non-captive market for phase-transfer catalysts is over $25 million per year. I estimate that the number of commercial PTC applications grew each year by 10-20%, although it should have been growing even faster due to industry trends for process requirements (described below). Ten years ago, Dr. Starks noted that Chemical Abstracts just

[1]E-mail address: halpern@phasetransfer.com

started to issue CA Selects - Phase-Transfer Catalysis, which has since been very successful and useful to PTC chemists and casual PTC browsers. During the past six years, specialized courses on industrial PTC have been offered and provided to chemists at well over 100 companies and a new journal "Phase-Transfer Catalysis Communications" was launched (1994) with 1200 subscribers and growing rapidly. Today, you can search Phase-Transfer Catalysis on the Internet (which hardly existed 10 years ago) and find references to the work of many of the authors in this book (there is even one web site dedicated to PTC, which will seem trivial to a reader of this book 10 years from now).

Indeed, PTC is alive, well and growing because it provides benefit to chemists and engineers. Much has been written over the years about the multitude of advantages which PTC offers,[1,2] including increasing yield, reducing reaction time and/or temperature, eliminating or solvent, using alternate raw materials, enhancing selectivity and safety and more. An informal non-scientific survey[3] of chemists and engineers at 38 industrial sites in 1995/6 suggested that the primary driving force for considering PTC at the outset of a project is usually productivity and sometimes replacement or elimination of solvent. The most desired PTC outcome in both lab and plant in 1985 was increasing or achieving high yield. In 1996, this has been supplemented by PTC driving forces for reducing cycle time (to squeeze more profit from a fixed plant) and eliminating, reducing or replacing solvent (to comply with ever increasing environmental demands). **These driving forces, which fit well with the capabilities of PTC, are timeless and fundamental to performing chemistry in the lab and in the plant**. Considering these driving forces and considering that PTC is applicable to such a diverse range of organic and polymer reactions, PTC is not likely to be totally replaced by some as yet unknown competing technology in the near future. The ability of PTC to deliver benefit in the lab and plant will probably assure that another ACS Symposium Series on PTC will be published within another decade.

The key barriers to executing PTC projects are usually cited to be (1) lack of time to develop/optimize choice of catalyst, catalyst separation and other reaction or process parameters or (2) lack of confidence (i.e., expertise) that the reaction or process (including related unit operations) will actually work. As a general observation, it may be stated that deadlines and project overload are responsible for an enormous amount of non-optimized processes being commercialized, whether they involve PTC or not.

The economic conditions and massive R&D layoffs in the chemical industry in the early 1990's have created very strong conflicting needs which are relevant to PTC development. Chemists have always felt project pressures, however the demands on chemists in the mid-1990's are quite different from the demands in the mid-1980's. The number of R&D scientists and engineers in the US chemical industry (non-pharmaceutical) dropped from 46,100 in 1990 to 41,200 in 1995.[4] In this work environment, chemists are required to develop much more cost efficient manufacturing processes than ever and have less time and more workload than ever. On the one hand, PTC excels in reducing cost of manufacture and pollution prevention. On the other hand, PTC often requires added time for development. Constrained R&D resources often result in managers consciously foregoing million dollar per year profit opportunities in favor of allocating several $10,000's of development resource for a "higher priority" non-PTC project.

Progressive companies recognize the potential profit opportunities which are likely to result from implementing PTC. Progressive companies have overcome the barriers to PTC development by cultivating internal PTC experts, contracting with industrial PTC expert consultants or bringing in industrial PTC training. All of these activities reduce the investment in development resource to a tolerable level while maximizing the actual benefit realized by the company. Review of the patent literature shows that most of the same companies which issued impressive PTC technology patents prior to 1985 (such as, but not limited to, General Electric, DuPont, Dow, Bayer, Sumitomo, ICI, Ciba-Geigy, Merck, Eli Lilly and others) continued to do so through the 1990's. The new PTC patents continue to cover polymers, pharmaceuticals, agricultural chemicals and a very wide range of intermediates, specialty and fine organic chemicals.[5] It is interesting to observe that not all large companies have embraced PTC and some small companies use PTC very effectively. One may assume that the growth of PTC in many large and small chemical companies is due to the profit enhancing virtues of PTC. Whereas, I would like to think that the lack of growth of PTC at certain large and small companies is a result of corporate culture and organizational resistance to change, despite the proven profit track record of PTC.

Although there have been several extremely impressive industrial contributions to fundamental PTC, the basis for most PTC knowledge has come from academic institutions. During the past decade, academia has also suffered funding shortages. Federal funding for basic and applied chemical research in the United States increased at annual rates of only 0.6% and 2.7%, respectively, during the period 1989-1996.[4] In constant dollars, these amounts actually represent decreases in funding. Despite the funding difficulties, major advances in PTC have come from academia, primarily as a result of the highly dedicated and very high quality work of a few groups around the world. Many, though not all of these outstanding dedicated groups, contributed papers to the Symposium reported in the chapters of this book.

The past decade provided major advances in almost all categories of theoretical and applied PTC. The major areas of PTC progress during this decade are represented well in this book and include better understanding of general PTC mechanisms (Starks, Chapter 2; Makosza, Chapter 4; Landini et al[6]) the critical role of water (landmark work by Liotta et al, Chapter 3; more landmark work by Sasson et al, Chapter 12), solvent effects (Yufit et al, Chapter 5), catalyst structure (Sirovski, Chapter 6; Dehmlow, Chapter 9; Halpern, Chapter 8), solvent-free PTC[7] (Diez-Barra et al, Chapter 14), chiral PTC (landmark work by O'Donnell et al, Chapters 7 and 10; Shioiri, Chapter 11), modification of polymers (wide application by Nishikubo, Chapter 17; Nakamura et al, Chapter 18; Iizawa et al[8]) and carbohydrates (vast application by Roy et al, Chapter 13), polymerization,[9,17] triphase catalysis (Ohtani et al, Chapter 19; Dutta et al, Chapter 20; Doriswamy et al[10], Svec et al[11]), kinetic characterizations[12], inverse phase-transfer catalysis,[13] new catalysts (among them Balakrishnan et al, Chapter 21; Ouchi et al, Chapter 22; plus too many others to cite here) and many synthetic improvements (such as Takido et al, Chapter 15; Jiang et al, Chapter 16). Almost all of these advances were initiated in academia and soon after found application in industry as witnessed in the growing patent literature (see below). Other notable advances originating in academia include the integration of PTC with supercritical CO_2[14] and transition metal co-catalysis[15] to name just two. It is surprising to this author that there are no known commercial applications of transition metal co-catalyzed PTC. It may be anticipated that during the next decade both transition metal co-catalyzed PTC as well as supercritical fluid PTC will be commercialized.

Industrial R&D groups have been exploiting PTC's virtues to enhance productivity, quality, safety and environmental performance. During the past decade, General Electric continued to pioneer the discovery and use of thermally stable guanidinium salts as phase-transfer catalysts[16] and apply PTC to greatly reduce phosgene usage in the manufacture of polycarbonates.[17] A new area of "fluorous media" was innovated, in which a catalyst mediates between a perfluoroalkane (which is immiscible with hydrocarbons) a hydrocarbon and a gaseous phase.[18] Other impressive patents include multi-Michael additions,[19] dehydrohalogenation with efficient catalyst recycle,[20] high temperature fluoride displacement[21] and nucleophilic aromatic substitution.[22] Ton quantities of crown ethers are being used in commercial industrial PTC processes, which was nearly unthinkable ten years ago. Other recent industrial PTC innovations are reviewed in reference 5.

Almost all of the mechanistic and significant synthetic *breakthroughs* in PTC originate from groups focused on PTC (with significant PTC publications and patents). However, it is interesting to note that, as in the past, a great percentage of the PTC publications and patents which describe *applications*, continue to originate from groups which produce only 1-3 PTC publications or patents. As opposed to other topics in chemistry, PTC has such wide applicability that many researchers who focus on non-PTC areas follow the PTC literature and apply PTC when the need arises. Indeed, one of the reasons for the popularity of PTC is that it is applicable to such a wide variety of organic chemical reactions. With increased time management constraints and dwindling libraries, chemists have to be selective in which technologies they choose to follow. PTC offers too much advantage for too wide a scope of reactions to be ignored. Thirty years of growth suggest that PTC is not a fad...PTC offers real benefits which stand the test of time.

Although statistics can be manipulated and are often misleading, following are some statistics relating to the development of PTC literature and patents.[23] By August 1996, abstracted PTC publications numbered 8000 and PTC patents numbered 1500. The most popular phase-transfer catalyst cited by authors and inventors was tetrabutyl ammonium bromide, "TBAB," which was cited in 3257 publications and 846 patents.

The number of PTC publications and patents issued per year has been growing steadily. Some measures indicate that the rate of growth of PTC publications/patent remains high and other measures suggest a slight decrease in the *rate* of growth, though year to year growth is still clearly observed. For example, the five year average of TBAB patents issued per year in 1985 was 38, in 1990 it was 46 and in 1995 it was 57. The five year averages of overall PTC publications (including patents) and PTC patents (separately) issued per year were, respectively, 389 and 70 in 1985, 467 and 90 in 1990 and 484 and 100 in 1995. It appears that in the late 1980's and the early 1990's, growth in PTC patents outpaced growth in PTC publications (excluding patents), though the overall rate of growth may have slowed slightly. It should be emphasized again that the absolute number of *new* PTC publications and patents continues to grow year to year.

Another interesting observation results from examining the country of origin of PTC publications. A random examination of 157 PTC publications in 1989 showed that 18% each originated from Japan and the former Soviet Union, 11% each from China and France, 8% from the US, 7% from Germany and 5% from India. A random examination of 182 PTC publications in 1996 showed 23% originating in China, 13% from the former Soviet Union, 12% from Japan, 11% from the US, 9% from India, 5% from Germany and 4% from France. The absolute number of PTC publications has been rather steady from Poland, Italy, the UK, Israel, Spain, Canada, Hungary, Romania and Korea. Despite the smaller quantities, the quality of the publications from the latter countries is quite high and some represent the new landmarks in PTC progress. Countries represented in the PTC literature in the 1996 sample which did not appear in the 1989 sample included Vietnam, Switzerland, Morocco, Pakistan, Iraq, Turkey, Bulgaria, Egypt and New Zealand.

It is usually easy to set up and analyze PTC reactions, though they can be difficult to optimize. Good quality academic PTC research can be performed on a relatively low budget (mostly for flasks and chemicals, assuming graduate students are "free"), as long as a GC or HPLC (and preferably but not necessarily an NMR instrument) is available. Industrial PTC R&D requires time allocation as do other industrial projects, plus time for special scale up issues such as heat transfer and agitation for the multiphase PTC systems and catalyst separation/recycle unit operations. As we all know, the ease at which research can be performed often has an impact on how much of it is chosen to be performed. The data here suggest that PTC seems to be gaining popularity in all regions (with constrained or less constrained resources). Again, the driving forces for PTC and the resulting achievable benefits are fundamental to performing chemistry and are often obvious to academic and industrial researchers alike, regardless of country of origin.

In summary, phase-transfer catalysis has progressed very well during the past decade, with patents and publications achieving nearly three times the cumulative level of the previous two decades. As the following chapters in this book suggest, even though tremendous progress has been made in elucidating the fundamentals of PTC mechanism and applying them for commercial gain, much progress needs yet to be made. Scarce academic resources pose a threat to this advancement in PTC as well as to science in general. In parallel, dwindling industrial R&D groups are narrowly missing great opportunity to generate self-sustaining profit by failing to evaluate, let alone implement, PTC in manufacturing processes. Many corporations have realized great profit from implementing PTC, whereas other corporations are missing great profit opportunities by PTC retrofit (potential short term gratification for profit & loss executives) and don't even know it! The opportunities to benefit from PTC in both academia and industry are great. The rewards will be reaped by those individuals, companies and even countries (China, Russia, India and Japan are leading the growth in 1996) who take the initiative to pursue the PTC opportunities, have the PTC expertise and awareness of benefits and who allocate the resources to execute the PTC programs.

Literature Cited

1. Starks, C. *ACS Symposium Series 326*, **1987**, American Chemical Society, Washington DC, Chapter 1
2. Starks, C.; Liotta, C.; Halpern, M. *"Phase-Transfer Catalysis: Fundamentals, Applications and Industrial Perspectives,"* **1994**, Chapman and Hall, New York
3. personal interviews conducted by this author (at 38 chemical company sites in the US, Europe, the Middle East and Asia from April 1995 to August 1996) without intent to conduct a statistically valid study
4. calculated from data shown in *Chemical and Engineering News*, Aug 26 **1996**, pp. 54, 55 and 60
5. Halpern, M.; *Phase Trans. Catal. Comm.*, **1996**, *2*, 17
6. see for example Gobbi, A.; Landini, D.; Maia, A.; Secci, D. *J. Org. Chem.*, **1995**, *60*, 5954 and references cited therein
7. a review of recent Solvent-Free PTC literature may be found in Halpern, M. *Phase Trans. Catal. Comm.*, **1994**, 1. For ongoing developments, follow also the work of Bram, Barry and Loupy.
8. see for example Iizawa, T.; Sueyoshi, T.; Hijikata, C.; Nishikubo, T.. *J. Polym. Sci. Part A: Polym. Chem.*, **1994**, *32*, 3091
9. see for example Percec, V.; Chu, P.; Ungar, G.; Cheng, S.; Yoon, Y. *J. Mater. Chem.*, **1994**, *4*, 719
10. see for example: Desikan, S.; Doriswamy, L. *Ind. Eng. Chem. Res.*, **1995**, *34*, 3524
11. see for example Trochimczuk, A.; Hradil, J.; Kolarz, B.; Svec, F. *Polym. Bull.* **1994**, *22*, 9

12. I recommend following the work of Wang et al; for example: Wang, M.; Chang, S. *Ind. Eng. Chem. Res.*, **1995**, *34*, 3696

13. see for example Wang, M.; Ou, C.; Jwo, J. *Ind. Eng. Chem. Res.*, **1994**, *33*, 2034

14. Dillow, A.; Yun, S,; Suleiman, D.; Boatright, D.; Liotta, C.; Eckert, C. *Ind. Eng. Chem. Res.*, **1996**, *35*, 1801

15. Alper, H. *J. Organomet. Chem.*, **1986**, *300*, 1

16. Brunelle, D. (General Electric), **1992**, US Patent 5,082,968

17. Boden, E.; Phelps, P.; Ramsey, D.; Sybert, P.; Flowers, L.; Odle, R. (General Electric) **1995**, US Patent 5, 391,692

18. Gladysz, J. *Science*, **1994**, *266*, 55 and Horvath, I.; Rabai, J. *Science*, **1994**, *266*, 72

19. Sabahi, M.; Irwin, R. (Albemarle) **1994**, US Patent 5,347,043

20. Reed, D.; Snedecor, T. (Dow Chemical) **1995**, WO 95/05352

21. Cantrell G. (Mallinckrodt) **1990**, US Patent 4,973,772

22. Grace H.; Wood, M. (Ciba-Geigy) **1994**, EP 0 648 755 A1

23. data were gathered by performing various types of searches with CAS Online and CA Selects - Phase-Transfer Catalysis

MECHANISMS AND RESEARCH GUIDELINES

Chapter 2

Modern Perspectives on the Mechanisms of Phase-Transfer Catalysis

Charles M. Starks

Cimarron Technology Associates, 73 Stoneridge, Ponca City, OK 74604

Phase Transfer Catalysis became a widely used tool for organic chemists and chemical engineers following the publication of basic mechanistic and physical chemical foundations in the 1960-1971 time period. Much sophisticated work and chemical evolution have now provided for an even better understanding of the various steps and principles that collectively define the many possible mechanisms of PTC. Many of these modern perspectives are discussed in this paper. This improved understanding allows development of even more sophisticated and complex PTC systems to come in the future.

Phase transfer catalysis has been used knowingly and unknowingly for many years. The catalytic activity of quaternary ammonium salts was evidently first patented[1] in 1912, and the first clear-cut example of its commercial use was production of butyl benzyl phthalate from sodium monobutyl phthalate and benzyl chloride in 1946,[2] as shown in the Figure 1. The phase transfer catalyst, benzyltriethylammonium chloride was formed *in situ* by addition of triethylamine.

Early publications on phase transfer catalysis saw many different theories proposed for the mechanisms of these systems. After twenty-five years and thousands of publications contributed to this field by many chemists, one must conclude that all the mechanisms offered are probably valid, given the right reactions and set of reaction conditions. This is true because within phase transfer catalysis systems many variables can be manipulated so as to control the relative rates of the steps involved in the process, and it is primarily only the number of steps and their relative rates that define the mechanism of any given PTC process. This great variability, which is certainly a strength of PTC, also means that PTC mechanisms are not fixed, in the sense that mechanisms have traditionally been fixed in organic chemistry. To properly understand the forces that affect phase transfer catalyzed reactions, the chemist must concentrate on understanding the fundamental steps involved in each chemical reaction sequence, rather than rely on stereotype mechanisms.

For many years we have heard that there are two mechanisms for phase transfer catalyzed reactions, i.e., the *extraction* and *interfacial* mechanisms. These terms are poorly defined but generally refer to whether the rate-determining step of the reaction sequence occurs in the organic phase with transferred reagent, or whether it occurs at the interface. These differences are confusing, since both extraction and interfacial reactions take place, usually at varying rates, during every PTC process, and it is the chemist's control of the balance of these rates that allows a process to be successful or not.

A simple example of the problem with these mechanisms can be illustrated with the cyanide displacement reaction as represented in Figure 2. This reaction is considered to be the very model of the extraction mechanism, since its rate is independent of stirring rate, presumably an identifying feature of the extraction mechanism. Indeed, in a given apparatus, its reaction rate is independent of stirring rate, when stirring is greater than 200 rpm, as in Figure 3. However, in the same apparatus at stirring speeds of less than 200 rpm the reaction rate is found to be proportional to the square of the stirring rate as shown in Figure 4. This is exactly what one would expect from the PTC reaction rate-limited because of slow cyanide transfer from the aqueous to the organic phase. Thus, the process now would presumably have the interfacial mechanism. The conclusion we draw from this data is that PTC cyanide displacement follows an extraction mechanism when stirring rate is greater than 200 rpm, but it follows an interfacial mechanism when stirring is at a lower rate. Obviously this conclusion provides little help in understanding and improving the process.

Modern Perspectives on PTC Mechanisms

The important thesis of this paper is that *chemists need to understand all of the separate chemical [kinetic] steps, as well as the physical chemistry, involved in a PTC process to do a complete job of mechanistic description and process optimization.* Then accurate predictions can be made on how to make the system more responsive and efficient. This paper will review several points of mechanistic interest.

To discuss phase transfer catalysis mechanisms in a logical way, it is useful to divide PTC reaction systems into three categories based on system complexity: (1) simple PTC reactions, (2) reaction with hydroxide or other bases, and (3) more complex reactions. The major mechanistic questions relating to the first two of these are outlined and briefly discussed in this chapter, but complex PTC reactions although noted, need much greater space for reasonable analysis. In addition to system complexity there are other important mechanistic points, such as catalyst stability and catalyst poisoning, catalyst location during reaction, the influence of water and organic solvents, simultaneous or concurrent reactions at the interface both catalyzed and non-catalyzed, and other features that may affect the mechanistic sequence of any phase transfer catalyzed reaction.

Please note in the discussion that follows that when the words *interface* and *interfacial* are used, these refer to the classical meanings of these words and not to mechanism names.

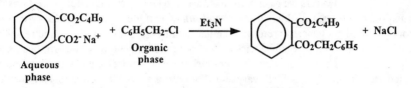

Figure 1: Early example of PTC with *in situ* catalyst formation.

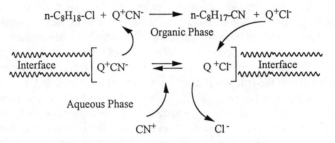

Figure 2: PTC Cyanide displacement reaction.

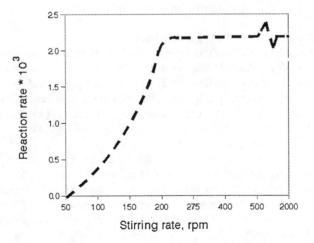

Figure 3: Effect of agitation rate on rate of PTC cyanide displacement reaction on 1-chlorooctane.

Simple Phase -Transfer-Catalyzed Reactions

Using cyanide displacement on 1-chlorooctane, Figure 2, as a model for simple phase transfer catalysis we clearly recognize that the reaction occurs in at least two steps: First, the catalyst transfers cyanide ion into the organic phase, *the transfer step*, and secondly the catalyst makes cyanide available to 1-chlorooctane in an activated form in the organic phase, so that cyanide displacement, the *intrinsic reaction step*, can proceed at a good rate. It is important to understand that the catalyst is required to function very well in *both* steps, otherwise the sequential nature of the process slows or stops reaction. Thus it is critical to choose a catalyst that can perform well in both steps. Each step has different chemical and physical characteristics that affect the overall reaction.

The Intrinsic Reaction Step: Anion   Activation and the organic phase reaction. See Fig. 5. The first requirement for a phase transfer catalyst is the elementary need of a sufficient quantity of reactant anion in the organic phase; i.e., the concentration of cyanide anion in the organic phase where it reacts with 1-chlorooctane must be large enough for the process to proceed at a satisfactory rate.

Once in the organic phase, cyanide anion must also be sufficiently reactive and available to allow displacement to proceed at a high rate. If uncomplexed sodium cyanide could be magically transferred into 1-chlorooctane, we know the cyanide would react extremely slowly, perhaps immeasurably slow. This poor reactivity is due to the tight ion pairs of NaCN, or large interaction energy binding the two ions together, as results in low-dielectric solvents. Unless activated by high temperatures or separated from sodium cations by use of special solvents such as dimethylsulfoxide or N,N-dimethylformamide, the cyanide anion is too closely held for displacement reaction with 1-chlorooctane to occur at an acceptable rate. However, if the Na^+ cation were replaced by Bu_4N^+ the reaction accelerates many thousand fold, perhaps even tens of thousands fold, due to the much lower interaction energy of the quat-cyanide ion pair, due in turn, to the much large size of the quaternary cation. This behavior of quaternary ammonium salts in nonpolar media has been experimentally explored and demonstrated by Uglestad and co-workers.[3] We may appreciate the difference in relative size of the sodium cation and tetrabutylammonium cation by the comparison shown in Figure 6. The larger sizes of quaternary cations, compared to alkali metal cations (as measured by electrochemical methods) are listed in Table 1.

Even the smallest quaternary cation has a larger ionic radius than any of the alkali metal cations. These differences in ionic radii can be translated into ionic interaction energies by simple Coulombic calculations.[4] Results from such calculations are listed for ion pairs with bromide in Table 1 (bromide anion is used because its anionic radius is accurately known). By subtraction, e.g., of tetrabutyl energy from potassium energy, one can compare the calculated differences between ion pairs. If these differences in ion-pair energies are translated into reduction of kinetic activation energies, then one can understand the extraordinary activating effect shown by large quaternary cations. For example a 5 kcal/mole difference in activation energy (approximately the difference in calculated binding energy between tetrabutyl-

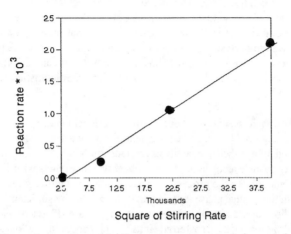

Figure 4: Dependence of cyanide displacement rate on stirring rate in transfer-limited range.

$$n\text{-}C_8H_{17}\text{-}Cl + Q^+CN^- \longrightarrow n\text{-}C_8H_{17}\text{-}CN + Q^+ CN^-$$

Figure 5: Intrinsic reaction step in PTC cyanide displacement.

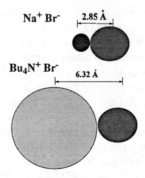

Figure 6: Comparative size of Na^+Br^- and $Bu4N^+$ Br^- ion pairs.

ammonium bromide and potassium bromide)is equivalent to a 4400-fold change in reaction rates. This ability of PTC catalysts to activate anions is often a critical feature of PTC reactions; otherwise, the intrinsic reaction rate may be too slow for the overall process to be practical.

Table 1: Cation Size and Coulombic Interaction Energies of Bromide Salts

Cation	Cation Radius, Å	Coulombic Interaction Energy with Br$^-$ (Kcal/mole)
Li$^+$	.6	12.8
Na$^+$	0.9	11.4
K$^+$	1.33	9.9
Rb$^+$	1.48	9.5
Cs$^+$	1.69	9
Me$_4$N$^+$	2.85	6.8
Et$_4$N$^+$	3.48	6.2
n-Pr$_4$N$^+$	3.98	5.5
n-Bu$_4$N$^+$	4.37	5.3

We might expect that by using larger and larger quaternary ammonium salts, the activating effect would become larger and larger; i.e., that the cationic radius of the quaternary cation would also become larger. However, the trend of increasing ionic radii with increasing chain length of the four alkyl chains of a tetraalkylammonium salt does not continue indefinitely. It levels off at larger chain lengths since the structure of tetraalkylammonium salts become quite open, as can be readily be seen from molecular models. As the structure becomes more open the anion is no longer pushed further away from the quaternary nitrogen, so the cation-anion interaction distances reach a limiting value, depending on the size of the anion. For most anions this leveling probably occurs with symmetrical quaternary cations having alkyl groups in the range of pentyl to octyl. It would be of great interest to specifically construct quaternary ammonium salts wherein the quaternary nitrogen was sterically hindered from approach by anion to no closer than say 10Å.

The Transfer Step. Turning to the much more complicated transfer step, it is often assumed that the reactions involved in the transfer step, being largely ionic in nature, are virtually instantaneous. This is not true. Many PTC reactions are transfer-limited. Slow transfer usually results from poor agitation; use of anions that are highly hydrated; exceptionally high interfacial tension between phases; or use of phase transfer catalysts that are inhibited from approaching the interface.

There are many subtleties and variations in the transfer step, but for simpleanion transfer two general pictures have evolved, as represented in Figure 7. The first view assumes that the whole quaternary salt with desired anion is extracted from the aqueous phase into the organic phase (extraction of ion pairs), while the second view assumes that the quaternary cation remains more or less permanently located in the organic phase with anions only crossing the aqueous-organic interface. Both transfer mechanisms are probably correct depending on the quaternary salt catalyst, with the first being more likely with small to medium sized quaternary cations, while the second is more correct for medium and large quaternary cations, also depending on the hydrophilicity of the associated anion. For simplicity in the present discussion we assume the use of medium to large quaternary salts that mostly reside in the organic phase, and therefore that transfer occurs mostly by extraction of anions across the interface.

A well-recognized aspect of the transfer step is that unlike the intrinsic step, it involves *at least* two reactions, i.e., transfer of two different anions, one from the aqueous to organic phase, another from the organic to the aqueous phase, all of which may or may not be in equilibrium. For simplicity here, we are viewing the transfer rate as the net rate of delivery of desired anion to the organic phase. More comprehensive analysis of these systems requires that both anion transfers (and other related reactions) be viewed as two kinetically distinct steps. (See reference 5.)

For a given anion, its rate of transfer across the interface is largely governed by four factors: (1) interfacial area, (2) anion activity and hydration at the interface, (3) bulkiness of the quaternary salt, and (4) sharpness of the interface. Some other factors might also be important in some situations, such as for example the distribution of droplet sizes, and unusually fast or slow rates of droplet coalescence.

Interfacial area: The overall opportunity for any kind of interfacial transfer in a two-phase system is governed by the amount of interfacial area available. Therefore, in two-phase reactions the dispersal of one phase as tiny droplets in the second phase is a critical feature. Three of the more important factors affecting the amount of interfacial area include interfacial tension, the presence of surfactants, and the degree of stirring or agitation.

> 1. *Interfacial tension:* Interfacial area under steady-state stirring conditions will increase with decreasing interfacial tension. The chemical natures of the organic and the aqueous phases determine the interfacial tension that exists between these two phases. Highly nonpolar solvents and highly concentrated aqueous solutions have the highest interfacial tension. Use of more polar solvents and more dilute solutions reduces interfacial tension.

> 2. *Presence of surfactants:* The quaternary salt, or other components present in the reaction mixture may lower interfacial tension because of its surfactant properties. Most quaternary salts are surface active to some degree. Those having a trimethyl- or triethyl-structure, such as dodecyltrimethylammonium chloride are powerful surfactants and are used commercially in hundreds of

millions of pounds per year for applications requiring cationic surfactants. Normally use of such powerful surfactants is avoided for PTC applications since these generate emulsions which cause handling problems. However, even small amounts of interfacial tension reduction can be helpful for speeding the rate of transfer-limited reactions.[5]

3. *Stirring or agitation energy*: Simplististically, and at low stirring rates, the interfacial area between two liquid phases varies with the square of the stirrer speed, (agitation energy = mass x velocity[2]) as was seen in Figure 4. More accurately, the steady-state interfacial area in a two-phase system depends on the balance between the rate of interfacial area generation from stirring (agitation power) compared to the rate of loss of interfacial area because of droplet coalescence, up to the point where cavitation begins.

Anion activity at the interface: Chemical characteristics of the anion and the reaction system, such as charge/volume ratio of the anion, polarity of the organic phase, and concentration of salts in the aqueous phase significantly affect the ease of transfer of an ion into an organic phase. While little actual kinetic information is available regarding the rates of anion transfer, we usually assume that the widely differing equilibrium tendencies of the anions to be extracted into an organic phase by a given oleophilic cation also reflect the rates of anion transfer. Thus, large weakly-hydrated or organic anions such as perchlorate, iodide and phenolate, are easily transferred while small highly hydrated anions such as fluoride or hydroxide are poorly transferred. [6]

In addition to the above characteristics of particular anions, there is also the question of how the PTC catalyst structure may affect concentrations and activities of anions actually at the interface. We chemists are in the habit of believing that when a salt is dissolved in water, it exists there as discrete separated cations and anions floating around largely independent of one another. However, the practical conditions most often used in PTC reactions call for concentrated aqueous solutions in which much or most of the anion and cation populations are in various kinds of aggregated forms such as ion pairs, triplet ions, etc., especially in the region near the interface. Since one side of the interface is an organic material having significantly lower polarity than water even greater ion aggregation results near the interface. To the extent that this ion aggregation occurs, a quaternary cation at the interface can activate anions much like it activates ions in the organic phase, such that it can be considered to function as a catalyst in yet another way.

Elegant ESCA studies [7] have recently begun to illuminate some interfacial chemistry of importance to PTC. This work on anion association behavior with tetrabutylammonium cations has shown that large poorly-hydrated anions such as perchlorate and iodide are closely associated as contact ion pairs at the interface, whereas more highly hydrated anions such as chloride and nitrate are much less strongly associated. This suggests that if the anion is highly hydrated it will not be tightly bound at the interface with the quaternary cation, but rather tend to be more dispersed in solution, removed from the interface.

Bulkiness of the Quaternary Cation: As the length of the alkyl groups of R_4N^+ becomes larger and larger, the rate of transfer of anions across the interface will become slower and slower, eventually causing the reaction rate of an intrinsic-rate-limited reaction to become transfer-rate-limited. These lower transfer rates due to bulkiness of the quaternary salt result from two factors. One factor, illustrated in Figure 8, is the increasing distance of the quaternary nitrogen cationic center from the interface. This concept is experimentally supported by results[8] from interfacial reactions that indicate the removal of a group about 10Å away from the interface causes a ten-fold reduction in reaction rate. In phase transfer catalyzed reactions with quaternary salt catalysts, we can of course fix this inhibition of transfer to some extent by using unsymmetrical quaternary cations such as trialkylmethyl or dialkyldimethylammonium catalysts, to allow much closer approach of the cationic center to the interface, as illustrated in Figure 8.

The second factor causing catalyst bulkiness to reduce transfer rates is due simply to lowering of the maximum possible concentration of quaternary cation at the interface, as shown in Figure 9. Fewer of the bulky quaternary cations can occupy available space at the interface than with small-footprint quaternary cations. Doubling of the quaternary salt radius (physical radius, not ionic radius) causes a four-fold reduction in maximum possible surface concentration of the quaternary cation within the interface, and presumably a four-fold reduction in transfer rates.

Sharpness of the interface: Interface sharpness seems to be a recognized, but poorly understood feature of interfacial chemistry, especially in its effect on kinetics. We expect, for example, a hydrocarbon mixed with a concentrated aqueous salt solution to have an extremely sharp interface (corresponding to high interfacial tension), where possibilities for interfacial mixing are minimized and interfacial reaction would be slow. On the other hand, a two-phase mixture of acetone and concentrated aqueous salt solution would be expected to have a diffuse interface where interfacial mixing is easy. Interfacial sharpness may be thought of as a strong or weak barrier or film through which one or both of the reagents need to penetrate for transfer to occur. Interfacial sharpness may also be thought of in terms of the thickness of the interfacial region, i.e., as the interface having depth and therefore volume, instead of just surface area, and having a concentration gradient from one side of the interface to the other. In some cases, notably with tetrabutylammonium PTC catalysts,[9] a third catalyst-rich layer can be formed.

The PTC Matrix. The confusion that surrounds the selection of a PTC catalyst and selection of appropriate reaction conditions for PTC reactions, is often due to unclear understanding of how the two steps in the PTC sequence interact with each other. For some reactions almost any catalyst and any set of reaction conditions seem to result in fast and convenient reactions. In other systems, the nature of the catalyst and the set of reaction conditions chosen are extremely important to obtain successful results.

Use of the PTC Matrix can help to sort out some important issues. The PTC

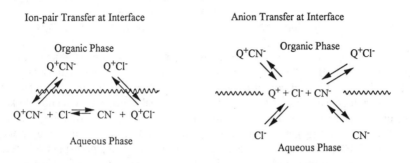

Figure 7: Two common views on anion transfer mechanisms in PTC reactions.

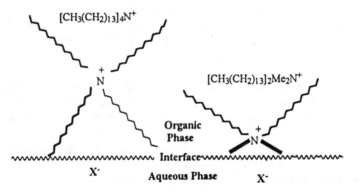

Figure 8: Hindrance of approach of quaternary nitrogen to the interface by long alkyl groups.

Figure 9: Effect of catalyst bulk at interface on interfacial concentrations.

Matrix brings the two PTC steps together in a plot of intrinsic rate vs. transfer rate as illustrated in Figure 10.We can look at this plot in several different ways, but hereit is arbitrarily divided into four quadrants, which allow us to see how different reaction patterns develop, depending on the relative rates of the two steps.

Fast quadrant. In the upper right quadrant both transfer and intrinsic reaction rates are fast. Such PTC processes are easy to run since almost any kind of catalyst can be used, and it is easy to find reaction conditions under which the reaction will go well. For these reactions our principal concern may be how to keep the process under control. Permanganate oxidations are a good example of such fast processes: permanganate ion is easy to transfer into organic solutions by a wide variety of PTC catalysts; and, once in the organic phase, permanganate reacts rapidly with most oxidizable groups, irrespective of the polarity of the solvent.

Intrinsic rate limited quadrant. Reactions with fast transfer rates but slow intrinsic rates, such as the cyanide displacement reactions with good agitation, are in this quadrant. For rate improvements in this quadrant one needs to select reactivity factors that enhance the rate of the intrinsic reaction, such as catalyst structure, temperature, organic solvent, etc.

Transfer rate limited quadrant. Opposite of intrinsic-rate-limited quadrant are processes where transfer is the slow step. Here, for example, is the cyanide reaction with poor agitation, or the cyanide reaction with an extremely reactive alkyl halide. If transfer is slow, and if the organic reaction is exceptionally fast, then the process will resemble a truly interfacial reaction, i.e., a reaction occurring only at the interface, since any transferred species will react almost as soon as it enters the organic phase. For reactions that are transfer limited we turn to bisphasic catalysts such as tetrabutylammonium, benzyltriethylammonium, etc. We would also be concerned to keep interfacial area high by careful selection of solvent, maybe use of some interfacial-tension lowering additive, and a high level of agitation.

Slow quadrant. Here, both steps of the PTC process are slow. An example of such a slow process in this quadrant is fluoride displacement on 2-chloroalkanes, since fluoride is difficult to transfer, and once transferred is slow to react. Reactions in this quadrant require the greatest amount of ingenuity to develop. One important tool exceedingly useful and economical for processes in this quadrant (and indeed for all PTC reactions) is the use of two PTC catalysts. This allows optimization of a separate catalyst for each step, usually resulting in synergistic performance. Use of dual PTC catalysts is a powerful tool to accelerate many PTC reactions, and to lower the costs of using PTC.

If we can analyze and sort out which quadrant a PTC reaction belongs, then we can begin to predict what reaction conditions would work best for that reaction, given the factors affecting rates such as listed in Figure 10. The PTC matrix shows that, depending on where one is in the matrix, the optimum catalyst may be different, e.g., a highly activating catalyst for a reaction with intrinsic rate limiting, but a biphasic catalyst for a transfer rate limited system.

Catalyst Structure

In the extensive literature on phase transfer catalysis it has been found that no one type of PTC catalyst is best for all reactions, but rather optimal catalyst structure depends on the chemical and physical nature of the reaction system, as indicated in the previous section. Generally the optimal catalyst structure can be identified as belonging to one of three solubility groups, (1) organic-soluble, (2) biphasic, or (3) aqueous-soluble. This classification based on solubility is necessarily imprecise since a quaternary ammonium salt or other PTC catalyst may change its phase residence depending on the particular solvent used, polarity changes in the organic phase as the reaction proceeds, and on the kind and concentration of inorganic salts in the aqueous phase. We would generally consider, for example, tetrahexylammonium or trioctylmethylammonium salts to be mostly organic-soluble. Typical of the biphasic type are benzyltriethyl ammonium and dodecyltrimethyl ammonium salts. Tetrabutylammonium may also be considered biphasic in many reaction systems, although it is clearly organic-soluble in high-polarity solvents, and with low-polarity solvents it may form a third liquid phase containing most of the catalyst.[10] Tetramethylammonium salts, which have found a surprising number of uses as catalysts for two-phase reactions, are clearly aqueous-soluble, functioning perhaps by activating anions at the interface or in the aqueous phase itself.

The effect of size of the quaternary salt catalyst, i.e., symmetrical R_4N^+, usually follows the pattern shown in Figure 11. The position of the maximum may move to lower or higher alkyl groups, depending on the reaction, the solvent, and reaction conditions. Likewise the slope of the curve on both sides of the maximum may vary significantly, depending on the reaction and physical properties of the system. But the pattern is a general one, reflecting all of the factors that we have discussed previously, and sometimes others. The more important structural characteristics affecting catalytic activity of quaternary ammonium salts as PTC catalysts are (1) the effect of quaternary cation size and shape on anion activation as discussed previously, (2) the ability of the catalyst to solubilize the aqueous-phase reagent into the nonaqueous phase, a function of total "organic structure" of the quaternary cation, the anion, and the particular organic solvent used, (3) decrease in reactivity due to quaternary salt cation bulkiness as discussed previously, (4) stability of the quaternary salt under the reaction conditions, and (5) the surfactant properties of the catalyst. Different reactions respond to these five characteristics in different ways so it should not be surprising that no one catalyst structure is the best for all reactions.

Note that the curve in Figure 10 only represents quaternary salts where all four groups in R_4N^+ are identical n-alkyl groups. Use of unsymmetrical quaternary salts such as $R_3NCH_3^+$ X^-, or alkyl groups containing branching, rings, etc., may significantly affect one of the five characteristics listed above without changing the others. Thus, for example, changing to an unsymmetrical quaternary salt that has greater surfactant characteristics may accelerate the rate of transfer-limited reactions but not of an intrinsic-rate-limited reaction.

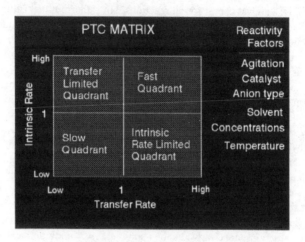

Figure 10: The PTC Matrix

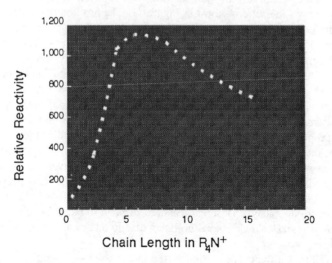

Figure 11: Typical rate pattern for simple PTC reactions on variation of quaternary cation size.

Many other features of the catalyst, such as cost, stability, availability, removability from reaction product, biodegradability, etc., aside from its influence on reactivity, are also exceedingly important for commercial processes.

PTC Reactions with Hydroxide or Carbonate

Reactions with hydroxide or carbonate (carbonate normally used only in liquid-solid PTC), constitute a very important sector in phase-transfer-catalysis chemistry, since it is estimated that more than 50% of all industrial PTC applications use sodium or potassium hydroxide, potassium carbonate or other organic-insoluble base.

Some few hydroxide reactions where hydroxide ion is transferred to the organic phase for reaction are similar to the simple PTC reactions previously discussed. For example, hydroxide displacement on benzyl chloride to form benzyl alcohol (although complicated by sequential formation of dibenzyl ether), probably requires hydroxide transfer into the organic phase. Dehydrohalogenations[11] also appear to require that hydroxide be transferred to the organic phase for reaction.

However, most PTC-hydroxide reactions, instead of hydroxide transfer, involve a new step, i.e., a *neutralization step,* in which the anion to be transferred is generated from the base and from an added acid. This neutralization constitutes a "third step" in the PTC sequence, in addition to transfer and intrinsic reaction steps, and the several chemical and physical aspects introduced by this step must be taken into account. In addition to the complications arising from this third step, PTC hydroxide reactions are also more complicated because of several additional issues including: (a) Hydroxide anion is much more difficult to extract into organic solutions than other anions, and most other anions produced during the reaction, such as halide, seriously retard hydroxide transfer. Large quaternary salts are needed for hydroxide transfer. (b) Quaternary ammonium hydroxides undergo decomposition readily at moderate temperatures. Caution should be used in interpreting mechanistic information of PTC hydroxide reactions based on yield data alone, because of different rates of catalyst decomposition. (c) Alcohols and other compounds may function as surrogates for transfer of hydroxide, i.e., by formation and transfer of alcoholate anions, and these may therefore act as co-catalysts that can significantly affect both transfer and reaction rates. (d) Hydroxide ion in water is highly hydrated, and from concentrated aqueous solutions is transferred in aggregation with various hydration levels, which cause varying reactivity. Although this hydration issue exists with other anions it is especially important with hydroxide.

For the present it is useful only to concentrate on the new reaction-sequence step, neutralization. Neutralization may occur at different locations in the reaction mixture: in the aqueous phase, at the interface, or, in the organic phase. For example, suppose that to a reaction mixture of benzyl chloride, quaternary ammonium salt PTC catalyst, and aqueous sodium hydroxide, hydrogen cyanide dissolved in benzyl chloride is added. Of course, the hydrogen cyanide would be almost instantaneously extracted into the aqueous phase and neutralized with NaOH. Then the displacement reaction with benzyl chloride that follows would be just like the simple type of PTC cyanide displacement reaction discussed previously.

Similarly, if instead of hydrogen cyanide, we used a phenol or an alcohol, then rapid neutralization of these compounds by sodium hydroxide will occur either at the interface because of the high affinity of the hydroxyl group for water, or perhaps in the water phase itself. The organic anion produced will be easily transferred into the organic phase by the quaternary cation. Because the reaction of phenol or alcohol with base at the interface is rapid, these reactions are also similar to the simple PTC reactions discussed previously.

Now, on the other hand, if we use a weak acid, such as phenylacetonitrile which has no special affinity for the interface, then its neutralization with NaOH at the interface will be slow. Because phenylacetonitrile is a weak acid the equilibrium concentration of phenylacetonitrile anion, PAN^-, will be small in the absence of PTC catalyst, especially in the interface where the water concentration is high. We can write this as an interfacial equilibrium, as shown in Figure 12.

If a biphasic quaternary salt is added to this reaction mixture, Fig. 13, two important results beyond those expected for simple PTC reactions are obtained: First, the quaternary salt, highly concentrated at the interface, can easily associate with hydroxide at the interface without ruinous competition with chloride or other anions. Because this hydroxide is activated by the quaternary cation and because the ion pair can easily approach phenylacetonitrile in the organic interface, the presence of the quaternary cation will greatly accelerate the neutralization of phenylacetonitrile to produce PAN^-. The PAN^- concentration in the interface will be governed by an equilibrium, as represented by Eq. 1:

$$(PAN^-)_i = \frac{K_p \, (PAN)_i \, (OH^-)_i}{(H_2O)_i} \tag{1}$$

Secondly, and also of great importance, the quaternary cation in association with PAN^- forms an ion pair, Q^+PAN^- with sufficient organic structure that the ion pair formed is overwhelmingly transferred into the organic phase, Figure 14, with associated equilibrium represented by Eq. 2.

$$(Q^+ \, PAN^-)_o = \frac{K_t \, (Q^+)_i \, (OH^-)_i}{H_2O} \tag{2}$$

Three points are notable here. The first is that transfer of PAN^- into the organic phase away from the high water concentration at the interface allows a great increase in the steady-state-concentration of reactive phenylacetonitrile anion in the reaction system, due to the low solubility of water in the organic phase. It is estimated that the water concentration in the organic phase is roughly 0.01 times its concentration at the interface. The second notable point is that because the two anions transferred are both highly organic, the equilibrium constant, K_t, will be large, probably on the order of 10^2 to 10^4. The third notable point from addition of PTC catalyst is that the PAN^- will be activated by the quaternary cation, compared to its activity as a sodium or potassium salt, probably 10^3 times more active.

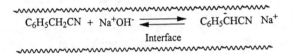

$$C_6H_5CH_2CN + Na^+OH^- \rightleftharpoons C_6H_5\overset{.}{C}HCN\ Na^+$$

Interface

Figure 12: Interfacial equilibrium for neutralization of phenylacetonitrile with OH^-.

$$(Q^+)_i + (OH^-)_i + (PAN)_i \xrightarrow{K_n} (Q^+)_i + (PAN^-)_i + H_2O$$

Figure 13: Interfacial neutralization equlibrium of phenylacetonitrile with quaternary participation.

$$(Q^+)_i + (PAN^-)_i \xrightarrow{K_p} (Q^+PAN^-)_o$$

Figure 14: Transfer equlibrium of phenylacetonitrile anion from interface to organic phase.

Taken together these three points suggest that use of PTC will vastly improve the probability that successful reactions with acids such as PAN can be readily accomplished with aqueous sodium hydroxide. We can estimate the huge effect of these three points, first by combining equations 1 and 2 to give an equilibrium relationship for the total concentration of reactive PAN anion:

$$(PAN)_{o \cdot i} = [1 + K_t (Q^+)_i] \frac{K_n (PAN)_i (OH^-)_i}{(H_2O)_i} \qquad (3)$$

If the concentration of Q^+ at the interface is estimated to be 1 to 5 m/L, and then one uses the values estimated in the previous paragraph, the active concentration of phenylacetonitrile anion in the system in the presence of a quaternary catalyst is three to four orders of magnitude greater than for simple interfacial concentrations. When this increase in reactant concentration is further multiplied by a 10^3 reactivity increase from activation by the quaternary cation, we can easily rationalize how the use of PTC with aqueous bases allows a million- to ten-million-fold increase in alkylation rates of even very weak organic acids.

Thus for these kinds of PTC hydroxide reactions, where interfacial neutralization of organic acids is slow, and where the anion generated by neutralization has appreciable organic structure, we can expect use of phase transfer catalysts, and especially biphasic quaternary catalysts, to increase both active concentration and activity of the conjugate anion by a very remarkable extent. Even acetone can be alkylated with using aqueous NaOH under PTC conditions.[12]

The stunning importance of the equilibrium constant, K_t is also a key to understanding some other behaviors observed in PTC alkylations, as for example the product distribution observed during PTC butylation of malononitrile with aqueous sodium hydroxide. This reaction gives almost exclusively dibutylmalononitrile and very little of monoalkylated product even in the presence of a significant excess of malononitrile.[13] Understanding that the monobutylated-malononitrile anion, $[BuCH(CN)_2]^-$ will be much more readily extracted into the organic phase (estimated more than 100 fold) than unalkylated malononitrile anion, $[CH(CN)_2]^-$ easily explains these results.

More Complex PTC Reactions. Many PTC reactions reported in the chemical literature involve more complex reaction sequences than the simple reactions and the PTC hydroxide reactions considered up to now. For example: liquid-solid PTC reactions; reactions involving co-catalysts such as metal ions of various kinds, alcohols for base-reactions, or traces of iodide for displacement reactions; insoluble catalysts such as polymer-bound or inorganic solid-bound PTC catalysts ("triphase catalysis"), third-liquid phase catalysis, and vapor phase reactions; electrochemical and photochemical reactions involving PTC; and complex reactions such as when dichlorocarbene is formed and consumed, all have added reaction steps. These result in increased complexity and difficulty of kinetic analysis of the steps and identification of the important factors involved. Some of these more complex systems, such as the use of polymer-bound PTC catalysts, have received intensive mechanistic study,[14] and the understanding which results can explain some unique observations. For example, the diffusion limitation of some reactions using polymer-bound PTC catalysts explains why these catalysts exhibit a much larger reactivity with 1-bromooctane compared to 1-bromohexadecane, whereas with ordinary PTC catalysts reaction rates are the samefor both bromoalkanes.

It is not possible in the space available here to review the fundamental features of these more complex reactions, but it is clear that their important features and basic sequence of steps can be sorted out, providing the understanding for improvements. In this day computers allows chemists to perform numerical integrations of the differential equations for kinetics of even extremely complex reaction systems such that it is possible to sort out the factors which affect these systems. When the chemist takes the time to do this kind of analysis it is found not only to be highly useful and practical, but it clearly illustrates the idea that understanding mechanisms for phase transfer catalyzed processes requires a level of sophistication well beyond that of simple solution-phase organic reactions.

Competition of PTC Reactions with Truly Interfacial Reactions

In PTC chemistry the fact is often ignored that truly interfacial reactions, or restricted-solution reactions (where reaction occurs, e.g., by dissolution of very small levels of the organic-phase reagent in the aqueous phase) can and do occur simultaneously with PTC reactions. Sometimes, reactions that seem to go well in the presence of a PTC catalyst go just as well in the absence of catalyst.

Sometimes these truly interfacial or restricted-solution reactions compete directly with PTC reactions to provide different products. This appears to be the case for example in recent work by Dehmlow and co-workers,[15] in O- vs. C-alkylation of 1,1-dimethyl-2-indanone, which gives predominantly C-alkylation in the absence of phase transfer catalyst, gradually reverses direction to yield predominantly O-alkylation as the size and organic-phase extractability of the quaternary salt catalyst increases. Also in the elegant studies by Dolling and co-workers [16] on chiral methylation of cyclic ketones, it was shown that use of toluene and benzene gave products with significantly greater enantiomeric efficiency than when more polar solvents such as methylene chloride and methyl t-butyl ether were employed, regardless of catalyst concentration. This suggests that the non-polar hydrocarbon solvents could inhibit truly-interfacial reactions which would have given predominantly racemic product.

References

1. Badische Anilin-und Soda Fabrik, Ger. Pat. 268,621(1912).

2. W.H.C. Ruggleberg, A. Ginsberg, R.K. Frantz, *Ind. Eng. Chem.*, **1946**, *38*, 207.

3. J. Uglestad, T. Ellingsen and A. Biege, *Acta Chem. Scand,* **1966**, *20*, 1593 .

4. C.M. Starks, C.L. Liotta, and M.E. Halpern, *Phase Transfer Catalysis, Fundamentals, Applications and Industrial Perspectives,* Chapman & Hall, New York, **1994**, p. 85.

5. D. Mason, S. Magdassi, and Y. Sasson, *J. Org. Chem.*, **1990**, *55*, 2714 .

6. See Reference 5, pp. 24-46.

7. Moberg, F. Bokman, O. Bohman, and H.O.G. Siegbahn, *J. Am. Chem. Soc.*, **1991**, *113*, 3663, and references contained therein.

8. Hughes and Rideal, *Proc. Roy. Soc.*, **1933**, *A140*, 253, as noted in N.K. Adam, *The Physics and Chemistry of Surfaces*, Oxford University Press, London, **1941**, p.95.

9. Der-Her Wang and Hung-Shan Weng, *Chem. Eng. Sci.*, **1988**, *43*, 2019 (1988).

10. D. Mason, S. Magdassi, and Y. Sasson, *J. Org. Chem.*, **1991**, *56*, 7229 .

11. V.A. Revyakin, S.V. Levanova, N.N. Semochkina, and F.S. Sirovskii, *Kinetika i Kataliz*, **1991**, *31*, 1336 (1990); V.A. Revyakin, A.I. Zabotin, and S.V. Levanova, *Kinetika i Kataliz*, **1991**, *32*, 191(1991).

12. S.S. Yufit, I.A. Esikova, *Izv Nauk SSSR Ser. Khim. Akad.*, **1981**, 1996; **1983**, 53.

13. C.M. Starks and C.L. Liotta, *Phase Transfer Catalysis, Principles and Techniques*, Academic Press, New York, **1978**, pp. 171,172.

14. See reference 5, Chapter 5.

15. E.C. Dehmlow and R. Richter, *Chem. Ber.*, **1993**, *126*, 2765.

16. U. H. Dolling, D.L. Hughes, A. Bhattacharya, K.M. Ryan, S. Karady, L.M. Weinstock and E.J.J. Grabowski, *ACS Symp. Ser.*, **1987**, *326*, 67 .

Chapter 3

Mechanisms and Applications of Solid–Liquid Phase-Transfer Catalysis

Charles L. Liotta, Joachim Berkner, James Wright, and Barbara Fair

School of Chemistry and Biochemistry, Georgia Institute of Technology, Atlanta, GA 30332

Historically, phase transfer catalysis (ptc) has been divided into two major classes: liquid-liquid ptc and solid-liquid ptc.[1] In their most primitive forms, the mechanisms employed to describe these two classes of ptc reactions are described in **Figure 1a** and **1b**. In each of the representations two phases were used to portray the systems. Liquid-liquid ptc is described by an organic phase in which the organic electrophile resides and an aqueous phase which contains the anionic nucleophile while solid-liquid ptc is described by a solid salt phase containing the nucleophilic anion and an organic phase containing the organic electrophile. In both cases the ptc catalysts were distributed between the two phases. The mechanism simply involved a ptc catalyst transporting the nucleophilic anion from the aqueous or solid phase into the organic phase where reaction with the organic electrophile ensued followed by ptc catalyst transport of the product anion from the organic phase into the liquid or solid phase. In their simplest form the original mechanisms always had the electrophilic reagent residing in the organic phase. This was not unreasonable since the electrophiles were usually organophilic.

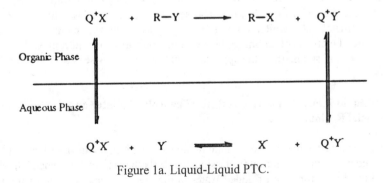

Figure 1a. Liquid-Liquid PTC.

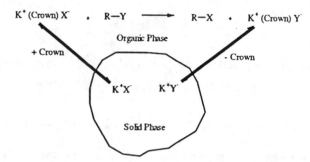

Figure 1b. Solid-Liquid PTC.

Since these early days many kinetic and thermodynamic studies have been reported which necessitated the modification and fine-tuning of the original mechanistic descriptions. In particular, it has been recognized that

(1) in both liquid-liquid and solid-liquid systems anions with large charge to volume ratios (i.e. hydroxide, fluoride) and multiply charged anions (i.e. carbonate) are not easily transferred from an aqueous or solid phase into an organic phase,[2]

(2) in liquid-liquid systems

(a) some effective ptc catalysts are highly hydrophobic and do not distribute into the aqueous phase,[3]

(b) addition of common ptc catalysts (quaternary ammonium salts) greatly reduced the interfacial tension,[4] and

(c) in certain cases a third phase has actually been observed upon addition of a ptc catalyst,[4] and

(3) in many solid-liquid systems small quantities of water play an important role in the catalytic process.[5]

As a consequence, the importance of interfacial regions and interfacial reactions in describing ptc mechanisms has developed.[1]

The first section of this presentation addresses kinetic measurements and mechanistic models dealing with the role of water and the omega phase in solid-liquid ptc reaction of benzyl halide with cyanide ion. The second section addresses the spectrum of inorganic metal salts which form an omega phase and the startling observation concerning the effects of an omega phase on the ionization of p-nitrophenol. Finally, the third section summarizes the application of the omega phase concept to the Wittig reaction.

Solid-Liquid Phase Transfer Catalysis: The Role of Water and the Solid-Liquid Interfacial Region.

In order to probe the mechanism of solid liquid ptc, the reaction kinetics of benzyl halide with solid potassium cyanide and potassium halide (both in excess) suspended in toluene was studied as a function of added water in the presence and absence of the phase transfer catalyst 18-crown-6 at 85°C. The results are summarized in **Table 1**. In the absence of crown no reaction takes place. In the presence of crown and in the absence of added water, reaction takes place but the reaction rate is independent of benzyl halide concentration; the reaction follows pseudo zero order kinetics. In contrast, in the presence of water, the pseudo first order behavior is observed. It can be seen that small quantities of added water result in a significant rate enhancement. However, as the amount of added water increases the pseudo first order rate constant decreases. An analogous kinetic investigation of the reaction of benzyl bromide with potassium cyanide was conducted at room temperature. The results are summarized in **Table 2**. In the absence of added water, pseudo zero order kinetics was again observed. Upon addition of water, the reaction kinetic profile followed pseudo first order kinetics.

Table 1. Pseudo Zero and First Order Rate Constants for Reaction of Benzyl Halide with Potassium Cyanide at 85°C in the Presence of 18-Crown-6 as a Function of Added Water

Added Water (mL)	$k \times 10^5$ (Benzyl Chloride)	$k \times 10^4$ (Benzyl Bromide)
0.00	3.2 M sec^{-1} [a]	0.02 M sec^{-1} [a]
0.36	9.2 sec^{-1}	
0.50	9.4 sec^{-1}	
1.00	11.6 sec^{-1}	13.9 sec^{-1}
2.00	14.7 sec^{-1}	
10.0	10.2 sec^{-1}	11.9 sec^{-1}
20.0	6.9 sec^{-1}	4.4 sec^{-1}
30.0	5.8 sec^{-1}	3.2 sec^{-1}

Reaction Conditions: 0.05 mol benzyl halide, 0.0025 mol 18-crown-6, 0.15 mol KBr, 0.15 mol KCN. (a) Zero-order kinetics (M sec^{-1})

In order to understand the mechanistic role of water, a series of experiments were designed to determine the location of 18-crown-6 in the heterogeneous reaction system. The results are summarized in **Table 3**. It was demonstrated that 92% of the 18-crown-6 added to a system composed of toluene and the inorganic salts KCN and KCl resided in the toluene phase. Upon addition of small quantities of water, and after equilibration of the system, approximately 1-2% of the crown was translocated onto the surface of the salt as determined from the gas-liquid chromatographic analysis of the organic phase. It is postulated that the initial water added to the salt-toluene system coats the surface of the salt particles to form a third phase consisting of water, "dissolved" salt, and toluene. It is this third phase, which we have termed the **omega phase**, that extracts the crown from the toluene phase. Since this new region of the reaction system is formed with the addition of small quantities of water, and since the pseudo first order rate constants for the reaction of benzyl halide with cyanide ion reaches a maximum upon addition of small quantities of water, it is conjectured that the **omega phase** is intimately involved in the observed rate enhancement.

Table 2. Rates of Reaction of Benzyl Bromide with Potassium Cyanide at 25° C in the Presence of 18-Crown-6 as a Function of Added Water

Water (mL)	$k \times 10^5$
0.0	2.06 M sec^{-1} [a]
0.5	1.71 sec^{-1}
1.2	2.54 sec^{-1}
1.4	2.84 sec^{-1}
1.5	9.40 sec^{-1}
1.6	5.10 sec^{-1}
1.75	5.34 sec^{-1}
2.0	5.12 sec^{-1}
2.5	3.36 sec^{-1}
3.0	3.50 sec^{-1}
3.5	3.26 sec^{-1}
4.0	2.20 sec^{-1}
6.0	1.49 sec^{-1}

Reaction Conditions: 0.05 mol benzyl bromide, 0.01 mol 18-crown-6, 0.15 mol KBr, 0.15 mol KCN. (a) Zero-order kinetics (M sec^{-1})

Figures 2a-c attempt to show the formation of the omega phase and **Figure 3** and **Figure 4** the consequences of this interfacial region in the observed kinetic order of the cyanide displacement reaction. **Figure 2a** shows a solid salt particle suspended in a non-polar organic phase. In this picture, the interface between the solid and the liquid is shown as a solid line. While this is a gross simplication, it provides a basis for the subsequent pictures in which additional chemical species are added to the reaction system. **Figure 2b** describes the result of adding small quantities of water to the system. The water is postulated to adsorb onto the surface of the salt to form a surface solvent phase containing the metal salt solvated by water molecules. In addition some of the non-polar solvent may also be present. This interfacial region is termed the **omega phase**. The picture described in **Figure 2b** is not unreasonable since the polar water molecules will be attracted more to the surface of the ionic solid than to the non-polar organic medium.

Table 3. The Effect of Added Water on the Concentration of 18-Crown-6 in Toluene at Ambient Temperatures

Water (μL)	Equiv. of Water [a]	% Crown in Toluene
0	0.00	91.5
10	0.14	81.4
15	0.21	77.3
21	0.29	72.7
22	0.31	50.0
23	0.32	40.0
25	0.35	34.6
30	0.42	17.7
36	0.50	5.8
45	1.25	2.5
50	1.39	2.0
80	2.22	1.0

Conditions: 0.0040 mol 18-crown-6, 0.027 mol KCl, 0.027 mol KCN, 10 mL toluene.
(a) Moles of H_2O/moles of crown

A further consequence of this representation is that the metal salt in the **omega phase** is weakly solvated and not part of the salt crystal lattice. **Figure 2c** describes the system upon addition of 18-crown-6. The crown initially dissolved in the non-polar solvent is extracted into the omega phase. It is postulated that the crown exists in the **omega phase** complexed with the metal cation of the salt. **Figure 3** describes the mechanistic sequence postulated for the 18-crown-6 catalyzed reaction of benzyl halide with cyanide ion in the absence of added water. In this case <u>no **omega phase** exists.</u> As a consequence, the mass transfer of potassium cyanide from the solid phase into the organic phase is slow compared to the organic phase reaction of cyanide ion with benzyl halide. This representation is consistent with the experimentally observed pseudo order kinetic behavior. **Figure 4** describes the mechanistic sequence postulated for the 18-crown-6 catalyzed reaction in the presence of added water. In this case, <u>the **omega phase** is an integral part of the reaction system.</u> Under these conditions, the steps involving the mass transfer of potassium cyanide from the solid phase to the **omega phase**, the complexation of the potassium cyanide with the crown, and the mass transfer of the complex into the organic phase are fast compared to organic phase reaction of cyanide with benzyl halide. This representation is consistent with the experimentally observed pseudo first order kinetic behavior.

Organic Phase

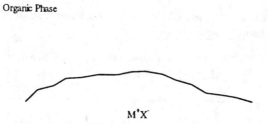

M^+X^-

Solid Phase

Figure 2a. Salt Suspended in Organic Phase.

Organic Phase

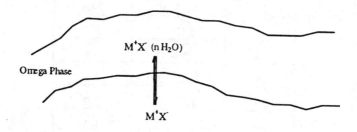

M^+X^- (n H$_2$O)

Omega Phase

M^+X^-

Solid Phase

Figure 2b. Formation of Omega Phase.

Organic Phase   Crown M^+ (Crown) X^- (nH$_2$O)

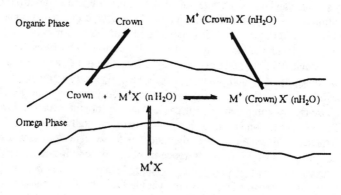

Crown + M^+X^- (n H$_2$O) → M^+ (Crown) X^- (nH$_2$O)

Omega Phase

M^+X^-

Solid Phase

Figure 2c. Addition of 18-Crown-6 and Extraction into the Omega Phase.

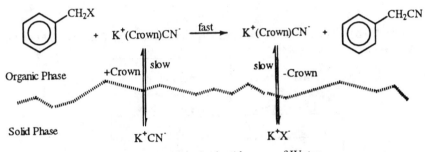

Figure 3. Reaction in the Absence of Water.

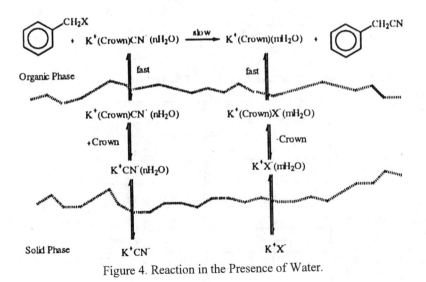

Figure 4. Reaction in the Presence of Water.

The results of distribution experiments of 18-crown-6 between the toluene phase and the **omega phase** as a function of added crown are summarized in **Table 4a.** While the quantity of crown varies from 3.50 to 11.45 mmoles, the amount of crown in the toluene phase remains low and essentially constant. The **omega phase** acts like a sponge and absorbs the added crown. In a companion series of experiments, the kinetics of reaction of benzyl bromide with cyanide were investigated as function of added crown. The pseudo first order rate constants are summarized in **Table 4b.** The results indicate that the rate constants remained relatively constant with increasing quantities of crown.

Table 4a. Distribution of Crown Between Organic and Omega Phases as a Function of Added 18-Crown-6

Total mmoles of 18-Crown-6	mmoles of 18-Crown-6 in Organic Phase
3.50	0.06
6.87	0.06
10.33	0.07
11.45	0.07

Conditions: 1.0 mL H_2O, 0.15 mole KBr, 0.15 mole KCN, 50 mL toluene, 25°C.

Table 4b. 18-Crown-6 Catalyzed Reactions of Benzyl Bromide with KCN as a Function of Added 18-Crown-6

mmoles 18-C-6	$k \times 10^5$ sec^{-1}
3.0	2.16, 2.47
5.0	3.97, 3.63
7.0	3.86, 3.99
10.0	3.75, 4.00
12.0	3.80, 3.60

Reaction Conditions: 1.0 mL H_2O, 0.15 mole KBr, 0.15 mole KCN, 50 mL, toluene, 25°C.

The experimental results presented so far suggest that (1) small quantities of water facilitate the distribution of crown between the toluene and the solid salt phase via the omega phase, (2) the displacement reaction takes place in the toluene phase, (3) in the absence of added water pseudo zero order kinetics is observed, and (4) in the presence of small quantities of water the kinetics become pseudo first order. The following mechanism is proposed to account for these experimental observations:

$$(K^{\cdot}\text{-}\,C\,\text{-}\,CN^{\cdot}\,\cdot nH_2O)_\omega \,\,+\,\, (K^{\cdot}\text{-}\,C\,\text{-}\,Br^{\cdot}\,\cdot mH_2O)_{org}$$

$$k_1 \,\,\downarrow\uparrow\,\, k_{-1}$$

$$(K^{\cdot}\text{-}\,C\,\text{-}\,CN^{\cdot}\,\cdot nH_2O)_{org} \,\,+\,\, (K^{\cdot}\text{-}\,C\,\text{-}\,Br^{\cdot}\,\cdot mH_2O)_\omega$$

$$(K^{\cdot}\text{-}\,C\,\text{-}\,CN^{\cdot}\,\cdot nH_2O)_{org} \,\,+\,\, (PhCH_2Br)_{org} \,\xrightarrow{k_2}\, products$$

where C represents the crown ether and the subscripts ω and org designate the **omega phase** and the organic phase, respectively. Step 1 represents the distribution of crown potassium salt complexes between the **omega** and the organic phases, and Step 2 represents the nucleophilic substitution reaction in the organic phase. In the absence of added water n and m are zero whereas in the presence of added water n and m are the degrees of hydration of the potassium salt complexes. Assuming the steady state approximation for $(K^+\text{-}C\text{-}CN^{\cdot}.nH_2O)_{org}$ the following differential expression for the rate of formation of products may be derived:

$$\frac{d(product)}{dt} = \frac{k_1k_2\,[C_T - C_\omega]\,(K^{\cdot}\text{-}\,C\,\text{-}\,CN^{\cdot}\,\cdot nH_2O)_\omega\,(PhCH_2Br)_{org}}{k_1\,(K^{\cdot}\text{-}\,C\,\text{-}\,CN^{\cdot}\,\cdot nH_2O)_\omega \cdot k_{-1}\,(K^{\cdot}\text{-}\,C\,\text{-}\,Br^{\cdot}\,\cdot mH_2O)_\omega \cdot k_2\,(PhCH_2Br)_{org}}$$

where C_T is the total concentration of crown ether (complexed and uncomplexed), C_F is the concentration of uncomplexed crown in the organic phase, and C_ω is the total crown in the **omega phase**. In the absence of added water the crown is primarily located in the organic phase. Under these conditions, the term $k_2(PhCH_2X)_{org}$ is greater than the other terms in the denominator and the above equation takes the form of a zero order expression. In the presence of added water, the crown is primarily located in the **omega phase**. Under these conditions, the omega terms in the denominator are greater than the term $k_2(PhCH_2X)_{org}$ and the above equation takes the form of a first order expression. These limiting cases are consistent with the experimental facts already discussed.

An attempt was made to determine which metal salts form an omega phase which is capable of extracting 18-crown-6 from the non-polar phase. **Table 5** summarizes some representative salts. In each of the experiments, known quantities of water were added to 20-40 mmoles of salt or salt mixture suspended in 50 mL of toluene containing approximately 3 mmoles of 18-crown-6. The quantities of water listed in **Table 5** are the amounts necessary to extract the maximum amount of crown from the toluene phase into the omega phase. In general, lithium, sodium and potassium salts of halides are effective in extracting crown into the omega phase. It is interesting to note that no water is necessary in the cases of lithium bromide and iodide.

With the exception of potassium acetate, metal salts of oxy anionic species show little ability to extract crown into the omega phase. In contrast, mixtures of potassium salts of oxy anions with potassium bromide or iodide are quite effective.

Table 5. Maximum Extraction of 18-Crown-6 as a Function of Metal Salt and Water

Salt	mL H$_2$O	% ω Phase	Salt	mL H$_2$O	% ω Phase
LiCl	1.1	96	KCl	0.3	90
LiBr	0	100	KBr	0.6	100
LiI	0	100	KCN	0.8	100
LiOH	1.7	0	KOAc	0.1	77
NaBr	0.8	97	K$_2$CO$_3$	0.7	7
NaOAc	0.8	12	K$_2$CO$_3$-KBr	1.4	100
KOH-KCl	0.3	8	KOAc-KCl	0.3	82
KOH-KBr	0.4	94			
KOH-KI	0.5	100			

Known quantities of water were added to 20-40 mmoles of salt or salt mixture suspended in 50 mL of toluene containing 3 mmoles of 18-crown-6.

The role of water in influencing the reactivity of metal salts suspended in non-polar solvents was dramatically illustrated by comparitive reactions of potassium carbonate with p-nitrophenol (pK$_a$ = 7.0) in a variety of organic solvent including toluene, acetone, methyl ethyl ketone and ethyl acetate. Unionized p-nitrophenol is an almost colorless solid while the corresponding anion is a distinct yellow. It was observed that addition of p-nitrophenol to oven-dried potassium carbonate suspended in the above orangic solvents resulted in no observable color change; the organic phase remained colorless and the solid phase remained essentially white. The same experiment, carried out with potassium carbonate treated with minute quantities of water, resulted in the solid becoming an intense yellow color. The organic phase remained colorless. It was concluded that, in
at least these cases, small quantities of water were essential for activating the surface of the solid base in order to promote a proton transfer reaction from an organic acid of moderate strength.

Table 5 shows that the even with rather large quantities of added water potassium carbonate suspended in toluene does not extract 18-crown-6 from the organic phase into this **surface solvent phase**. In the presence of potassium bromide or potassium iodide (not shown in **Table 5**), however, only small quantities of water are necessary to facilitate this transfer. These observations provided the basis for developing a simple method for conducting Wittig reactions. The method involves (1) the addition of the appropriate phosphonium salt to a reaction system consisting of toluene, potassium carbonate, potassium bromide, water, and 18-crown-6 to form the corresponding ylide and (2) the addition of the appropriate aldehyde to produce the desired alkene. In all cases studied, the reaction system was stirred with either a magnetic or mechanical stirring device. Some representative results are shown in **Figure 5** and **Figure 6**.

In the examples outlined in **Figure 5**, the isolated yields are as follows: (a) methyl cinnamate, 98%, (b) 1-phenyl-1,3-butadiene, 40%, (c) 1,4-distyrylbenzene, 53%, and (d) 1,4-bis(o-cyanostyryl)benzene, 63%. **Figure 6** shows the synthesis of vitamin A acetate. When KOH/KBr was employed as the solid base, large quantities the hydrocarbon shown at the bottom of the figure was obtained. In the presence of K_2CO_3/KBr, vitamin A acetate was obtained in 86% isolated yield with no hydrocarbon byproduct.

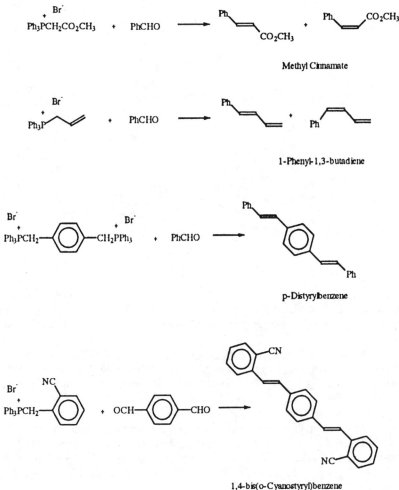

Figure 5. Representative Omega Phase Reactions.

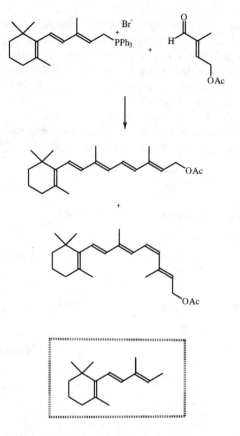

Figure 6. Synthesis of Vitamin A Acetate.

References.

1. C. M. Starks, C. L. Liotta, M. Halpern, "Phase Transfer Catalysis: Fundamentals, Applications, and Industrial Perspectives, Chapman & Hall Inc., New York-London, 1994.

2. (a) A. W. Herriott, D. Picker, J. Am. Chem. Soc., **97**, 2345 (1975); (b) J. de la Zerdo, Y. Sasson, J. Chem Soc. Perkin Trans. II, 1147 (1987); (c)D. Landini, A. Maia, A. Rampoldi, J. Am. Chem. Soc., **51**, 5476 (1986); E. V. Dehmlow, R. Thieser, Y. Sasson, E. Pross, Tetrahedron, 41, 2927 (1985).

3. (a) A. Brandstrom, "Principles of Phase Transfer Catalysis by Quaternary Ammonium Salts," in Advances in Physical Organic Chemistry, Academic Press, London-New York, 1977; (b) D. Landini, A. Maia, F. Montanari, J. Chem. Soc. Chem. Comm., 112 (1977).

4. (a) D. Mason. S. Magdassi, Y. Sasson, J. Org. Chem., **55**. 2714 (1990); (b) Y. Sela, S. Magdassi, Tenside, **27**, 179 (1990); (c) D. Mason, S. Magdassi, Y. Sasson, J. Org. Chem., **56**, 7229 (1991).

5. (a) S. Deremik, Y. Sasson, J. Org. Chem., **47**, 2264 (1985); (b) I. A. Esikova, S. S. Yufit, J. Phys. Org. Chem., **4**, 149, 341 (1991).

Chapter 4

New Results in the Mechanism and Application of Phase-Transfer Catalysis

M. Makosza

Institute of Organic Chemistry, Polish Academy of Sciences,
01–224 Warsaw, Poland

After a short historical and mechanistic introduction of phase-transfer catalysis new solutions of some synthetic and mechanistic problems are discussed, namely: PTC generation and reactions of difluorocarbene, PTC base induced ß-elimination processes, and effects of produced anions on PTC reactions.

In the begining of 1960, on the basis of a short note by Jarrousse (*1*) we started a systematic investigation of reactions of carbanions generated in the presence of conc. aqueous NaOH and catalysts - tetraalkylammonium salts, Q^+X^-. Initially, alkylation of arylaceto- nitriles was studied because of the great practical value of this reaction to pharmaceutical industry. Results of this work were immediately implemented in industrial processes (*2*) and subsequently published in 1965 in a series of papers in Roczniki Chemii (*3*).

$$PhCH_2CN \ + \ R-X \ \xrightarrow[NaOH_{aq}]{0.01 \ Q^+Cl^-} \ PhCHCN \atop R \qquad (1)$$

These experiments can be considered as the first examples of a new methodology in organic chemistry presently known as the Phase-Transfer Catalysis. Thus, in this Symposium, we celebrate the 30th anniversary of the creation of this powerful technique.

Our initial reports and the following papers reporting the alkylation of indene (*3*), Reissert compound (*5*), nitroarylation reaction (*6*), etc. did not, at first, attract much attention within the chemical community. Only our paper on the generation and reactions of dichlorocarbene in the presence of aqueous NaOH and a tetraalkylammonium catalyst, published in 1969 (*7*), precipitated great interest, and soon this method of carbene generation become widely applied.

$$CHCl_3 + \,>\!C\!=\!C\!<\, \xrightarrow[\text{NaOH}_{aq}]{Q^+Cl^-} \; >\!\underset{Cl}{C}\!-\!\underset{Cl}{C}\!<$$

$$Q^+Cl^-_{org} + NaOH_{aq} \rightleftharpoons Q^+OH^-_{org} + NaCl_{aq} \qquad\qquad (2)$$

$$Q^+OH^-_{org} + CHCl_{3\,org} \rightleftharpoons Q^+CCl^-_{3\,org} + H_2O$$

$$Q^+CCl^-_3 \rightleftharpoons CCl_2 + Q^+Cl^-$$

In our first reports mechanistic questions were not discussed. It was assumed that carbanions reacted in the form of the corresponding tetraalkylammonium salts, and it was shown that lipophilic anions such as iodide, perchlorate, etc. inhibited the formation of these salts. It was also shown that the catalytic process was arrested when products were stronger CH-acids than the starting materials (6). However, the question of how the tetraalkylammo- nium derivatives of carbanions were formed was not addressed. The first mechanistic hypothesis was formulated in our 1969 dichlorocarbene paper: It was postulated that the ion exchange between tetraalkylammonium (TAA) chloride and aqueous NaOH resulted in the formation of TAA hydroxide which acted as the base inside the organic phase (7).

In 1971 an important paper was published by C.M. Starks in which the concept of the phase-transfer catalysis (PTC) was formulated for reactions of inorganic anions with lipophilic organic compounds (8). This concept is best exemplified with nucleophilic aliphatic substitution e.g., cyanation of alkyl halides.

$$R\!-\!Cl + Q^+CN^- \longrightarrow R\!-\!CN + Q^+Cl^-$$

$$NaCl + Q^+CN^- \rightleftharpoons Na^+CN^- + Q^+Cl^- \qquad\qquad (3)$$

In this paper generation the of some carbanions and of dichlorocarbene was also described and was practically identical to that reported in our papers. The use of the term PTC in application to these reactions was implying the mechanistic picture analogous to that of the cyanation process: namely, extraction of OH^- anions from the aqueous NaOH into the organic phase, where they act as a base, as it was proposed in our 1969 paper.

Further studies of PTC reactions revealed that they depend on the extraction equilibrium (4b) which in turn is a function of the charge density of anions, so in row (4c) the rightside partner always expels from the organic phase its left side located anions.

$$R\!-\!X_{org} + Q^+Y^-_{org} \longrightarrow R\!-\!Y_{org} + Q^+X^-_{org} \qquad\qquad a$$

$$Q^+X^-_{org} + Na^+Y^-_{aq} \rightleftharpoons Q^+Y^-_{org} + Na^+X^-_{aq} \qquad\qquad b \qquad\qquad (4)$$

$$SO_4^{2-}, F^-, OH^-, Cl^-, CN^-, Br^-, I^-, SCN^-, ClO_4^- \qquad\qquad c$$

Disussion of The Mechanism

A more detailed analysis of the reactions of carbanions and dihalocarbenes generated in the presence of aqueous NaOH and a tetraalkylammonium catalyst raised, therefore, substantial doubts whether the OH^- anions could, indeed, be extracted into the organic phase in sufficient quantity. In fact, the extraction equilibrium is shifted to the left, moreover, the formation of large amounts of the halogen anions during the reaction should have resulted in autoinhibition. Additionally, in the case of dichlorocarbene, the presence of water and of OH^- anions in the organic phase should have resulted in a substantial hydrolysis of this reactive intermediate, nevertheless such hydrolysis was not observed. On this basis, shortly thereafter we abandoned our extractive mechanistic concept presented in 1969 and formulated an alternative mechanistic hypothesis for generation and reactions of carbanions and dihalocarbenes in the presence of aqueous NaOH. According to this hypothesis, the crucial step - deprotonation of the starting CH acid to form the carbanion takes place at the interface between the organic and the aqueous phases. However, the carbanion generated in this way cannot enter the organic phase as the Na^+ salt, nor can it penetrate into the aqueous phase because of the strong salt-out effect (note that the NaOH solution is saturated), so it stays at the phase boundary in an adsorbed-like state. The catalyst, i.e., the TAA salt, dissolved in the organic phase enters into an ion exchange process with the carbanion adsorbed at the phase boundary to form a lipophilic ion pair, which can migrate into the organic phase. All further reactions: alkylation, dissociation to dichlorocarbene, reactions of the carbene, etc., take place inside the organic phase.

$$\underset{\displaystyle \text{NaOH} \rightleftharpoons \text{Na}^+ + \text{H}_2\text{O}}{\overset{\displaystyle {\ge}\text{C-H} \rightleftharpoons {\ge}\text{C}^- + \text{Q}^+\text{Cl}^- \rightleftharpoons {\ge}\text{C}^-\text{Q}^+ + \text{Cl}^-}{\wwwwwwwwwwwww}} \qquad \text{Na}^+ \tag{5}$$

This mechanistic concept was first presented in the plenary lecture: "Reactions of Carbanions and Dihalocarbenes in Two-Phase Systems: Are they Examples of Phase-Transfer Catalysis?" at the Conference of the American Chemical Society Organic Division, Cape Cod, 1973, where it was met with a strong criticism and opposition.

In our further work we accumulated numerous experimental results supporting this interfacial mechanism which were then presented in a plenary lecture at the First IUPAC International Symposium on Organic Synthesis, in Belgium in 1974 and subsequently described in the Pure and Applied Chemistry (9).

Although the term Phase-Transfer Catalysis being short and convenient, was immediately generally accepted, it implies a well-defined mechanistic picture - namely catalytic transfer of the reacting species from phase to phase. Because of that implication we had for long time avoided the use of this term and used the descriptive term Catalytic Two-Phase Reactions in our papers on reactions of carbanions and carbenes in which such transfer was not operating. However, when the mechanistic questions were fully clarified and the operation of the two different mechanisms become widely accepted, there were no more reasons to avoid the use of the convenient term Phase-Transfer Catalysis.

Whereas it was found that the extraction mechanism was not operating for the majority of reactions of carbanions and carbenes generated in a two phase system in the presence of aqueous NaOH and a TAA catalyst, this mechanism was fully confirmed for reactions of inorganic anions and also some organic anions. The major correction, which was introduced into the original scheme, was based on the observation that highly lipophilic TAA salts, which were practically insoluble in the aqueous phase (aqueous solutions), were active as PT catalysts (10). Thus, it did not appear to be necessary for the

TAA salts to migrate into the aqueous phase in order to enter the ion exchange process, which can obviously also occur at the interface. Numerous mechanistic studies disclosed the fine features of the process, such as the influence of the charge density of the reacting ions on the extraction coefficient, solvation effects, and effect of concentration of the aqueous phase on the PT catalyzed reactions.

Thus, it is presently commonly accepted that there are two fundamental general mechanisms operating in the PT catalyzed reactions: (i) The extraction mechanism, in which the reacting anions (ion-pairs) are transferred from the aqueous phase into the organic phase as lipophilic TAA salts. The ion pairs can be formed via the ion exchange in the aqueous phase or at the phase boundary (ii). The interfacial mechanism which involves an organic substrate, typically a carbanion precursor, being initially deprotonated with conc. aqueous NaOH at the phase boundary. The resulting carbanion subsequently enters into an ion exchange with a TAA salt to form a lipophilic ion pair, which migrates into the organic phase. Since in both these mechanisms the ion exchange takes place at the boundary between the phases, the unambiguous criterion differentiating between these mechanisms is that in the former mechanism the matter (reacting anions) is transferred from the aqueous phase into the organic phase and the produced anions are going back into the aqueous phase, whereas in the interfacial mechanism there is no such transfer, and the only matter which migrates between the phases is the two-step transfer of HX from the organic phase into the aqueous phase. Of course, as is the case in all processes in Nature, there is no sharp border between these two mechanisms and in some cases both of them might operate concurrently.

Although the mechanistic studies of the PTC process and related phenomena formed a fascinating chapter of organic chemistry, the major research efforts were directed towards practical application of this new methodology (*11,12,13,14*). Discussion of these issues is beyond the scope of this paper; it is sufficient to mention that PTC is presently one of the most general and powerful techniques in organic synthesis. Although PTC is presently widely used both in research laboratories and in industrial processes, and the majority of the mechanistic questions are already solved, there are still some problems which await clarification, processes which are not sufficiently understood and phenomena which should be better rationalized. Some of such problems will be addressed below, namely: PTC generation of difluorocarbene, base-induced ß-elimination of HX from alkyl halides, and the effect which anions produced during PTC reactions exert on the catalytic process.

PTC Generation and Reactions of Difluorocarbene. PTC offers the most efficient methodology for generation of dihalocarbenes such as CCl_2, CBr_2, $CFCl$, etc. via the α-elimination of hydrogen halides from the appropriate trihalomethanes. However, this simple and efficient technique cannot be applied in syntheses of difluorocyclopropanes via generation of CF_2 in an α-elimination reaction of HX from $CHXF_2$, followed by its addition to alkenes. It is, probably, because rates of deprotonation of difluorohalomethanes and dissociation of the corresponding carbanions to difluorocarbene are similar (*15*). Because of this, the transfer of the carbanions from the interface into the organic phase does not occur, and the difluorocarbene molecules formed in the interfacial region undergo a fast hydrolysis. In order to bypass this problem it appears necessary to generate the short-lived CXF_2^- carbanion, not at the interface, where carbanions are generated under the PTC conditions, but in the organic phase. An ingenious solution to this problem consists of the generation of the intermediate CXF_2^- carbanions, not via deprotonation, but via the PTC halophilic reaction.

It is well known that PTC generated carbanions react with CCl_4 to produce the CCl_3^- anions and chlorinated products. These active compounds can subsequently enter a variety of reactions as shown below (scheme 6) (*16*).

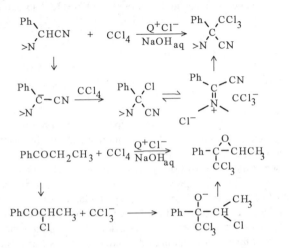

(6)

It was found that di- and tribromomethyl carbanions generated in the two-phase system via deprotonation of methylene dibromide and bromoform underwent halophilic bromination with dibromodifluoromethane to produce bromodifluoromethyl carbanions which immediately dissociated to difluorocarbenes. When the organic phase contained alkenes, the addition took place to produce difluorocyclopropanes. This simple process can be recommended as a method of choice for synthesis of difluorocyclopropanes.

$$CHBr_{3 \, org} + Q^+X^-_{org} + KOH_{aq} \rightleftharpoons Q^+CBr_3^-{}_{org} + NaX_{aq}$$

$$Q^+CBr_3^-{}_{org} + CBr_2F_{2 \, org} \rightarrow \left| Q^+CBrF_2^- \right|_{org} + CBr_4 \rightarrow Q^+Br^-_{org} + CF_{2 \, org}$$

$$CF_2 + {>}C{=}C{<} \longrightarrow {>}\underset{\underset{F \quad F}{X}}{C}{-}C{<}$$

(7)

$$PhCH{-}CHMe \; 80\%; \quad Me_2C{-}CHMe \; 72\%; \quad MeOCH{-}CHMe \; 62\%$$
$$\underset{F \quad F}{X} \qquad\qquad\qquad \underset{F \quad F}{X} \qquad\qquad\qquad \underset{F \quad F}{X}$$

Base-induced ß-Elimination. Base-induced ß-elimination of haloalkanes to alkenes is an important process of organic synthesis and the possibility of using the PTC in this reaction is of great practical interest. However, as can be seen in scheme 8, the very nature of the catalytic process makes it rather unsuitable for the ß-elimination reaction.

$$Q^+X^-_{org} + Na^+OH^-_{aq} \rightleftharpoons Q^+OH^-_{org} + Na^+X^-_{aq} \qquad\qquad a$$

$${>}\underset{\underset{H \quad X}{C}{-}C{<}}{}_{org} + Q^+OH^-_{org} \rightarrow {>}C{=}C{<}{}_{org} + Q^+X^-_{org} + H_2O_{org} \qquad b$$

(8)

The difficulties arise because equilibrium (8a) is strongly shifted to the left; hence, taking into account that halide anions (X^-) are produced in the reaction, transfer of the OH^- anions into the organic phase is negligible and the catalytic process is hindered. These difficulties could be eliminated if an equimolar amount of $Q^+HSO_4^-$, able to extract OH^- anions into the organic phase, was used, as in scheme 9 (*19*).

$$Q^+HSO_4^-{}_{org} + NaOH_{aq} \longrightarrow Q^+OH^-{}_{org} + Na_2SO_4{}_{aq}$$

$$Q^+OH^-{}_{org} + \underset{H\ \ X}{>\!C\!-\!C\!<}{}_{org} \longrightarrow >\!\!C\!=\!C\!\!< + Q^+X^- + H_2O$$

$$(9)$$

Indeed, this ion-pair extraction procedure is very efficient for ß-elimination; unfortunately it does not offer the main practical advantage of PTC - namely the use of the ammonium salt in only small (catalytic) amounts.

Nevertheless, in spite of the problems with catalytic transfer of OH^- anions into the organic phase, there are many papers reporting PTC ß-elimination reactions in the presence of aqueous NaOH (*20,21*). It was also observed that addition of small amounts of alcohols to such two-phase systems promoted the PTC elimination reactions. Although some mechanistic rationalizations of these phenomena were doubtful (*22*), the solid results reported by Dehmlow indicated that the formation of the alkoxide anions was responsible for this co-catalytic effect (*23*).

A general solution of this important and difficult ß-elimination problem could be based on the following reasoning: since the strongly basic OH^- anions, necessary to promote the ß-elimination, are *hydrophilic* and, thus, cannot be transferred efficiently into the organic phase by the catalyst, one should use other strongly basic, but *lipophilic* anions generated by the action of NaOH on the appropriate weak Y-H acid. A weak organic acid Y-H should be selected as the precursor of the Y^- anions, which should be strongly basic and lipophilic, but weakly nucleophilic. The latter quality is necessary in order to avoid undesired S_N2 type substitution reactions. As shown in scheme 10 which presents this concept, compound Y–H should behave as the second catalyst so that the anions X^-, produced in the elimination step, do not exert the inhibitory effect (*24*).

$$Y\!-\!H_{org} + NaOH_{aq} \rightleftharpoons Y^-Na^+{}_{int} + H_2O_{aq}$$

$$Y^-Na^+{}_{int} + Q^+X^-{}_{org} \rightleftharpoons Y^-Q^+{}_{org} + NaX_{aq}$$

$$(10)$$

$$Q^+Y^-{}_{org} + \underset{H\ \ X}{>\!C\!-\!C\!<}{}_{org} \longrightarrow >\!\!C\!=\!C\!\!<{}_{org} + Q^+X^-{}_{org} + YH_{org}$$

This idea was not new, we had observed already in 1966 that attempts to alkylate phenylacetonitrile in a PTC system with ethylene dihalides resulted in ß-elimination of HX which was catalyzed by both the TAA salts and phenylacetonitrile acting as YH (scheme 11) (*25*).

$$0.1\ PhCH_2CN + 1\ X\!\diagdown\!\diagup\!\diagdown\!\diagup^X \xrightarrow[NaOH_{aq}]{Q^+X^-} 0.1\ PhCH_2CN + X\!\diagdown\!\diagup\!\diagdown\!\!\!\diagdown$$

$$(11)$$

It was recently shown that the cyclopropanation of this CH acid proceeds satisfactorily when 1-bromo-2-chloroethane was used as alkylating agent (*26*).

As a rule carbanions are, however strong nucleophiles and thus Y^- should rather be $-O^-$ or $>N^-$ anions. Indeed we showed that some alcohols and indoles acted as efficient co-catalysts in the ß-elimination reactions (scheme 12) (*24*).

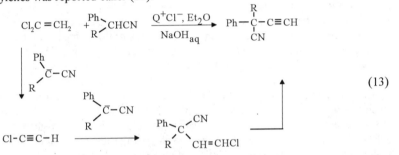

$$Y-H \; = \; CF_3CH_2OH, \;\; 98\% \; ; \; no \; Y-X \; , 4\%$$

In recently reported reactions with 1,1 dichloroethylene producing ethynyl derivatives, arylalkanenitrile carbanions act as cocatalysts promoting ß-elimination and as nucleophiles adding to chloroacetylene (scheme 13) (*27*). PTC addition of such carbanions to acetylenes was reported ealier (*28*).

On the basis of the co-catalytic action of Y-H acids it is possible to rationalize the early examples of unexpectedly efficient PTC ß-elimination reactions carried out without intentionally added co-catalyst. Since haloalkanes, if not thoroughly purified, often contain small amounts of alcohols, these impurities could act as the co-catalysts. Indeed, we observed that meticulously purified haloalkanes are much more resistant towards PTC ß-elimination reactions than the crude reagents containing traces of alcohols.

This rationalization of unexpectedly efficient PTC ß-elimination reactions appeared still insufficient. Our attempts to explain it and also the high effectiveness of some other base promoted PTC reactions lead us to conclusion that position of the equilibrium governing transfer of OH^- anions or carbanions into the organic phase is a function of the size of the interface, in other words - rate of stirring (*29*). This conclusion, although unusual, is not unprecedented (*30*).

Effects of Anions Produced in PTC Reactions on The Catalytic Process.
It had been observed already by Jarousse that alkylation of phenylacetonitrile in the two-phase system proceeded efficiently with ethyl chloride, ethyl bromide was less efficient, and the reaction with ethyl iodide did not proceed at all. We had partially confirmed these observations and found that the chemical reactivity of ethyl halides follows the normal pattern: EtI>EtBr>EtCl. However, Br^- and, particularly, I^- anions produced during the reaction inhibited the catalytic process (*3*). This behavior was due to the high lipophilicity of the I^- anions, which stayed as tetraalkylammonium iodide in the organic phase, thus hindering the transfer of the carbanions into the organic phase. This inhibitory effect of

the iodide anions on the PTC reactions was confirmed in many subsequent reports; however, there were also some papers describing the successful PTC alkylation of various CH, OH and NH acids with alkyl iodides (*31-33*). Explanation of this apparent discrepancy is necessary, and it should be based on an analysis of the extraction equilibrium between the iodide anions and the carbanions.

$$R_3C^- Q^+_{org} + I^-_{int} \rightleftharpoons I^- Q^+_{org} + R_3C^-_{int} \tag{14}$$

The position of this equilibrium, and, hence the concentration of R_3C^- in the organic phase, should be a function of two parameters: the relative lipophilicity of R_3C^- and I^- and the fraction of the interfacial area covered by each of these anions, which in turn should depend on acidity of the carbanion precursor. This hypothesis was verified experimentally using substituted phenylacetonitriles. It seemed reasonable to assume that changing the ring substituents from H to Cl or CH_3 should not affect the relative lipophilicities of R_3C^- and I^-, whereas it would affect the acidity of the nitrile CH acids. According to the interfacial mechanism, deprotonation of the carbanion precursors takes place at the interface to produce carbanions in adsorbed state. The extent of the deprotonation and consequently the fraction of the interfacial area occupied by the particular carbanions should be, therefore, a function of acidity of the CH acids. Consequently, one can expect that the degree of the inhibitory effect on this catalytic process of the I^- anions produced during the alkylation of carbanions with alkyl iodides should depend on the acidity of the carbanion precursors. Scheme 15 shows the degree of PTC alkylation of substituted phenylacetonitriles with propyl iodide depending on their acidity (Lasek,W., Ph.D. Theses, 1995) and confirms this hypothesis.

$$X{-}PhCH_2CN + Pr{-}I \xrightarrow[\text{NaOH}_{aq}]{Q^+X^-} X{-}Ph\underset{\underset{Pr}{|}}{C}HCN \tag{15}$$

X	4-MeO	4-Me	H	4-Cl	3,4-Cl$_2$	2,4-Cl$_2$
pKa	23.8	22.9	21.9	20.5	18.7	17.5
conversion %	4	10	17	45	55	90

From the data shown in scheme 15 it is possible to conclude that the inhibition of PTC alkylation with the iodide anions is, for the most part, connected with the competition between carbanions and iodide anions for the interfacial area. Thus, PTC alkylation of carbanions derived from CH acids of higher acidity should proceed smoothly.

An important problem in alkylation of carbanions concerns the selectivity of mono-*versus* dialkylation, particularly with methyl halides. Since inhibition of the PTC alkylation with I^- depends on the acidity of the carbanion precursors, and since the introduction of the Me group by alkylation should decrease this acidity, one could expect that the PTC alkylation with MeI should be more selective (towards the mono-alkylated product) than with MeCl, and also more selective than the reaction in a homogeneous system. Indeed, the results presented in scheme 16 (Lasek, W., Ph.D. Theses 1995) support this hypothesis.

$$PhCH_2CN + MeX \xrightarrow[\text{solv.}]{B^-} PhCH_2CN + Ph\overset{Me}{\underset{}{C}}HCN + Ph-\overset{Me}{\underset{Me}{C}}-CN$$

$NaOH_{aq}$, $Bu_4N^+Cl^-$	MeCl	10.5	78	11.5	(16)
	MeI	7.3	85	7.7	
t-BuOK, THF	MeCl	12.1	73	14.9	
	MeI	15	69	16	

The PTC mono-methylation of phenylacetonitrile with MeI proceeds more selectively than with MeCl. This effect is, however, weaker than expected, apparently because the extent of the dialkylation is governed not only by the ratio of the carbanions generated at the interface but also by the equilibrium in the organic phase. The latter problem was discussed in one of our early papers (*34*).

Another important feature of the PTC reactions of methylenic carbanions is connected with the effect of introduced substituents on the reaction course. In the case of alkylation with alkyl halides, monoalkylated products are usually weaker CH acids than the starting compounds, thus they can be obtained in high yield. There are, however, cases when the products are stronger CH acids than the starting materials. In such cases the products are preferentially deprotonated to form the product carbanions which enter into further reactions to form disubstitution products or, because of weak nucleophilicity, do not react further. In the latter case the catalytic process is arrested at the very beginning. This situation is exemplified by the PTC nitroarylation of phenylacetonitrile and its derivatives (scheme 17) (*6, 35*).

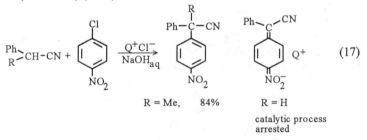

$$R = Me, \quad 84\% \qquad R = H$$

catalytic process
arrested

This obvious rule is, perhaps, insufficiently understood and not always recognized. Recently, a publication has appeared in which the PTC nitroarylation of phenylacetonitrile with 4-chloro-3-trifluoromethylnitrobenzene was reported to occur in high yield (*36*). Since this result violated the well-established rule described above, and, if confirmed, would contribute to the better understanding of the mechanistic features of PTC, we have studied this reaction and found that the results reported in the paper were incorrect. In the PTC reaction between 4-chloro-3-trifluoromethylnitrobenzene and phenylacetonitrile three products were formed: the major product was 5-chloro-7-phenyl-3-trifluoromethyl-benzisoxazole. On the other hand, when the tetraalkylammonium salt was used in an equimolar amount, following the ion-pair extraction procedure, the nitroarylation occured in high yield. The nitroarylation did also occur also with 2-phenyl propionitrile, because the product in this case was not a CH acid (*37*).

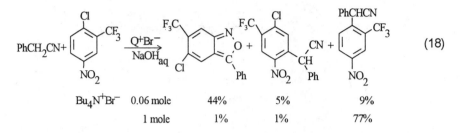

| | Bu$_4$N$^+$Br$^-$ | 0.06 mole | 44% | 5% | 9% |
| | | 1 mole | 1% | 1% | 77% |

Thus, the rule already formulated in 1969, stating that the PTC reactions of carbanions do not occur when the products are stronger CH acids than the starting materials, was once more confirmed. It should also be mentioned that chlorotrifluoromethyl nitrobenzenes behave similarly to other nitroarenes in the reactions with carbanions. The initial addition always occurs at the position occupied by a hydrogen atom (*38*). In the case of the phenylacetonitrile carbanion this addition results in the formation of oxidative products or benzisoxazole, whereas with α-halocarbanions vicarious nucleophilic substitution takes place (*39*).

This short paper shows how during the 30 years PTC made a long way from specific conditions for some reactions to a general technique widely applied in laboratories and industrial processes where it brings great economical advantages. Its complicated mechanistic features are now fairly well understood so its application become very efficient in many fields. There is no doubt that studies on mechanism and application of PTC will continue.

Literature Cited.

1. Jarrousse, M. J., *C. R. Acad. Sci.* Paris, **1951**, *232*, 1424.
2. Urbański, T.; Serafinowa, B.; Bełzecki, C.; Lange, J.; Makarukowa, H., Mąkosza, M. Polish Patent Nr. 46030, filed 1960, 47902 filed 1961, CA, **1964**, *61*, 4279g.
3. Mąkosza, M.; Serafinowa, B., *Rocz. Chem.*, **1965**, *39*, 1223, 1401, 1595, 1799, 1805; C. A. **1966**, *64*, 12595h, 17474g, 17475c, 17475e, 17475gs.
4. Mąkosza, M., *Tetrahedron Lett.*, **1966**, 4621.
5. Mąkosza, M., *Tetrahedron Lett.*, **1969**, 677.
6. Mąkosza, M., *Tetrahedron Lett.*, **1969**, 673.
7. Mąkosza, M.; Wawrzyniewicz, M., *Tetrahedron Lett.*, **1969**, 4659.
8. Starks, C. M., *J. Am. Chem. Soc.*, **1971**, *93*, 195.
9. Mąkosza, M., *Pure Appl. Chem.*, **1975**, *43*, 439.
10. Landini, D.; Maia, A. M.; Montanari, F., *Chem. Comm.*, **1977**, 112.
11. Dehmlov, E. V.; Dehmlov, S. S., Phase Transfer Catalysis, 3rd ed. Verlag Chemie, Weinheim 1993.
12. Mąkosza, M.; Fedoryński, M., *Advances in Catalysis*, **1987**, *35*, 375.
13. Starks, C. M.; Liotta, C. L.; Halpern, M., Phase-Transfer Catalysis, Chapman & Hall, New York, London 1994.
14. Goldberg, Y., Phase-Transfer Catalysis, Gordon and Breach, Amsterdam 1992.
15. Hine, J.; Langford, P. B. J., *J. Am. Chem. Soc.*, **1957**, *79*, 5497.
16. Mąkosza, M.; Kwast, A.; Jończyk, A., *J. Org. Chem.*, **1985**, *50*, 3722.
17. Balcerzak, P.; Fedoryński, M.; Jończyk, A., *Chem. Comm.*, **1991**, 826.
18. Balcerzak, P.; Jończyk, A., *J. Chem. Res. (5)*, **1994**, 200.

19. Le Coq, A.; Gorgues, A., *Org. Synthesis*, **1979**, *59*, 10.
20. Halpern, M.; Sasson, Y.; Rabinowitz, M., J. Org. Chem., **1984**, *49*, 2011.
21. Dehmlov, E. V.; Lissel, M., *Tetrahedron*, **1981**, *37*, 1653.
22. Shavanov, S. S.; Tolstikov, G. A.; Shutienkova, T. V.; Riabova, N. A., *Zh. Org. Khim.*, **1989**, *9*, 1868.
23. Dehmlov, E. V.; Thieser, R.; Sasson, Y.; Pross, E., *Tetrahedron*, **1985**, *41*, 2927.
24. Mąkosza, M.; Lasek, W., *Tetrahedron*, **1991**, *47*, 2843.
25. Mąkosza, M.; Serafinowa, B., *Rocz. Chem.*, **1966**, *40*, 1647, CA **1967**, *66*, 94792.
26. Fedoryński, M.; Jończyk, A., *Org. Prep. Proc. Int.*, **1995**, *27*, 355.
27. Jończyk, A.; Kuliński, T., *Synthesis*, **1992**, 757.
28. Mąkosza, M.; Czyżewski, J.; Jawdosiuk, M., *Organic Synth.*, **1976**, *55*, 99.
29. Lasek, W.; Mąkosza, M., *J. Phys. Org. Chem.*, **1993**, *6*, 412.
30. Watari, H.; Freiser, H., *J. Am. Chem. Soc.*, **1983**, *105*, 191.
31. Dehmlov, E. E.; Schrader, S., *Z. Naturforschung*, **1990**, *456*, 409.
32. Sato, R.; Senzaki, T.; Goto, T.; Saito, M., *Bull Chem. Soc. Japan*, **1986**, *59*, 2950.
33. DiCezare, P.; Gross, B., *Carbohydr. Res.*, **1976**, *48*, 271.
34. Mąkosza, M., *Tetrahedron*, **1968**, *24*, 175.
35. Mąkosza, M.; Jagusztyn-Grochowska, M.; Ludwikow, M.; Jawdosiuk, M., *Tetrahedron*, **1974**, *30*, 3723.
36. Durantini, E. N.; Chiacchiera, S. M.; Silber, J. J., *J. Org. Chem.*, **1993**, *58*, 7115.
37. Mąkosza, M.; Tomashewskij, A., *J. Org. Chem.*, **1995**, *60*, 5425.
38. Mąkosza, M., *Polish J. Chem.*, **1992**, *66*, 3.
39. Mąkosza, M.; Winiarski, J., *Acc. Chem. Res.*, **1987**, *20*, 272.

Chapter 5

Adsorption on Interfaces and Solvation in Phase-Transfer Catalysis

S. S. Yufit[1], G. V. Kryshtal[1], and Irina A. Esikova[2]

[1]N. D. Zelinsky Institute of Organic Chemistry, Russian Academy of Sciences, 47 Leninsky prospekt, 117913 Moscow, Russia
[2]Chiron Corporation, 4560 Horton Street, Emeryville, CA 94608

It was shown that PTC reactions both in solid—liquid and liquid—liquid systems can proceed at interface via the formation of cyclic adsorption complexes. The conditions for the formation of the latter were formulated and the kinetic and thermodynamic parameters of the reaction with solid ionophoric salts were estimated. A linear relationship between the energy of crystalline lattice of the salt and its melting point was estimated. The latter was shown to be useful as the simplified estimation of the reactivity of solid salts in the reactions where crystalline lattice is destroyed. The connection between solvent characteristics and the regio- and stereoselectivity in the aldol condensation and Horner — Wadsworth — Emmons reaction of aldehydes with derivatives of 3-methyl-4-phosphono-2-butenoic acid was studied. In solid/liquid systems the "benzene effect" of solvents was observed. The role of hydration in liquid/liquid systems was discussed.

It is well known that the interface in a liquid—liquid system is an intricate voluminous formation (1). The calculations showed (2) that the effective thickness of the phase boundary in the water—CCl_4 system is up to 9—10 Å due to the formation of capillary waves. The comparison of activities of two phase transfer catalysts (tetraoctylammonium (Oct_4N^+) and methyltrioctylammonium ($MeOct_3N^+$) chlorides) in the S_N2-substitution reaction (1) showed that the first-order rate constants are $1 \cdot 10^{-5} \, sec^{-1}$ and $3.2 \cdot 10^{-5} \, sec^{-1}$ for $MeOct_3N^+$ and Oct_4N^+, respectively.

$$RBr + MC1 \xrightarrow{\; Q^+Cl^- \;} RCl + MBr,$$

$$R = C_6H_{13}, \; Q^+ = MeOct_3N^+, \; Oct_4N^+ \qquad (1)$$

The calculations of solvation and energy parameters for the reaction by the Monte-Carlo procedure demonstrated (3) that the decrease in the number of carbon atoms in one of the radicals of Q^+ (Q^+ is the quaternary ammonium cation) result⁻

in the increase in the hydration degree of the tight ion pair (the hydration number of $MeOct_3N^+Cl^-$ is $n = 3+2$ as compared with $Oct_4N^+Cl^-$ ($n = 3+1$)).

The fact that decrease in the hydration degree results in increase in reaction rate is well known in the PTC. This increase could be attained both by binding of water molecules with adding alkali to the aqueous phase (4, 5) as well as by the complete elimination of the aqueous phase (6). So the interaction of the catalyst adsorbed at the interface with the aqueous phase exerts a strong influence over reaction rate thus demonstrating the significant role of the catalyst hydration in the PTC.

Adsorption at the interface (4, 7, 8). Adsorption of the biphyle molecule at the interface (nearly all phase transfer catalysts are biphyle) results in the formation of a double electrical layer typical for electrochemical systems. The strong electrostatic field of the ion at the interface pulls molecules together, thus promoting the elementary act of reaction (pull-together catalysis) and determines the mechanism of the PTC reaction.

The role of phase boundary is clearly shown in process of aldol condensation. The conventional condensation of aromatic aldehydes with n-alkanals in alcohols or aqueous alcohol solutions of alkali bases (9-11) proved to be of little effectiveness. The yields of the required product did not exceed 15 %. The application of the PTC procedures recommended for the aldol condensation (12,13) was also disappointing: in the system aqueous NaOH—benzene—benzyltriethylammonium chloride (BTEA-Cl) the reaction proceeded slowly to give **2** in no more than 20% yields, whereas in the systems solid K_2CO_3 (or Na_2CO_3)—benzene (or CH_2Cl_2)—BTEA-Cl compound **2** was not formed.

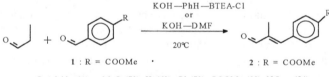

1 : R = COOMe   •   2 : R = COOMe

R (yield , %) = MeO (70), H (62), Cl (73), COOMe (90), NO$_2$, (84)

Far better results were obtained in two other heterogeneous systems, i.e., solid KOH (or NaOH)—benzene—BTEA-Cl and solid KOH (or NaOH)—DMF where the yield of the enal **2** (R = COOMe) amounted to 90%. Only catalytic amounts of the bases are necessary for a successful synthesis (14). The efficiency of the biphasic system solid KOH(cat) — DMF was demonstrated on a series of related reactions. Benzaldehyde and its derivatives (R = Cl, OMe, NO$_2$) reacted with propanal to give the respective enals **2**. Electron-withdrawing substituents (CO$_2$Me, NO$_2$) accelerated the reaction while the electron-donating MeO group reduced its effectiveness. Similar results were obtained upon condensing n-pentanal with benzaldehydes.

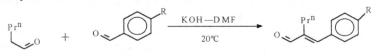

R (yield, %) = H (62), Cl (64), COOMe (62)

This procedure of aldol condensation was successfully extended to various heterocyclic aldehydes (furfural, 2-thiophene- and 2-pyridinecarboxaldehyde), which

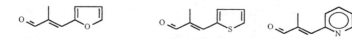

readily reacted with propanal to afford the respective 3-heterylsubstituted 2-methyl-2-propenals (*14*) in yields of ~ 69-71%:

According to the GLC and NMR data, in all these cases the reaction proceeded stereospecifically with the formation of geometrically pure *E* isomers (*14*).

The apparent advantage of the solid-liquid system could be explained by the possible co-ordination of the reagents at the interface. Another reason of this is the absence of water influence on the second step of the aldol condensation - water elimination.

In order to understand the role of the interface in the PTC reaction it is necessary to give a some theoretical consideration of the substitution. Such a consideration is based on the quantum chemical restrictions, values of dissociation constants of ion pairs in the solvents usually used in PTC reactions as well as phase electroneutrality and steric demands. Taking these factors into account one can formulate the five requirements that should be satisfied for all proposed mechanisms of PTC S_N2 reaction (*15*):

1) the retention of the stereochemistry of the intermediate complex according to the quantum chemistry rules, i.e. the linear arrangement of reacting molecules;

2) providing for the leaving group solvation;

3) conforming to the phase electroneutrality during ion transfer through the interface (*i. e.* the synchronous transfer of similarly charged ions);

4) no dissociation of ion pairs in organic phase due to energy restrictions;

5) mutual compensation of dehydration (desolvation) and hydration (solvation) heat effects in the course of the sole elementary act of reaction.

These demands are completely met in case of the substitution that includes formation of the intermediate adsorbed at the interface (Scheme 1).

Scheme 1

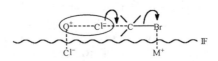

(IF — interface)

Such a mechanism of S_N2 substitution has been shown in the solid/liquid PTC systems (*24-26*). Reactions in the liquid/liquid systems can proceed more complicatedly due to strong influence of the aqueous phase. However, the rules formulated above are necessary in all the cases.

A look at the stereochemistry of conventional scheme of the S_N2 substitution in a homogeneous solution that conforms to quantum chemical restrictions (requirement 1) shows that the demand 4 will be violated in case of such organic solvents as hexane or toluene because the reaction results in the dissociation of the ion pair that is unlikely.

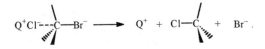

It is evident that the role of water is very important in the liquid/liquid PTC substitution. It is connected with the possibility of the solvation of the leaving group (complying with the requirement 2).

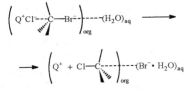

However, the act of reaction will be in agreement with the requirement 3 only if there is compensating transfer of the anion from the aqueous phase (aq) to the organic one (org).

All these "problems" of the above mechanistic restrictions are abolished if the binary complex resides at the interface thus giving rise to the formation of the ternary absorbed complex.   Absorption mechanism conforms to all the five requirements to the PTC substitution mechanism.   We believe that the adsorption reaction mechanism can be also realized in the liquid—liquid systems if the lifetime of the substrate adsorbed at the interface is large enough for the reaction act to proceed. The discussion of this problem is given in (16).

All the set of structural, energetic, quantum-chemical and thermodynamic factors together with clearly expressed biphylity and complexing properties of all phase transfer catalysts make it possible for the absorption mechanism to take place in all PTC reactions.

Reactions inside ion pairs. The different consequences of proton oscillations in hydrate shells of OH^- and F^- ions have been discussed elsewhere (cf. ref.(4)).

$$HO-H---^=OH \quad \longleftrightarrow \quad HO^=---H-OH$$

$$HO-H---^=F \quad \longleftrightarrow \quad HO^=---H-F$$

In the first case the number and composition of the participating species are not changed, while in the second case two new species are formed, i.e. a chemical reaction takes place there. However, in both of the cases the superposition of the states due to proton oscillations inside the ion pair should result in strengthening the separated ion pair. This was confirmed in (17), where it was shown that at certain arrangement of H_2O molecule inside the separated ion pair t-Bu$^+$//H$_2$O// Cl$^-$ there appears a minimum on the potential energy surface. The stabilization of separated ion pair is proved in this manner.

Under certain conditions the stabilization of the separated ion pair could result in the acceleration of its reactions. This can explain the acceleration of PTC reactions already observed on adding a trace of water to anhydrous systems (*18, 19*), as water would hydrate the anion including itself inside the separated ion pair.

The formation of so-called quadruplets in substitution reactions was suggested long ago by C.K.Ingold (*20*). These quadruplets are really the binary reactant complexes. Ingold believed that in the act of reaction the quadruplet unfolds itself, and the reaction proceeds according to quantum-chemical restrictions. However, the reaction can proceed in an another way that we call the reaction inside an ion pair. Actually, if the solvent-separated ion pair is separated by the reacting molecule, then the reaction may proceed inside the ion pair (Scheme 2).

Scheme 2 illustrates the aldol condensation of acetone inside the ion pair consisting of onium cation and acetonide anion. The first step is the formation of a typical Ingold quadruplet. Then the second solvating acetone molecule stumbles into the ion pair making it solvent-separated. Finally, the electron transfer results in the formation of the reaction product i.e. alkoxide of the diacetone alcohol. Just the same explanation can be applied for many other reactions that include intermediate ion pairs.

Scheme 2

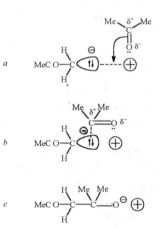

a. Coordination of the ion pair $MeCOCH_2^-...Q^+$ with acetone molecule;
b. Quasi-separated ion pair $MeCOCH_2^-//(Me)_2CO//Q^+$;
c. Alkoxide Q^+ (tight ion pair) $MeCOCH_2(Me)_2CO^-...Q^+$.

Reactions on solid surface (solid—liquid systems) (*21-29*). Earlier (*21-23*) it was shown that PTC substitution reactions of the type (1) in solid—liquid systems occur at the surface of the ionophoric salt. The absence of the exchange shown below served as the direct evidence of this fact:

$$Q^+Br^-_{org} + KC \ l_{solid} \xrightarrow{\quad\times\quad} Q^+Cl^-_{org} + KBr_{\ solid}$$

$$(2)$$

The above reaction is the base of the commonly recognized PTC mechanism for liquid—liquid systems. Because the substitution reaction (1) proceeds in the absence of the ion exchange (2) it is clear that the process does not occur according to the traditional "extraction" mechanism.

The analysis of the kinetic data of the substitution (*24*) together with thermodynamic and activation parameters (*25*) allowed us to suggest a new mechanism of solid-liquid PTC. The mechanism includes step-by-step formation of the cyclic ternary complex (*26*). The resulting data are consistent with formation of two pairs of binary and ternary complexes obtained from RY, MX and QX adsorbed on solid salt surface (Scheme 3; where K is the equilibrium constant, k is the rate constant, BC_1, BC_2 are binary complexes, and TC_1 and TC_2 are the ternary ones).

Scheme 3

$$RBr + MCl + QX \underset{\longleftarrow}{\overset{K_{TC1}}{\longrightarrow}} TC_1 \underset{k_-}{\overset{k_+}{\rightleftarrows}} TC_2 \underset{\longleftarrow}{\overset{K_{TC2}}{\longrightarrow}} RCl + MBr + QX$$

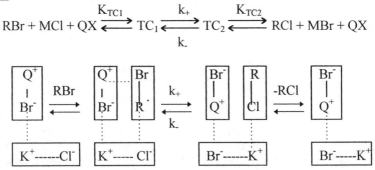

$$(BC_1) \qquad\qquad (TC_1) \qquad\qquad (TC_2) \qquad\qquad (BC_2)$$

Analysis of energetics of the substitutions with solid salts of different strength of M—Cl-bonds showed that the rate-limiting step is the rearrangement of the ternary complexes. The possible structures of these ternary complexes are expected to provide for adherence to the detailed balancing principle in this reversible reaction. In design of the ternary complex models the following three factors should be taken into account: 1) according to S_N2 mechanism it is assumed that the rear part of the carbon atom associated with the leaving group Y^- is attacked by the X^- ion; 2) the quantum-chemical study of these systems (*27*) indicates that during such an attack the X^- ion is capable of being diverted from the X^-...C—Y axis by up to 20° without a noticeable rise in the system's energy (10% maximum); 3) the $Cl^-...Q^+...Cl^-...M^+$ and $Br^-...M^+...Cl^-...Q^+$ electrostatic interactions do not exhibit any predominant direction so that the arrangement of the component ions is controlled only by the strength of these interactions.

It should be also noted that in the formation of ternary complexes the ion—ion and ion—dipole interactions impose some requirements on the geometry of the entering species. In this context, the effect of the geometric factors will be significant, and the reaction rates depend on the size of the catalyst Q^+ cation. The use of bulkier Q^+ cations results in a sharp deceleration of the substitution reaction. The rate constant in the presence of Bu_4NBr is four-fold that in the case of Oct_4NBr, although the latter is completely dissolved in toluene. The reaction rates in the presence of different QX have no connection with the solubility of the catalyst in toluene, since the addition of RBr to the system brings about complete dissolution of all the QX salts at 70°C.

The volume of the catalyst anion also effects the reaction rate, and a satisfactory linear relationship between $\log k_{\pm}$ and reciprocal anion radius $1/R$ (Å^{-1}) both for the forward and the reverse reaction were established.

The kinetic analysis of different reaction mechanisms (25) allowed to elucidate the single mechanism that reflects all kinetic peculiarities of the reaction (3):

$$RX + MY + QX \; \underset{K_\Sigma}{\xrightarrow{\hspace{2cm}}} \; TC \; \underset{k_-}{\overset{k_+}{\xrightarrow{\hspace{2cm}}}} \; RY + MX + QX \qquad (3)$$

$$k_{obs} = \frac{k_+ K_\Sigma [MY]_o [QX]_o}{1 + K_\Sigma [RX]_o ([MY]_o + [QX]_o)} \qquad (4)$$

According to the equation (4), a limiting value of the observed rate constant (k_{obs}) will be observed at certain initial concentrations of reagents. Actually, the invertion of both sides of equation (4) results in subsequent linear plots ($1/k_{obs}$ vs. $1/[Q^+]$).

The comparison of enthalpy changes during formation of ternary complexes including different solid salts MCl and MBr (M = Li, Na, K, Rb, Cs) showed (25) that the energy of their formation is linked with the crystalline lattice energy of the salt. It was found that the less is the latter the weaker is the former. In the case of CsCl the formation of a ternary complex is even endothermic corresponding thus to its high reactivity (23). This hypothesis is confirmed by the analysis of ΔH_{TC2} values for the reverse reaction: the lower values of crystal lattice energies of bromides result in a zero values of ΔH_{TC2}. It allows one to believe that in case of salts with the «weak» crystalline lattice (little values of crystalline lattice energy ΔH) the reaction scheme may be transformed in such a way that TC_1 and TC_2 would not constitute kinetically independent species any more, but would become "diffusion" pre-equilibrium complexes.

Starting with the earlier published data (22) on the dependence of k_{obs} on $[QBr]_0$ and $[RBr]_0$ in the reaction with KCl, we determined the formation constant of TC_1, $K_{TC1} = 20.48$ M^{-2}, and the rate constant of its rearrangement to TC_2, $k_+ = 0.75$ s^{-1} M^{-2} at 84°C. Using the equation $\Delta G = \Delta H - T\Delta S$, where $\Delta H_{TC} = -11.7$ kcal·mol^{-1} (see (24)), we estimated the entropy of formation for TC_1 and that of activation involved in the step-by-step rearrangement between TCs. It was found that $\Delta S_{TC1} = -26.8$ e.u., and $\Delta S^{\neq} = +47$ e.u. A comparison of these values suggests that the structure of the TCs is more rigid than that of the transition state during the complex rearrangement.   This agrees well with the idea that the solid-phase molecule forms a part of the TCs. It is believed that the transition between the TC_1 and TC_2 occurs via a cyclic state with a delocalized bond system (Scheme 4; arrows show the direction of acceptor (A) or donor (D) forces of solvent molecules (S) and their polarisation component (Sp)).

Scheme 4

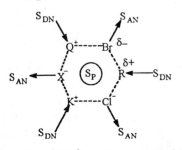

The large positive value of $\Delta S^{\neq}$ may be explained by the destruction of the crystalline lattice of the solid salt, resulting in the formation of lattice-free M^+ and Cl^- (or Br^-) ions and, consequently, in a sharp rise in the entropy of the system. Such a substitution mechanism with participation of TC_1 and TC_2 involves the destruction of crystalline lattice of the ionophore by a catalyst.   The fact was demonstrated earlier also for other reactions (*28, 29*).

It seems that the desorption of QBr from the surface of MCl is not necessary for the reaction to proceed, the movement of QBr on the crystal surface is more probable. This phenomenon is well known in the heterogeneous catalysis. In this case we have a modification of PTC that should be called the "true inter-phase catalysis", and not the "phase transfer catalysis", because a catalyst does not transfer any solid phase.

"Strength" of the crystalline lattice. As it was indicated above, reaction with solid salt results in the destruction of the crystalline lattice because anion of the M^+Y^- salt (KCl for example) is included in the reaction products:

$$RX + M\,^+Y_{solid}^- \longrightarrow RY + M\,^+X_{solid}^- \tag{5}$$

It is found in the several cases (*25,28,29,43*) that there is a correlation between log k_{obs} and crystalline lattice energy. That is why the crystalline lattice energy should be taken into account in consideration of any processes in solid— liquid systems. The problem is that the determination of the crystalline lattice energy is fairly difficult, and sometimes there are no published data. Because the crystalline lattice energy ($\Delta H_{cr.l.}$ or $U_{cr.l.}$) is connected with most different properties of the crystal one may suppose that such a relationship exists also for the crystal melting point. Actually, such a link does exist. Fig. 1 shows the relationship between the crystalline lattice energy and the salt melting point.

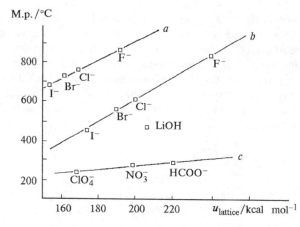

Fig. 1. Crystalline lattice "strength": melting point—crystalline lattice energy relationship for some salts: potassium halides (*a*), lithium halides (*b*), other lithium salts (*c*). (Reproduced with permission from reference 8. Copyright 1995 Russian Academy of Sciences.)

It is clearly seen that salt melting points linearly depend on the crystalline lattice energies and form several series, concerned with cation and anion types. One should note the difference between simple symmetrical halide anions and more complex anions of the formate type. It is useful to introduce the term: the crystalline lattice "strength" of solid salts. It is evident that the smaller crystalline lattice "strength" would facilitate the reaction accelerating as it was observed actually. The comparison of the crystalline lattice "strength" of salts lying on one line in Fig. 1, with that for salts on the other lines can sometimes provide more information than the comparison of the salt $\Delta H_{cr.l.}$. Table 1 shows some parameters of lithium formate and chloride.

Table 1.   The comparison of some anion parameters in formates and chlorides

Parameter	Chlorides	Formates
Anion radii[a]/ Å	1.81	2.4-3
Crystalline lattice energy[a] /kcal•mole^{-1}	201.4	201.7
Lattice "strength"[a] (m.p.)/°C	610	273
Nucleophility[b]/DN	1.21	(0.95)
Basicity	− 3.0	(+6.46)
Symmetry[c]	Spherical	Planar

[a] See ref.(*30*); [b] see ref.(*31*); [c] see ref.(*32*).

Source: Reproduced with permission from reference 8. Copyright 1995 Russian Academy of Sciences.

The analysis of these data reveals three main distinctions, i.e., ones in the type of symmetry, in basicity and in the crystalline lattice "strength". These distinctions can explain the observed differences in kinetics and mechanisms in S_N2-substitution (reaction 1). First of all these relates to the different influence of water. For planar ions of the formate type with the low crystalline lattice "strength" there was observed a sharp acceleration of the reaction rate (and yield) on addition of small amounts of water (*18*). For spherical halide ions with the high crystalline lattice "strength" small amounts of added water exert no influence on the reaction rate (*26*).

The symmetrical halide ion and planar formate ion differ also in their "strength" (i.e. melting point of lithium salts) on 337°C, and in their basicity nearly on 9 units of pK. While for the cubic crystalline lattice of chlorides free unshielded ions as centres of formation adsorption complexes are typical, the formate lattice is flaky (*32*), and the ion itself is planar. That makes difficulty to approach to oxygen atom of formate from any side. These differences result in different reaction mechanisms for formates and halides.

The spherical symmetry of Cl^- ion that provides the freedom of approach from any side permits in turn the formation of the most favourable cyclic complex. In this process a catalyst does not only co-ordinate the reagents, but also solvates the leaving group. Thus, there is no need in its hydration (Scheme 5). Such reactions are low sensitive to water addition.

In case of formate the formation of such complex is impossible and the reaction can proceed by two pathways. One of those is the reaction inside an ion pair (path A), and the second one is the reaction of the unpaired formate ion (path B). In the second case the participation of the solvating agent is necessary.

Scheme 5

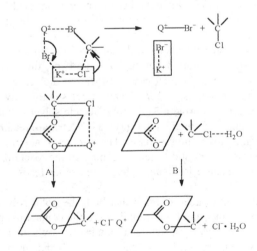

This role can be played by water, that is why the reaction would become strongly water-sensitive.

The influence of solvents upon reactions in solid—liquid system (*34, 35, 37*). The problem of solvent effect on chemical reactions is as old as the chemistry. There was proposed a multitude of methods for estimation of solvent properties. In the book (*33*) it was shown that six main scales: E_T ,Y, $logk_{ion}$, β, Ω, S, are linearly related with the seventh one (Z) and, consequently related with each other. In ref. (*31*) it is proved that all other scales can be reduced to the donor (*DN*) and acceptor (*AN*) numbers. However, it is a purely chemical approach, and some physical properties of solvents such as dielectric constants and Hildebrand solubility parameters and others are not reflected by these scales.

It is well known that solvent can be used as convenient tool to change both the rate and selectivity of the reaction. The stereoselectivity in Horner — Wadsworth — Emmons reaction of aldehydes with derivatives of 3-methyl-4-phosphono-2-butenoic acid was investigated in various solvents with and without crown ether (*34, 35*).

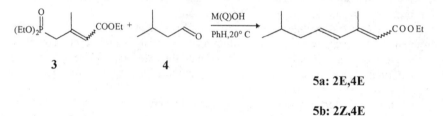

3   **4**

5a: 2E,4E

5b: 2Z,4E

This reaction produced a mixture of isomers. An addition of 18-crown-6 (10 mol %) resulted in an increase in the ratio **5a/ 5b**. In both cases the fraction of the 2*E*,4*E*-isomer in the product **5** increased on passing from hydrocarbons to polar aprotic solvents (Table 2). This trend became even more evident in the presence of crown ether: in this case in all of the tested solvents the 2*E*,4*E* product predominated in a binary mixture.

In the low-polar solvents (from hexane to CH$_2$Cl$_2$) without the crown ether, the 2*E*,4*E* : 2*Z*,4*E* ratios varied but slightly. Perhaps, the reaction occurs on the surface of the solid base. In such case, neither the dielectric constant (ϵ) nor the acceptor and donor numbers (*AN, DN*) can affect significantly the stereochemical result of the interaction of **4** with **3**. In polar aprotic solvents (PhCN, MeCN, DMF, DMSO) the carbanions, formed on the surface of KOH, are readily solvated and thus capable of migrating to the liquid phase. The transfer of anion is enhanced by using PTC. Thus, we can say about different directions of the reaction, one of which leads to 2E,4E isomer (reaction in organic phase) and other leads to 2Z,4E (reaction of interface).

It is important to keep in mind that both the polar solvent and crown ether facilitate the transformation of nucleophile from the contact ion pairs to the solvent-separated ones. Analysis of the data obtained shows that there is no linear correlation between the ratio of 2*E*,4*E*-**5a** to 2*Z*,4*E*-**5b** and any of the polarity criteria (ϵ, *AN,DN*) taken alone.

Table 2.   Effect of the solvent and crown ether additive (10 mol.%) on
the stereochemistry of the reaction between aldehyde **4** and
phosphonate **3**

| Solvent | Polarity parametrs | | | 2E,4E : 2Z,4E ratio in 5 | |
	ε	AN	DN	solidKOH	solid KOH-18-C-6
Hexane	1.9	0.0	0.0	41 : 59	66 :34
Dioxane	2.2	10.8	14.8	46 : 54	58 : 42
Benzene	2.3	8.2	0.0	44 : 56	62 : 38
THF	7.4	8.0	20.0	43 : 57	58 : 42
CH_2Cl_2	8.9	20.4	0.0	40 : 60	71 : 29
Pyridine	12.3	14.2	33.1	49 : 51	62 : 38
PhCN	25.2	15.5	11.9	51 : 49	71 : 29
MeCN	36.0	18.9	14.1	57 : 43	68 : 32
DMF	36.7	16.0	26.6	65 : 35	71 : 29
DMSO	46.7	19.3	29.8	71 : 29	79 : 21

Source: Reproduced with permission from reference 35. Copyright
1995 Russian Academy of Sciences.

AN and *DN* are empirical, nondimentional parametres characterizing the ability of
a solvent to solvate the anions (or the negative poles of molecules) and cations (or
positive poles of molecules), respectively (*36*). SOURCE: Adapted from ref. (*35*).

However, the existence of relationship between the solvent polarity
parameters and the 2E,4E stereoselectivity of the reaction was demonstrated by
employing an equation that takes into account the effects of both *AN* and *DN*:

$$\log(\%2E,4E) = 1.573 + 0.0018(DN) + 0.0088(AN)$$
$$(n = 9; \; r = 0.908)$$

The coefficients in this equation were calculated by applying the equation
borrowed from (*31*) to nine of the ten solvents (except for CH_2Cl_2). They show that
the stabilisation of the anions by solvent is important for the stereochemical outcome
of the reaction.

Presently there are a few systematic studies of solvent influence under PTC.
French investigators (*5*) have studied 15 different solvents in the liquid—liquid
substitution but did not come to any certain conclusion. For the solid—liquid system
with onium salt as the catalyst we have studied the solvent effect in the pseudo first-
order displacement (1) (*37*). The results are listed in Table 3.

In opposite to the liquid—liquid system (*5*), we did not obtained any link
with solvent acceptor numbers. However, the correlation between rate constants of
the displacement and donor numbers was observed. These results confirm our
hypothesis that there is no any transfer of anions in the solid-liquid system with
onium salt and the ions do not effect the course of a reaction.

Table 3.   Pseudo first-order rate constants of the reaction (1) in various solvents and associated paramerers.

No Solvent	$k/10^{-5}\,s^{-1}$	AN	DN	$\log\varepsilon$	P*
1. Pyridine	11.22	11,5	43,0	1.09	0.299
2. Hexametapol	9.33	12.3	38.1	1.471	0,273
3. 1-Methylpyrrolidin-2-one	8.3	14.0	27.3	1.518	0.277
4. N,N-Dimethylformamide	4.95	16.1	25.1	1.565	0.257
5. Acetophenone	1.85	12.9	14.1	1.24	0.311
6. Pentan-2-one	1.39	10.3	12.6	1.23	0.231
7. Benzonitrile	1.15	13.8	12.0	1.401	0,308
8. Tetrahydrofuran	1.11	7.8	20.6	0.869	0.246
9. Nitrobenzene	0.91	13.8	3.7	1.541	0.321
10. Acetonitrile	0.876	18.9	13.2	1.574	0.21
11. Toluene	0.709	3.3	3.9	0.376	0.293
12. Nitromethane	0.667	19.3	4.8	1.555	0.232
13. Ethyl acetate	0.657	8.7	10.9	0.779	0.227
14. Anisole	0.61	7.6	7.9	0.636	0.302
15. Carbon tetrachloride	0.44	8.6	0	0.35	0.274
16. Methanol	0.297	41.3	12.0	1.513	0.203
17. Ethanol	0.15	37.1	6.5	1.386	0.221

* $P = (n^2 - 1)/(n^2 + 2)$.
Source: Reproduced with permission from reference 37. Copyright 1993 Royal Society of Chemistry.

    In the several polar solvents (for example, 1-4) the reaction proceeded with a high rate. In these solvents the less was the accelerating solvent effect (zero rate) the more pronounced was the catalyst influence. Thus, the reaction rate in pyridine (DN = 33.1) did not depend on the catalyst, but there was observed unexplained lag-time not related with the possible solvent quaternizing (Fig. 2).

    In all other solvents the reaction proceeded only in the presence of the catalyst. The dissolution of the solid salt in solvent did not result in the reaction acceleration. The reaction is very slow in alcohols (solvents 16, 17) though these alcohols provide rather high salt solubility.

    It can be seen from Table 3 that there is no certain relationship between the reaction rate and majority of solvent "physical" parameters. Aprotic solvents of low polarity usually provide moderated rates in the displacement.   Sometimes the reaction is faster in solvents with lower dielectric constant thus confirming the hypothesis that PTC reactions are hindered by the dissociation of ion pairs. At the same time there seems to be correlation between $\log k_{obs}$ and polarizability (and subsequently various functions incorporating refraction index).   The analysis of $\log k_{obs}$ via $(n^2 - 1)/(n^2 + 2)$ gives two series: one of which includes all aromatic solvents and CCl_4, and the other one — all the remaining (Fig. 3). "Unusual" influence of CCl_4 is possibly the result of the formation of charge-transfer complexes (41,42).

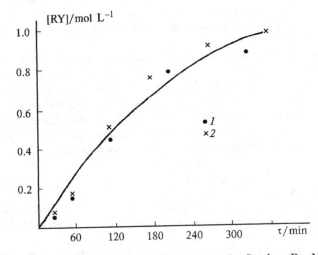

Fig. 2. Kinetic curve of the displacement (reaction 1). Catalyst Bu$_4$NCl, RX = C$_6$H$_{13}$Br, RY = C$_6$H$_{13}$Cl, M$^+$Y$^-$ = KCl, M$^+$X$^-$ = KBr. Solvent pyridine. The curve corresponds to first-order kinetics. *1* — reaction without a catalyst; *2* — catalytic reaction. (Reproduced with permission from reference 8. Copyright 1995 Russian Academy of Sciences.)

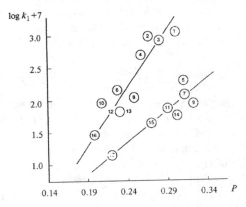

Fig. 3. The effect of the polarizability (P) of the solvents (Table 2) on logk_{obs} in reaction (1). Q$^+$ = Bu$_4$N$^+$. (Adapted from (*37*))

Special effect of aromatic solvent, so-called "benzene effect", was observed earlier (38). Its essence is that benzene in spite of its low dielectric constant effects many chemical reactions just as strongly as polar solvents. The high polarizability of the aromatic ring is believed (39, 40) to be the cause of the "benzene effect". It explains why pyridine that has high donor number DN, along with its aromaticity is the best solvent.

We think that there is observed some dualism in the effect of aromatic solvents on the displacement. On one hand, a solvent acts as a donor-acceptor agent polarizing respective reagents' bonds. On the other hand, the interaction of the aromatic p-electron system with reagents and intermediates is stronger in case of the intermediate. It is connected with the greater charge spreading in the intermediate complex (Scheme 4). Thus, donor and acceptor properties as well as its polarizability should be taken into account in choice of solvent.

Literature Cited

1. Frank-Kamenetsky, D.A. *Diffusia i teploperedacha v khimicheskoi kinetike*, [*Diffusion and heat transfer in chemical kinetics*] Nauka Publ., Moscow, **1987**, 491.
2. Varnek, A. A.; Gol'dberg, A.V. in *Mezhfaznyi kataliz. Novye idei i metody. II Moskovskaya konf. po mezhfaznomu katalizu* [*Phase transfer catalysis. New ideas and methods. II Moscow conference on phase transfer catalysis*], IOCh RAS, Moscow, **1994**, 9
3. Varnek, A. A.; Gol'dberg, A.V.; Danilova, O. I.; Yufit, S. S. in *Mezhfaznyi kataliz. Novye idei i metody. II Moskovskaya konf. po mezhfaznomu katalizu* [*Phase transfer catalysis. New ideas and methods. II Moscow conference on phase transfer catalysis*], IOCh RAS, Moscow, **1994**, 11
4. Yufit, S. S., *Mekhanizm mezhfaznogo kataliza* [*Phase transfer catalysis mechanism*], Nauka Publ., Moscow, **1984**, 265.
5. Antoin, J. P.; de Aguirre, I.; Janssens, F.; Thyrion, F. *Bull. Soc. Chim. Fr.* **1980**, *5—6*, 11.
6. Danilova O. I.; Esikova I. A.; Yufit S. S.; *Kinetika i kataliz* **1990**, *31*, 1484 [*Kinet. Catal.*, **1990**, *31* (Engl. Transl.)].
7. Mason D.; Magdassi S.; Sasson Y. *J. Org. Chem.*, **1990**, *55*, 2714.
8. Yufit, S.S. *Izv. Akad. Nauk, Ser. Khim.*, **1995**, *No.11*, 2126 [*Russ. Chem. Bull.*, **1995**, *44*, 1989 (Engl. Transl.)].
9. Nielsen, A.T.; Houlihan, W.J. *Org. React.*, **1968**, *16*, 1.
10. Lipp, M.;Dallaker, F. *Chem. Ber.*, 1957, *90*, 1730.
11. Wattanasin, S.; Murphy, W.S. *Synthesis*, **1980**, 647.
12. Sarkar, A.; Dey, P.K.; Datta, K. *Indian J. Chem.*, Sect. B, **1986**, *25B*, 656.
13. Dryanska, V.; Ivanov, C. *Tetrahedron Lett.*, **1975**, 3519.
14. Kryshtal, G.V.; Zhdankina, G.M.; Serebryakov, E.P. *Zh. Org. Khim.*, **1994**, *30*, 732 [*J. Org. Chem. USSR*, **1994**, 30 (Engl. Transl.)].
15. Yufit, S.S. *Zhurn. Vsesoyuz. khim. obshch. im. D. I. Mendeleeva*, **1986**, *31*, 134. [*J. of D.I. Mendeleev All-Union Chem. Soc.*, **1986**, *31*, 134 (in Russian)].

16. Afon'kin, A. A.; Shumeiko, A. E.; Popov, A. F. *Izv. AN, Ser. Khim.*, **1995**, No 11 [*Russ. Chem. Bull.*, **1995**, *44*, No. 11 (Engl. Transl.)].
17. Jorgensen, W.; Buckner, J.; Tlinston, Sh.; Rossy, P. *J. Am. Chem. Soc.*, **1987**, *109*, 1891.
18. Arrad, O.; Sasson, Y. *J. Am. Chem. Soc.*, **1988**, *110*, 185.
19. Zahlka, H. A. ; Sasson, Y. *Chem. Comm.*, **1984**, 1652.
20. Ingold, C. K. *Structure and Mechanism in Organic Chemistry*, Cornell University Press, Ithaca, **1969**.
21. Danilova, O. I.; Esikova, I.; Yufit, S. S., *Izv. AN SSSR, Ser. Khim.*, **1986**, 2422 [*Bull. Acad. Sci. USSR, Div. Chem. Sci.*, **1986**, *35*, 2215 (Engl. Transl.)]; **1988**, 314 [*Bull. Acad. Sci. USSR, Div. Chem. Sci.*, **1988**, *37*, 239 (Engl. Transl.)].
22. Esikova, I. A.; Yufit, S. S. *Izv.AN SSSR, Ser. Khim.*, **1988**, 1520 [*Bull. Acad. Sci. USSR, Div. Chem. Sci.*, **1988**, *37*, 1342 (Engl. Transl.)].
23. Yufit, S. S.; Esikova, I. A. *Dokl. AN SSSR*, **1987**, *295*, 621 [*Dokl. Chem.*, **1987**, *295* (Engl. Transl.)].
24. Esikova, I. A.; Yufit, S. S. *J. Phys. Org. Chem.*, **1991**, *4*, 149.
25. Esikova, I. A.; Yufit, S. S. *J. Phys. Org. Chem.*, **1991**, *4*, 341.
26. Yufit, S. S.; Esikova, I. A. *J. Phys. Org. Chem.*, **1991**, *4*, 336.
27. Faustov, V. I.; Yufit, S. S. *Zhurn. fiz. khim.*, **1982**, 2226 [*J. Phys. Chem. USSR*, **1982** (Engl. Transl.)].
28. Yufit, S. S.; Esikova, I. A. *Dokl. AN SSSR*, **1982**, *266*, 358 [*Dokl. Chem.*, **1982**, *266* (Engl. Transl.)].
29. Yufit, S. S.; Esikova, I. A. *Izv. AN SSSR, Ser. Khim.*, **1983**, 47 [*Bull. Acad. Sci. USSR, Div. Chem. Sci.*, **1983**, *32*, 36 (Engl. Transl.)].
30. Mischenko, K. P.; Ravdel', A. A. *Kratkii spravochnik fiziko-khimicheskih velichin* [*Brief handbook of physico-chemical parameters*], Khimiya, Leningrad, **1967**.
31. Schmid, R.; Sapunov, V. N. *Non-Formal Kinetics*, Verlag Chemie, Weinheim, **1982**.
32. Fuess, H. *Acta Cryst.*, **1982**, *B38*, 736.
33. Gordon, A. J.; Ford, R. A. *The Chemist's Companion*, Wiley, New York, **1972**.
34. Kryshtal, G.V.; Serebryakov, E.P.; Suslova, L.M.; Yanovskaya, L.A. *Izv. Akad. Nauk SSSR, Ser. Khim.*, **1990**, 2544 [*Bull. Acad. Sci. USSR, Div. Chem. Sci.*, **1990**, *39*, 2301 (Engl. Transl.)].
35. Kryshtal, G.V.; Serebryakov, E.P. *Izv. Akad. Nauk, Ser. Khim.*, **1995**, No. 10, 2126 [*Russ. Chem. Bull.*, **1995**, *44*, 1785 (Engl. Transl.)].
36. Mayer, U.; Gitmann, V.; Gerger, W. Monatsh. Chem., **1975**, *106*, 1099.
37. Danilova, O. I.; Yufit, S. S. *Mendeleev Commun.*, **1993**, 165.
38. Wada, A. *J. Chem. Phys.*, **1954**, *22*, 198.
39. Eliel, E. L. *Angew. Chem. Int. Ed.*, **1972**, *11*, 739.
40. Eliel, E. L.; Hofer, O. *J. Am. Chem. Soc.*, **1973**, *95*, 8041.
41. Abraham, R. J.; Griffiths, L. *Tetrahedron*, **1981**, *37*, 575.
42. Abraham, M. H.; Abraham, R. J. *J. Chem. Soc. Perkin Trans. II.*, **1975**, *15*, 1677.
43. Esikova, I. A. *Izv. Akad. Nauk SSSR, Ser. Khim.*, **1989**, 2690-2696.

Chapter 6

Phase-Transfer and Micellar Catalysis in Two-Phase Systems

F. S. Sirovski

N. D. Zelinsky Institute of Organic Chemistry, Russian Academy of Sciences, 47 Leninsky prospekt, 117913 Moscow, Russia

Quaternary ammonium salts with a long hydrocarbon chain are a kind of bifunctional catalysts as the reactions in their presence proceed by two catalytic pathways, *i.e.*, phase-transfer and micellar one. The former is inhibited by lipophilic anions such as chlorate etc. The structure — activity relationship for quaternary salts can be described quantitatively using Hansch π-hydrophobicity constants.

The method of phase transfer catalysis (PTC) is applied in organic synthesis for 30 years. The micellar catalysis is much older and more developed, in spite of its more narrow application field. Till present these two types of catalysis were thought to be quite independent ones, although the onium salts are used as catalysts in both cases. Nevertheless, it is not only synthetic organic chemists, usually quite indifferent to the meanders of high theory, who do not realize the existence of the close enough link between these two types of catalysis, but also physical chemists. However, the existence of such link allows one to make a number of theoretical as well as practical conclusions. We intend to demonstrate this link and its practical consequences taking as an example some practically important reactions. First of them are the dehydrochlorination reactions, the kinetics of which we have studied for a number of years. These reactions are not very well investigated under PTC conditions in spite of their practical importance and also are interesting as a model ones. Organic compounds that are able to eliminate hydrogen halide are to a some extent also CH-acids. The deprotonation stage that is the first step of the elimination reaction (for E2H- and E1cb-mechanisms) (*1*) is a common one also for such important reactions as the addition of CH-acids and dihalocarbenes to double bonds. The kinetics of these latter reactions under PTC conditions are not investigated altogether (*2*).

The problems of catalyst activity—quaternary cation structure relationship and PTC—micellar catalysis link are adjoined quite closely. There are but a few attempts to link the catalytic activity of the quaternary ammonium cation (quat) with its structure. At the same time there are data on quat activity as extragents and physiologically active substances. The presence of two immiscible liquid phases is common for these two types of activity. Also there are quite a few statistics on structure—activity relationship for physiologically active substances, and a developed mathematical formalism exists for the structure description (*3—5*). It seemed very interesting to use data from those different fields for obtaining the structure—activity model for phase-transfer catalysts.

1. Noncatalyzed elimination in liquid-liquid systems

For a catalyzed dehydrochlorination in a liquid—liquid system (organochlorine compound — aqueous NaOH) it is quite common to proceed together with an noncatalyzed one. We have studied its kinetics in order to assess its contribution into the overall reaction (*6*). The model compound was 1,1,2,2,3-pentachloropropane (PCP).

In spite of non-homogeneity of the reaction system elimination proceeds not at the interface (as one can suppose on general consideration) but in the aqueous phase, that is confirmed by the first-order kinetics of the reaction in spite of its bimolecularity. Also, as shown in Figure 1 (*6, 7*), the observed rate constants change concurrently with the substrate concentration in the aqueous phase, the latter being determined according to the Sechenov equation (*8*).

$$s = s_0 \cdot \exp[-\lambda C],$$

where s and s_0 are the solubilities of a non-electrolyte in an electrolyte solution and in pure water respectively, C is the concentration of the electrolyte (NaOH in our case), and λ is a salting-out constant.

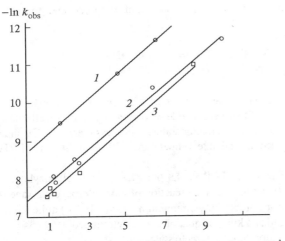

$$[\text{NaOH}]_0 + [\text{NaCl}]_0 / \text{mol L}^{-1}$$

Figure. 1. The effect of the initial concentrations of alkali and sodium chloride on the observed first-order rate constant at *1* — 279 K; *2* — 293 K; *3* — 302 K. (Reproduced with permission from ref. 7. Copyright 1995 Izvestiya Akademii Nauk, Ser. Khim. (Russ. Chem. Bull.).)

In order to determine the activation parameters the observed rate constants were extrapolated to the indefinite dilution of the aqueous phase. The second-order rate constants were determined by dividing these extrapolated constants on the substrate solubility in water. The authenticity of the obtained thus rate constant of the PCP dehydrochlorination was confirmed by the fact that it is satisfactorily correlated (6, 7) together with rate constants for the dehydrochlorination of a number of chloroethanes (9) with Taft's σ^*-constants ($\rho^*=5.73$; correlation coefficient 0.975). It must be underlined that in the latter case data were obtained in a homogeneous system (the reaction was studied in water at substrate concentrations below the solubility limit) (9).

2. Catalyzed dehydrochlorination of chlorinated hydrocarbons in the two-phase systems

The introduction of the catalytic onium salt (QX) into two-phase system results in a sharp change of the observed kinetics. We shall discuss the data on PTC-dehydrochlorination in the presence of short- and long-chained quats separately, because of the essential differences in the observed kinetics. The already published studies of elimination kinetics are somewhat fragmentary. Thus, kinetics of dehydrobromination of bromoethylbenzenes were studied by Sasson and Rabinovitz (10, 11). In the first work (10) the authors came to the conclusion that this reaction under PTC conditions is diffusion-controlled. It is without any doubt the result of the insufficient mixing, because the experiments were performed at evidently insufficient stirring rates of 750 and 1450 rpm, and does not reflect upon the inherent properties of the reaction. The authors themselves admitted that the observed reaction rates strongly depended on stirring speed. In their second work on the topic (11) the authors supposed that the quaternary salt itself may play the role of a base and proposed the following mechanism:

$$RCH_2CH_2X + QX \rightarrow RCH=CH_2 + QX \cdot HX,$$
$$QX \cdot HX_{org} + NaOH_{aq} \rightarrow QX_{org} + NaX_{aq} + H_2O$$

(the subscripts *org* and *aq* relate to the reagents in organic and aqueous phases).

Although the elimination of hydrogen halide by quaternary salts is well known (12), the reaction proceeds at fairly elevated temperatures (not less than 150 °C), and it is quite doubtful for it to proceed at moderate temperatures (70 °C in the case of PTC elimination).

For the dehydrochlorination of PCP in the presence of triethylbenzylammonium chloride (TEBA-Cl) (13) and for the same reaction of hexachlorocyclohexane α-isomer (α-HCCH) in the presence of tetrabutylammonium hydroxide (14) quite similar reaction schemes were proposed and kinetic equations were determined.

In the first case the reaction scheme is as follows:

$$Q^+_{aq} + OH^-_{aq} \xrightleftharpoons{K_I} (Q^+OH^-)_{if} \quad (1)$$

$$Q^+_{aq} + Cl^-_{aq} \xrightleftharpoons{K_{II}} (Q^+Cl^-)_{if} \quad (2)$$

$$(Q^+OH^-)_{if} + CH_2Cl-CCl_2-CHCl_2 \xrightarrow{k_{cat}} CH_2Cl-CCl_2=CCl_2 + (Q^+Cl^-)_{if} \quad (3)$$

(the subscript *if* relates to the interface).

Then the rate of the catalytic reaction would be

$$r_{cat} = k_{cat} [Q^+OH^-][PCP].$$

It is known that the value of K_{II} is bigger than the same of K_I about 10^4 times (15). The following equation was obtained by combined solution of rate and material balance equations:

$$r_{cat} = \frac{k_{cat} K_I \alpha^2 / (1-\alpha)^2 q([PCP]+[PCP]_0 \Delta n)[PCP]}{1 + K_{II}\alpha^2/(1-\alpha)^2([PCP]_0 - [PCP])} \tag{4}$$

where α is the fraction of the organic phase relative to the volume of reaction mass,

$$\Delta n = ([NaOH]_0 - [PCP]_0)/V_{org}[PCP]_0, \quad q = [Q]/\alpha,$$

[Q] is the catalyst concentration in the reaction volume, [PCP] is the PCP concentration in the organic phase. The value of the distribution constant K_{II}, determined independently by the spectrophotometric method was in a good accord with the one obtained from the kinetic measurements, thus supporting the proposed reaction scheme.

For the dehydrochlorination of α-HCCH the scheme of the reaction is

$$QOH_{aq} \underset{k_{-1}}{\overset{k_1}{\rightleftharpoons}} QOH_{org/if} \quad K_1 = k_1/k_{-1} \tag{5}$$

$$QOH_{org/if} C_6H_6Cl_6 \xrightarrow{k} C_6H_5Cl_5 + QCl_{org} \tag{6}$$

$$QCl_{org} \underset{k_{-2}}{\overset{k_2}{\rightleftharpoons}} QCl_{aq}; \quad K_2 = k_2/k_{-2} \tag{7}$$

$$QCl_{aq} + NaOH \underset{k_{-3}}{\overset{k_3}{\rightleftharpoons}} QOH_{aq} + NaCl; \quad K_3 = k_3/k_{-3} \tag{8}$$

where K_1, K_2, K_3, are equilibrium constants.
(The form of the equation does not change if the catalyst is not transferred to organic phase and is fixed at the interface.) So

$$-dC_6H_6Cl_6/dt = K_1 QOH_{aq}C_6H_6Cl_6/[1+(k/k_{-1})\cdot C_6H_6Cl_6], \tag{9}$$

where QOH_{aq} and QOH_{org} are the molar quantities of the quaternary ammonium base in the aqueous and organic phases; $C_6H_6Cl_6$ is the molar quantity of α-HCCH. The shape of the kinetic curves shows that up to high conversion levels $(k/k_{-1})\cdot C_6H_6Cl_6 \gg 1$. Integration gives the equation

$$-\frac{X}{K_3} - \frac{[NaOH]_0(1-\alpha)}{3K_3[HCCH]_0\alpha} \ln\left(\frac{[NaOH]_0(1-\alpha) - 3[HCCH]_0\alpha X]}{[NaOH]_0(1-\alpha)}\right) + X = \frac{k_{obs}q\tau}{[HCCH]_0},$$

$$\tag{10}$$

where X is the conversion of α-HCCH, k_{obs} — is the observed zero-order rate constant, mole/(l·s); $[NaOH]_0$ and $[HCCH]_0$ are the initial concentrations of alkali and α-HCCH in aqueous and organic phases respectively, mole/l; q is the concentration of the catalyst, referred to the organic phase volume, mole/l; α is the fraction of the organic phase relative to the reaction mass volume. Converting this equation, we obtain its linear form (in coordinates X/τ—$X/\tau+(a/\tau)\ln(X/a)$)

$$\frac{X}{\tau} = \frac{1}{K_3}\left(\frac{X}{\tau} + \frac{a}{\tau}\ln\left(\frac{X}{a}\right)\right) + \ln\frac{k_{obs}q}{[HCCH]_0} \tag{11}$$

where $a = [[NaOH]_0(1-\alpha)]/3[HCCH]_0\alpha.$ (12)

But the dependencies of the observed reaction rate constants on initial NaOH concentration in cases of PCP and α-HCCH differ dramatically (see Figs 2. and 3).

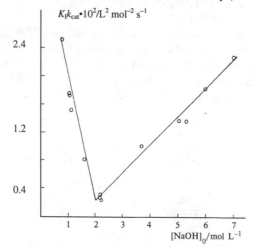

Fig. 2. Dependence of $k_{cat}\cdot K_I$ in the dehydrochlorination of PCP in the presence of TEBA-Cl on the initial NaOH concentration ($T = 293$ K). (Reproduced with permission from ref. 7. Copyright 1995 Izvestiya Akademii Nauk, Ser. Khim. (Russ. Chem. Bull).)

In case of PCP there exists an extremum dependence of $k_{cat}K_I$ (or initial rate) on initial concentration of NaOH (see Fig. 2). The increase in $k_{cat}K_I$ on the left branch of this dependence is explained by the salting-out of the catalyst out of the aqueous phase at high NaOH concentrations. On the opposite, the decrease in NaOH concentration results in the rise of PCP concentration in the aqueous phase, caused by the salting-in effect of the quaternary ammonium salt (8). This in turn results in the increase in overall rate constant, thanks to the increase in the fraction of the reaction in the aqueous phase. The same dependence of the initial rate on initial concentration of NaOH was obtained also in the dehydrochlorination of dichlorobutenes (16) and trichloroethane (17). In Fig. 4 the influence of alkali and quats on solubility of organic reagent in the aqueous phase is shown. It is clearly seen that the small addition of the quaternary salt winds up in the sharp rise of the concentration of this reagent in water, thus leading to the increase in the fraction the reaction, which proceeds in the aqueous phase. In the case of α-HCCH there are no noncatalyzed reaction (in the aqueous phase), so the obtained dependence of reaction rate on initial NaOH concentration reflects the salting-out effect of the latter (Fig. 3).

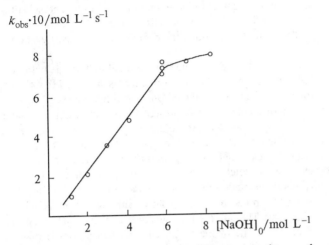

Fig. 3. The effect of the initial NaOH concentration on the observed zero-order rate constant of PTC dehydrochlorination of α-HCCH catalyzed by Bu_4NOH (345.5 K, $[HCCH]_0 = 5.01 \cdot 10^{-2}$ mole/l, $q = 7.7 \cdot 10^{-5}$ mole/l). (Reproduced with permission from ref. 7. Copyright 1995 Izvestiya Akademii Nauk, Ser. Khim. (Russ. Chem. Bull).)

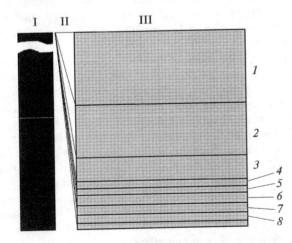

Fig.4 The solubility of the organochlorine compound in water in the presence of alkali and quaternary ammonium salt. $[Bu_4NBr^+]$/mole·l^{-1}: 0.25 (*1*), 0.1 (2), 0.05 (*3*), 0.01 (*4*); H_2O (*5*); [NaOH]/mole·l^{-1}: 2 (*6*), 5 (*7*), 10 (*8*). I — organic phase, II — interface, III — aqueous phase. (Reproduced with permission from ref. 7. Copyright 1995 Izvestiya Akademii Nauk, Ser. Khim. (Russ. Chem. Bull).)

So one can make a conclusion that the extremum dependence of the reaction rate on the initial alkali concentration should be common for all compounds that are reactive enough to eliminate the hydrogen chloride without the catalyst. The substitution of the activity of the alkali instead of its concentration did not change the appearance of the obtained dependence, thus confirming the deciding role played by salting-out effect.

In the recent work of Makosza (18), devoted to the alkaline PTC dehydrobromination of alkyl bromides in the presence of co-catalysts, it was supposed that the lypophilic alkoxide RO$^-$, formed as the result of the by-process of hydrolysis of the substrate, or from the admixtures of alcohol already present in the latter, plays a major role in this reaction. It is a fact that the promoting effect of alcohols in PTC elimination was observed by a number of investigators (19, 20). The authors (18) have supposed that having in view the extremely low lipophility of quaternary ammonium bases the above-mentioned effect is solely responsible for the proceeding of the PTC elimination. But this hypothesis is undermined by the fact that the PTC alkaline dehydrochlorination of α-HCCH proceeds quite rapidly, although there is no by-process of hydrolysis in this case. The authors themselves further have admitted that their explanation is valid only in some of the cases (21) and had proposed that the intensive stirring, resulting in sharp increase in the area of the interface where the quaternary ammonium hydroxide is absorbed, can shift the position of the extraction equilibrium in which the latter participates. So the extractability of such hydroxides under static and dynamic conditions can differ considerably. The data obtained have confirmed this hypothesis only in part. So, under static conditions Bu$_4$NBr does not practically change its anion for the hydroxide, but under dynamic conditions this exchange does proceed for 3—4 %. The authors (21) came to the conclusion that this effect is also insufficient for the explanation of the possibility of PTC elimination but did not offer any other explanation.

The dependencies observed in the course of the PTC elimination, catalyzed by quats with surfactant properties, completely differ from the ones discussed above. This can be readily demonstrated by carrying out the alkaline elimination in the presence of inorganic salts with the anions that "poison" the phase-transfer catalyst, i.e. sodium chlorate. This salt is the most common impurity in the aqueous alkali (especially in the technical grade). If the catalyst is TEBA-Cl, then the reaction stops completely even at moderate concentrations of the chlorate. But if the catalyst possesses the properties of the surfactant, such as, PhCH$_2$N$^+$ (CH$_2$CH$_2$OH)$_2$(C$_{12}$H$_{25}$—C$_{16}$H$_{37}$)Cl$^-$ (Katanol), than the reaction rate does not drop to zero (16, 22). In Fig. 5 are shown the kinetic curves of 3,4-dichlorobutene-1 dehydrochlorination catalyzed by Katanol in the presence of ions, that poison usual phase-transfer catalysts. It is seen that even at high chlorate concentrations the inhibition of the catalyst is not complete. This fact indicates the existence of the second catalytic pathway, i.e. the micellar one.

It is known that surfactants, unlike the phase transfer catalysts, can catalyze reactions in strongly diluted solutions that occur, as a rule, in the absence of the bulk organic phase in so-called pseudo-homogeneous systems. The already published works(23, 24) on elimination catalyzed by the long-chain quats, were performed at exactly same conditions, i.e. in the absence of the bulk organic phase. Thus there was investigated

(23) the reaction of 1-bromo-2-phenyl propane with NaOH catalyzed by $C_{16}H_{33}N^+(CH_3)_3$ Br^-:

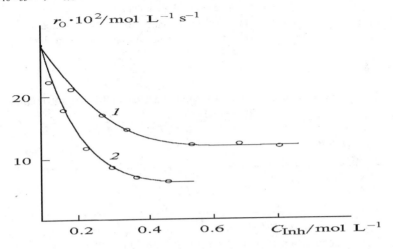

Fig. 5. Dependence of the initial rate of the dehydrochlorination of 3,4-dichlorobutene on the concentration of the inhibitor (C_{inh}) (*16, 22*) (catalyst — Katanol, $4.24 \cdot 10^{-3}$ mole·l^{-1}, 303 K): *1* — Cl$^-$; *2* — ClO$_3^-$. (Reproduced with permission from ref. 7. Copyright 1995 Copyright 1995 Izvestiya Akademii Nauk, Ser. Khim. (Russ. Chem. Bull).)

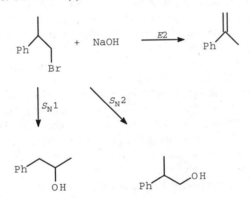

The catalytic and noncatalytic reaction rate constants on each pathway were determined. It is interesting to note that in the presence of the catalyst with the hydroxy group in a β-position (*24*) the reaction rate is much higher. The obtained results (*23, 24*) are listed in Table I.

These data show that cationic micelles are selective catalysts of the *E2*-reactions. This effect was explained (*23*) by the stabilization of the negative charge of the *E2*-transition state by high positive charge of the cationic micelle. The growth of *E2*-reaction rate can also be explained by the decrease in the microscopic dielectric constant inside the micelle. The dependence of the reaction rate on the catalyst concentration is the characteristic S-like curve. In the second work (24) of the same authors it was shown that the change of the methyl group for the β-hydroxyethyl one in the catalyst molecule results in the notable increase in the catalytic effect together with the inhibition of the by-processes (see run 3 in Table I). The authors (24) assumed that it is the result of the decrease in the activity of the substrate and increase in hydroxide activity.

Table I. The Rate Constants of The Reaction of 1-Bromo-2-Phenyl Propane with NaOH (23, 24)

Run	Catalyst Concentration, $mole \cdot l^{-1}$	$k_{E2} \cdot 10^4$ s^{-1}	$k_{SN2} \cdot 10^4$ s^{-1}	$k_{SN1} \cdot 10^4$ s^{-1}
1	0	0.55	0.10	0.35
2	10^{-2}	2.55	0.22	0.22
3[a]	10^{-2}	16.20	0.60	0.25

[a] Catalyst —$C_{16}H_{33}N^+(CH_3)_2CH_2CH_2OHBr^-$ (24)[1].
[1] SOURCE: (Reprinted with permission from ref. 7. Copyright 1995 Izvestiya Akademii Nauk, Ser. Khim. (Russ. Chem. Bull).)

It was interesting to investigate whether a long-chain quat— $PhCH_2N^+Me_2(C_{12}H_{25}—C_{16}H_{33})Cl^-$ (Katamin AB) would catalyze the dehydrochlorination reaction at low alkali concentrations in a true two-phase system and not in a pseudo-homogeneous one. This is important also from the practical viewpoint.

The investigation of the α-HCCH dehydrochlorination (25) revealed that unlike TEBA-Cl Katamin AB can catalyze the reaction even at pH 8—10, when the noncatalyzed reaction does not proceed. The reaction rate at the constant pH fell in time parallel to the increase in Cl⁻ concentration in the aqueous phase. The likewise lowering of solubility of non-electrolytes in water with the increase in NaCl concentration was observed elsewhere (8). The reaction proceeds in the micelles of the catalyst in the aqueous phase, the α-HCCH being solubilized in its trichlorobenzene (TCB) solution. The fall in reaction rate is caused by salting-out of the latter from the aqueous phase by NaCl. The micellar type of the catalysis is also indicated by the extremum dependence of the rate constant on the catalyst concentration (Fig. 6).

The rate of the micellar catalyzed second-order reaction is described by the equation (26—28)

$$r = k_m [C_6H_6Cl_6]_m [OH]_m (C_{cat} - CMC)V, \qquad (13)$$

where k_m is the second-order reaction rate constant in the micellar pseudo-phase, $[C_6H_6Cl_6]_m$, $[OH]_m$ are the concentrations of the reagents in the same phase, V is the molar volume of the catalyst, CMC is the critical micellar concentration.

One can usually determine the concentrations of the reagents in the micellar pseudo-phase, starting from the hypothesis of their distribution between the micellar and aqueous phases. In our case the problem is additionally complicated by the presence of the bulk organic phase. Assuming that the concentration of α-HCCH in TCB on transition from the organic phase to the micellar one does not change, we obtain:

$$P_{OH^-} = [OH]_m/[OH]_{aq}, \tag{14}$$
$$P_{TCB} = [TCB]_m/[TCB]_{aq}, \tag{15}$$
$$P_{NaOH} = [NaOH]_m/[NaOH]_{aq}, \tag{16}$$

where P is the coefficient of the distribution of the reagents between the micellar and aqueous phases.

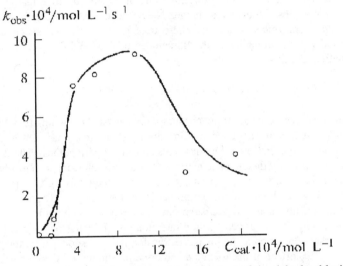

Fig. 6. Dependence of the observed rate constant of the dehydrochlorination of α-HCCH on the concentration of the catalyst. (Reproduced with permission from ref. 7. Copyright 1995 Izvestiya Akademii Nauk, Ser. Khim. (Russ. Chem. Bull).)

The reagents' material balance is described by the following equations:

$$[NaOH] = (C_{cat} - CMC)V]([NaOH]_{aq} + [OH^-]_{aq}) + V_c (C_{cat} - CMC)(NaOH]_m + [OH^-]_m \tag{17}$$
$$[C_6H_6Cl_6] = [1 - (C_{cat} - CMC)V][C_6H_6Cl_6]_{aq} + V (C_{cat} - CMC)[C_6H_6Cl_6]_m \tag{18}$$

Considering that $(P - 1)V = K$, where K is the constant of the solubilization, we obtain:

$$r = \frac{k_m K_{TCB} K_{OH} (C_{cat} - CMC)[NaOH][HCCH]\exp[-\lambda C]}{\left\{1 + K_{TCB}(C_{cat} - CMC)\right\}\left\{1 + K_{OH}(C_{cat} - CMC)\right\}}. \tag{19}$$

If [HCCH] = const. and [NaOH] = const., than

$$r = k_{obs}\exp[-\lambda C], \ \lambda=4.3,$$

where C is the concentration of NaCl in the aqueous phase.

The value of K_{TCB} determined by the independent procedure is in quite satisfactorily agreement with the one obtained from the kinetic calculations.

The obtained results demonstrate that the kinetic model of micellar catalysis that was developed for the pseudo-homogeneous systems is valid also for the heterogeneous ones. It is quite possible that catalytic effect of long-chain onium salts in alkaline dehydrochlorination at low concentrations of NaOH (up to 5 mass %) is brought about by micelles. Evidently at higher NaOH concentrations there are two catalytic pathways —the micellar and the phase transfer one. We had successfully applied this model for rationalizing the kinetic results of 3,4-dichlorobutene-1 dehydrochlorination (*29, 30*).

$$CH_2ClCHClCH=CH_2 + NaOH \rightarrow CH_2=CClCH=CH_2$$

The kinetic curves of the substrate consumption consisted of two parts. At first there was a sharp fall in substrate concentration during short time, than it was slow consumed according to the first-order kinetics. This confirms our hypothesis about two catalytic pathways caused by the dualistic nature of the onium salt catalyst. The phase-transfer pathway is blocked in the course of the reaction by the evolving Cl$^-$ ion. The micellar pathway is caused by the surfactant properties of the catalysts. The rate of the reaction neglecting the noncatalyzed one is the sum of rates along both pathways.

The catalysts providing for the larger reaction rate along the micellar pathway are the most active. The rate of reaction along the phase transfer pathway shortly after the start sharply falls due to the effect of the so-called "chloride poisoning" of the catalyst. So the substrate is consumed mainly thanks to the reaction along the micellar pathway.

Thus one can conclude that the long-chain quaternary onium salts are a kind of a missing link, which can bind PTC and micellar catalysis thanks to their "bifunctionality". In other words, they can be both micellar and phase transfer catalysts simultaneously.

The obtained results allow us to propose a kind of phase diagram showing the different fields of catalysis by various quaternary ammonium salts (Fig. 7). The realm of pure PTC is confined between chain length of 2 to 8 carbon atoms and NaOH concentration higher 1 mole/l. The field of the purely micellar catalysis extends from the chain length of 8 carbon atoms and is bordered by NaOH concentrations from 0.01 to 1.5 mole/l. Between these two fields the scarcely investigated wilderness extends where two types of catalysis coexist. Surely, this map is very rough, but nevertheless it can be helpful in the choice of the catalyst and reaction conditions. This diagram also makes it clear that the long-chain quats are more active catalysts, as they display catalytic activity in a wider range of alkali concentrations due to their dualistic catalytic properties. So, in general, the elimination proceeds by three pathways: noncatalyzed (in water), PTC-catalyzed and micellar-catalyzed.

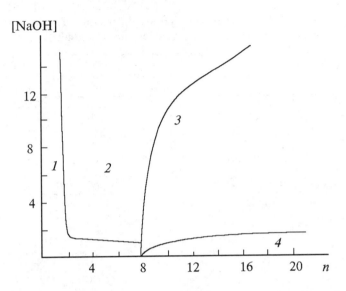

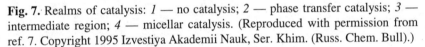

Fig. 7. Realms of catalysis: *1* — no catalysis; *2* — phase transfer catalysis; *3* — intermediate region; *4* — micellar catalysis. (Reproduced with permission from ref. 7. Copyright 1995 Izvestiya Akademii Nauk, Ser. Khim. (Russ. Chem. Bull).)

3. Some Practical Applications of the Catalytic Properties of the Long-Chain Quats.

The dual catalytic nature of the long-chain quats sometimes leads to very peculiar consequences. Thus in the following reaction

that proceeds in the liquid—liquid system it was not TEBA-Cl that was the best catalyst (Yagniukova, Z., Sirovski F., Kalinina L., L'vova N., unpublished results) but surfactant quats.

It is seen from Table II that compounds with surfactant properties are better catalysts than a conventional quat.

The specific properties of long-chain quats were exposed also in the liquid/solid system. Investigation of kinetics of two-phase catalyzed acetylation of vitanin A half-product showed that different results are obtained with TEBA-Cl and Katamin AB as catalysts.

Table II. Activity of Different Catalysts in Phosphorylation (Yagniukova, Z., Sirovski F., Kalinina L., L'vova N., unpublished results)

Catalyst	Catalyst concentration, % mass	Initial rate, $mole \cdot (l \cdot min)^{-1}$	Catalyst activ in relation to PEG-5
$[(C_8H_{17}-C_{10}H_{21})_3N^+-Me]_2SO_4^{2-}$	1.0	2.83	17.1
$PhCH_2N^+Me_2(C_{12}H_{25}-C_{16}H_{33})Cl^-$	1.0	3.90	21.0
$PhCH_2N^+ (CH_2CH_2OH)_2(C_{12}H_{25}-C_{16}H_{37})Cl^-$	3.0	3.25	12.2
$Et_3N^+CH_2PhCl^-$	1.0	1.19	2.0
$C_9H_{19}C_6H_4O(CH_2CH_2O)_nH$ ($n=9$—10)	0.5	0.66	5.5
$HO(CH_2CH_2O)_5$	3.0	1.73	1.0
dicyclohexano-18-crown-6	1.0	1.90	5.1

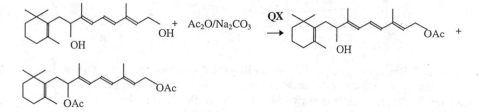

The dependence in case of the surfactant quat is shown in Figure 8. In case of TEBA the usual linear dependence was observed (Sirovski, F., Bobrova, E. unpublished results).

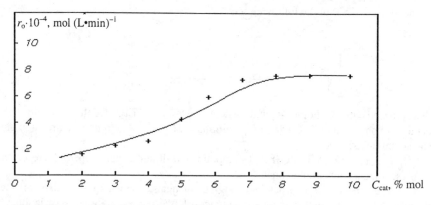

Fig. 8. The catalyst influence in solid/liquid acetylation

The surfactant quats also displayed different properties in the reaction of *p*-xylene with NaOCl (*31*).

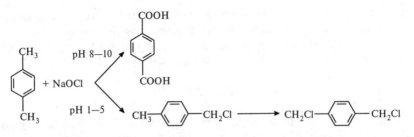

As shown in the above scheme the reaction proceeded along two pathways, depending on pH. But this effect was observed only in the presence of Katamin AB. On the opposite, Bu₄NOH weakly catalyzed chlorination and did not at all effect oxydation. Possibly the micellar effects also influence the reaction direction as well as its rate.

4. Catalysts' Structure—Activity Relationship in PTC

The problem of catalysts' structure—activity relationship is a central one in the theory of PTC. We would leave out the also very important question of the counter-ion influence and are going to concentrate the attention on the influence of a cation structure. Some investigators (2, 32) have operated with the coefficient of the distribution of the quat between the phases of the reaction mass using it for the estimation of the catalyst hydrophilic-lipophilic balance (33). Fukunaga and co-workers (34) proposed a fresh but arguable approach to the problem using for this aim the Hildebrand (35) solubility parameter δ. For the calculation of the catalyst hydrophilic-lipophilic balance $D(\delta)$ they used the equation (34)

$$D(\delta) = (\delta - \delta_{aq})^2/(\delta - \delta_{org})^2, \qquad (22)$$

where δ, δ_{aq}, and δ_{org} are respectively the solubility parameters of the catalyst (quat), water and the organic solvent. It is the matter of fact that $D(\delta)$ is somewhat linked with the change in the standard free energy of the catalyst on its transition from one phase to another. This change is the natural and quite sensitive parameter that reflects the alteration of the catalyst structure. The partial free energy of the solute in terms of the theory of regular solutions is expressed by the following equation (35):

$$\Delta G_2 = RT \ln X_2 + V_2 \varphi_1^2 (\delta_2 - \delta_1)^2, \qquad (23)$$

where X_2, V_2, and δ_2 are respectively the molar fraction, the molar volume, and the solubility parameter of the solute; δ_1 and φ_1 are respectively the solubility parameter and the volume fraction of the solvent.

However the link between the values of $D(\delta)$ and reaction rate constants is more qualitative than quantitative. The scarcity of values of δ for quats also restricts the possibilities of the wide application of Fukunaga's equation (23) for the picking of the catalyst. It is also quite doubtful that the theory of the regular solutions, that was destined for the solutions of nonelectrolytes, would correctly describe the interactions with the participation of the quats, which are usually strong electrolytes.

The parameter reflecting the change in the standard free energy of the catalyst on its phase transfer is the distribution coefficient. It is known (36), that

$$\mu^0_{org} - \mu^0_{aq} = RT \ln P', \qquad (24)$$

where μ^0_{org} and μ^0_{aq} are the standard chemical potentials of the solute in the organic and the aqueous phase; P' is the «thermodynamic» distribution coefficient (the ratio of the molar fractions of the solute in organic and aqueous phases).

Nevertheless the distribution coefficient is determined by the whole quat, although it is known that isomeric quats display different catalytic activity. So this integral parameter is not always fitting for the description of the catalyst structure. Beside this, the distribution coefficients of the same quat determined in different solvents vary strongly. In some solvents due to their ion-selective properties the distribution coefficient strongly depends on the anion. In order to eliminate the vagueness caused by this effects we had used the Hansch' π-constants of hydrophobicity for the description of the quat structure. These constants are defined analogously to Hammett and Taft constants (36):

$$\pi_X = \log P_X - \log P_H, \qquad (25)$$

where P_H is the distribution coefficient for the standard compound; P_X is the same for its derivative with the X substitutient in the standard 1-octanol—water system that is low ion-selective in relation to halide and hydroxide ions.

The cation of the quat can be pictured as consisting of two fragments, *i.e.* the hydrophilic ammonium "head" and the hydrophobic hydrocarbon "tail". This predetermines the choice of the two p-constants (π_{R_3N+} for the 'head" and $\pi_{R'} = n \cdot \pi_{CH_2}$ for the "tail") for the description of the quat structure. The values of the π-constants were calculated according to the equation (25) from the known in the literature (36—38) distribution coefficients for quats. As the result we obtained the values of the slope (π_{CH_2}), and the intercept (π_{R_3N+}) (fig. 9). Table III lists the values of the hydrophobicity constants for the various ammonium "heads" (39).

The determined by us values of π_{CH_2} somewhat differ from the most common value of 0.5 found by other investigators (π_{CH_2}).(8, 36) This can be explained by the conformation effect or the «folding» of the long hydrocarbon chain (36). In order to validate our approach we have treated the kinetic data published in a few works that compare the activity of different quats.

In the known work (32) of Herriott and Picker such a comparison was made in the following model reaction:

$$C_8H_{17}Br + PhSNa \rightarrow PhS\overset{\shortmid}{C}_8H_{17} + NaBr,$$

which was carried out in the benzene—water system. The treatment of their results by the method of the multi-parameter correlation yielded the following equation (39):

$$\ln k_2 = 1.41 \pm 0.64 + (1.33 \pm 0.18)\,\pi_{R_3N+} + (0.94 \pm 0.29)\,\pi_{R'} - (0.21 \pm 0.04)\,\pi^2_{R_3N+}, \qquad (26)$$

where k_2 is the second-order reaction rate constant, $l \cdot (\text{mole} \cdot \text{s})^{-1}$; the corrected for the degrees of freedom correlation coefficient is 0.969; the root mean square is 0.76.

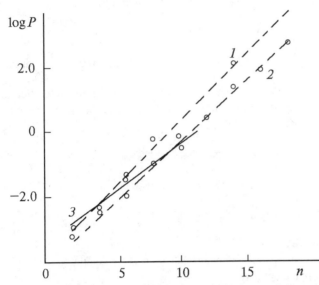

Fig. 9. Dependence of the distribution coefficients of the quaternary ammonium salts on the length of the alkyl radical in 1-octanol—water system: *1* — salts of the $C_6H_5CH_2N(Me)_2C_nH_{2n+1}$ type; *2* — salts of the PyC_nH_{2n+1} type; *3* — salts of the $Me_3NC_nH_{2n+1}$ type. (Reproduced with permission from ref. 7. Copyright 1995 Izvestiya Akademii Nauk, Ser. Khim. (Russ. Chem. Bull).)

Table III. Hydrophobicity constants of different functional groups

Ammonium «head»	π_{R_3N+}	π_{CH_2}
Me_3N^+	−3.91	0.35
Et_3N^+	−3.00	0.35
Bu_3N^+	−0.80	0.35
$(C_8H_{17})_3N^+$	4.49	0.35
$PhCH_2N^+Me_2$	−4.09	0.45
	−4.28	0.39
	−4.66	0.35^1

[1]   SOURCE: (Reprinted with permission from ref. 7. Copyright 1995 Izvestiya Akademii Nauk, Ser. Khim. (Russ. Chem. Bull).)

Other investigators (40) have studied the reaction in the solid—liquid system:

$$AcONa_{solid} + PhCH_2Cl_{org} \xrightarrow{\text{Q+X-}} AcOCH_2Ph + NaCl$$

The treatment of their data yielded other correlation equation (39):

$$\ln k_2 = (1.15\pm0.20)\, \pi_{R_3N+} + (1.05\pm0.15)\, \pi_{R'}, \tag{27}$$

(the corrected for the degrees of freedom correlation coefficient is 0.986, the root mean square is 0.55).

It is interesting to note that if the same reaction is carried out in the liquid—liquid (41) system and not in a solid—liquid one than the quadric member $\pi^2_{R_3N+}$ appears again (comp. eqs. (27) and (28)).

$$\ln k_2 = -2.85 + 0.19(\pi_{R'} + \pi_{R_3N+}) - 0.11\, \pi^2_{R_3N+} \tag{28}$$

The equation of the same type was obtained for the dehydrochlorination of PCP (42).

$$\log k_{cat} K_I = -(1.80\pm0.47) + (0.11\pm0.06)\, \pi_{R_3N+} + (0.53\pm0.16)\, \pi_{R'} - (0.049\pm0.016)\, \pi^2_{R_3N+}, \tag{29}$$

the root mean square is 0.47.

The presence of the quadric member also means that for the given reaction there exists the quat with the optimal π_{R_3N+} value.

The same very interesting parabolic dependence of the catalyst activity on the length of its hydrocarbon chain was obtained by Sasson, Halpern and co-workers (43) in the study of allylbenzene isomerisation in the liquid—liquid system catalyzed by the quaternary salts of R_4NX type. This dependence confirm our conclusion that there exist quaternary salts with the optimal value of the distribution coefficient P.

Halpern had also suggested another parameters for the rough estimation of the catalyst efficiency (44, 45). One of them is simply C# — i.e., the number of C atoms in the longest hydrocarbon chain. One can see that is proportional to $\pi_{R'}$. The other one is q, defined as

$$q = \Sigma 1/C\#$$

for all hydrocarbon chains. There were obtained interesting dependencies of conversion degrees and rate constants on this parameter. They would be discussed in more detail in Halpern's chapter. (46).

The generality of our «thermodynamic» approach was authenticated by investigation of the quats catalytic activity in the reaction in the protic solvent (47). The rate of the reaction that proceeds in an aprotic solvent rises with the increase in the catalyst hydrophobicity (2). We supposed that in a protic solvent it would be just the opposite, i.e. the quat catalytic activity would rise with the increase in its hydrophilicity. As the model we choose the oxidation of durene to pyromellitic acid on cobalt bromide catalyst in acetic acid in the presence of various quaternary ammonium bromides as co-catalysts. It is known from the literature that amines and ammonium salts accelerate the oxidation of alkyl aromatics (48).

The formal scheme of the reaction is as follows:

$$1 \xrightarrow{k_1} 2 \xrightarrow{k_2} 3 \xrightarrow{k_3} 5,$$

$$4 \xrightarrow{k_4} 3,$$

where **1** is the sum of isomeric 4,6-dimethylisophthalic and 2,5-dimethylterephthalic acids, **2** is the corresponding methylcarboxyphthalide, **3** is 5-methyl trimellitic acid, **4** is 4,5-dimethylphthalic acid, **5** is pyromellitic acid.

The rate constant of the rate-limiting step of the formation of 5-methyltrimellitic acid was used as the measure of the catalytic activity. Also log P was used as the measure of the hydrophobicity (hydrophilicity) because there was no sense in dividing the cation of symmetric quats, used as catalysts, on fragments. The obtained values of log k_2 / k_2^{NaBr} were plotted against the calculated values of log P (k_2^{NaBr} is the same constant of the formation of 5-methyltrimellitic acid in the presence of NaBr instead of the quat). The k_2 / k_2^{NaBr} ratio is the measure of the quaternary cation catalytic activity relative to sodium ion.

The obtained dependence was satisfactorily described by the equation (47)

$$\log(k_2 / k_2^{NaBr}) = -(4.88 \pm 0.85) \cdot 10^{-2} \log P + (1.59 \pm 0.43) \cdot (\log P)^2 \qquad (30)$$

with the correlation coefficient equal to 0.95. That confirms the assumption that catalytic activity of the quats in protic solvents rises with the increase in the hydrophilicity of the former. The presence of the quadric member in the obtained relationship also demonstrates the importance of the hydrophobic interactions in this case. The similar oxydation of *p*-xylene under PTC conditions was also studied by Sasson et al. (49).

The obtained data show that the relationships of the quat catalytic activity—lipophilicity in the protic and the aprotic solvents are the mirror reflections of each other (compare signs of the quadric members in eqs. 26, 28, 29, and 30).
It should be noted that the like parabolic relationships link the biologic activity of the quats with their structure (3—5).

$$\log(1/C) = a_1 (\log P)^2 + a_2 \log P + a_3, \qquad (31)$$

This activity was expressed by $1/C$, where C is the molar concentration of the active substance that is necessary for obtaining a certain physiological response (3—5). There were also determined the values of $(\log P)_{max}$ for the maximum biologically active quats. The comparison of these values with the ones for $(\pi_{R_3N+})_{max}$ in the PTC reactions demonstrated their closeness.

The necessity of the introduction of the quadric member was shown in the works using the multi-cell approximation (50—52). The authors have used the following transport model:

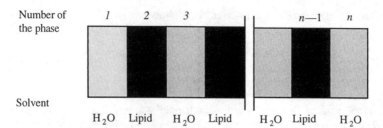

Number of the phase: *1* *2* *3* ... *n—1* *n*

Solvent: H_2O Lipid H_2O Lipid H_2O Lipid H_2O

It was shown (*50—52*) that the solution of the system of rate equations describing the transfer of the active molecule form any aqueous phase to the neighboring lipid phase yields the parabolic function.

What is the physical cause of the obtained parabolic relationships? In all probability they are the result of the interplay of the two opposite trends. Small highly hydrophilic cations are located mainly in the aqueous phase. On the opposite, large cations are located in the organic phase. The phase transfer is the most easy at some optimal intermediate size of the cation, thus determining the parabolic form of the structure—activity relationship.

The similitude of the mathematical model of PTC in liquid—liquid systems and the same of the quaternary salts interaction with the biological substrates reflects, in our opinion, the likeness of the mechanisms of these processes. The transfer of the quat molecule from the water (protein) phase to there lipid one accompanies its interaction with the biological substrates. Thus the PTC reactions in the liquid—liquid systems in a very approximate degree mimic the processes in biological membranes.

In the conclusion, the author sincerely hopes that this work would serve to attract attention to the following problems.

The first of them is the interrelation of the micellar and phase-transfer catalysis and the observed duality of properties of the long-chain quats.

The second problem is the frequently obtained extremum (most often parabolic) dependence of the catalyst effectiveness (alias the catalytic rate constant) on the size of the quat cation. The proposed method of the preliminary estimation of the extractability of the catalytic quat using its π-constants of hydrophobicity (we mean just the extractability neglecting the inherent catalytic activity, which is called intrinsic activity) is simple enough to be used by synthetic organic chemists.

The third problem is the connection of the PTC reactions with processes in biological membranes. The existence of this link again shows us that everything in the nature is interrelated.

Acknowledgements

The author wishes to express his sincere gratitude to all his coworkers, especially to Dr. S. Velichko, Dr. O. Grechishkina, Dr. Z. Yagniukova, Dr. V. Revyakin, Mrs. E. Bobrova, Mrs. A. Kamyshanova (Voronkina), Mrs. M. Panova, Miss S. Pilygaitseva for their conscientious work and friendly attitude. Special thanks to Prof. Yu. Treger who 20 years ago turned the attention of the author to quaternary salts.

References

1. Banthorpe, D. V. *Elimination reactions*; Elsevior: London, 1973.
2. Dehmlow, E. V.; Dehmlow, S. S. *Phase Transfer Catalysis*; 2nd rev. ed.; Verlag Chemie: Weinheim, 1987.
3. van Rossum, J. M.; Ariens, L. J. *Arch. Int. pharmacodyn.*, **1959**, *118*, 447.
4. Hansch, C.; Clayton, J. M. *J. Pharm. Sci.*, **1973**, *6*, 1.
5. Lien, E. J.; Ariens, E. J.; Beld, A. J. *Europ. J. Pharm.*, **1976**, *35*, 245.
6. Sirovsky, F. S.; Velichko, S. M.; Treger, Yu. A.; Chimishkyan, A. L.; Koptelov V. D. *Kinet. Catal.*, **1985**, *26*, 8 (Engl. Transl.).
7. Sirovsky, F. S.. *Russ. Chem. Bull.*, **1995**, *44*.
8. Sergeeva, V. F., *Russ. Chem. Rev.*, **1965**, *34*, 309 (Engl. Transl.).
9. Walraevens, R.; Trouillet, P.; Devos, A., *Int. J. Chem. Kinetics*, **1974**, *6*, 777.
10. Halpern, M.; Sasson, Y.; Rabinovitz, M. *J. Org. Chem.*, **1984**, *49*, 2011.
11. Halpern, M.; Zahalka, H.; Sasson, Y.; Rabinovitz, M. *J. Org. Chem.*, **1985**, *50*, 5088.
12. Lutz, E. F.; Kelly, J. T.; Hall, D. W. Brit. pat. 1112068, 1968.
13. Sirovsky, F. S.; Velichko, S. M.; Neimark, A. V.; Treger Yu. A.; Chimishkyan, A. L.; Kheifets, L. I., *Kinet. Catal.*, **1985**, *26*, 11 (Engl. Transl.).
14. Sirovsky, F. S.; Velichko, S. M.; Panova, M. V.; Treger, Yu. A.; Chimishkyan, A. L. *Kinet. Catal.*, **1985**, *26*, 731 (Engl. Transl.).
15. Dehmlow, E.; Slopianka, M., Heider, J. *Tetrahedron Lett.*, **1977**, 2361.
16. Levanova, S. V.; Revyakin, V. A.; Sirovsky, F. S.; Rodova, R. M.; Martirosyan, G. T.; Asatryan, E. M.; Malkhasyan A. Ts. *Zhurn. Vsesoyz. Khim. o-va im. D. I. Mendeleeva*, **1986**, *31*, 237.
17. Sirovsky, F. S.; Treger, Yu. A.; Panova, M. V.; Voronkina, A. V. *Zhurn. Vsesoyz. Khim. o-va im. D. I. Mendeleeva*, **1985**, *30*, 580.
18. Makosza, M.; Lasek W. *Tetrahedron*, **1991**, *47*, 2843.
19. Dehmlow, E.; Thieser, R.; Sasson, Y.; Neumann, R., *Tetrahedron*, **1986**, *13*, 3569.
20. Shavanov, S. S.; Tolstikov, G. A.; Shutenkova, T. V.; Viktorov, G. A. *J. Gen. Chem.*, **1987**, *7* (Engl. Transl.).
21. Lasek, W.; Makosza M. *J. Phys. Org. Chem.*, **1993**, *6*, 412.
22. Revyakin, V. A.; Zabotin, A. L.; Levanova, S. V. *Kinet. Catal.*, **1991**, *32* (Engl. Transl.).
23. Lapinte, C.; Viout, P. *Tetrahedron Lett.*, **1973**, 1113.
24. Gani, V.; Lapinte, C.; Viout, P. *Tetrahedron Lett.*, **1973**, 4435.
25. Sirovsky, F. S.; Skibinskaya, M. B.; Berlin, E. R.; Treger, Yu. A.; Stepanova, N. N.; Molodchikov, S. I. *Kinet. Catal.*, **1986**, *27* (Engl. Transl.).
26. Bunton, C. A., *Catal. Rev. Sci. Eng.*, **1979**, *20*, 56.
27. Romsted, L. S., in *Micellization, solubilization, and microemulsions*, 1, 2, Editor Mittal, K. L. Plenum Press: New York, London, 1977.
28. Berezin, I. V.; Martinek, K.; Yatsimirsky, A. K. *Russ. Chem. Rev.*, **1973**, *42*, 787 (Engl. Transl.).
29. Revyakin, V. A.; Levanova, S. V.; Sirovsky, F. S. *Kinet. Catal.*, **1988**, *28* (Engl. Transl.).
30. Revyakin, V. A.; Levanova, S. V.; Semochkina, N. N.; Sirovsky, F. S. *Kinet. Catal.*, **1990**, *31* (Engl. Transl.).
31. Sirovsky, F. S.; Kamyshanova, A.V. *Khim. Prom-st*, **1990**, 330.
32. Herriott, A.; Picker, W. *J. Am. Chem. Soc.*, **1975**, *97*, 2345.1
33. Griffin, W. G. *J. Soc. Cosmetic Chemists*, **1943**, *1*, 311.
34. Fukunaga, K.; Shirai, M.; Ide, S.; Kimura M. *J. Chem. Soc. Jap., Chem. and Ind. Chem.*, **1980**, 1148.
35. Hildebrand, H.; Scott R. L. *Regular solutions*, Prentice Hall: New York, 1962.

36. Leo A.; Hansch, C.; Elkins, D. *Chem. Revs.*, **1971**, *71*, 525.
37. Czapkiewicz-Tutaj, B.; Czapkiewicz, J. *Rocz. Chem.*, **1975**, *49*, 1353.
38. Zapior, B.; Czapkiewicz-Tutaj, B. *Zesz. nauk. Univ. Jagellonskiego, Chemia*, **1972**, *17*, 87.
39. Sirovsky, F. S.; Velichko, S. M.; Treger, Yu. A.; Chimishkyan, A. L. *Russ. J. Phys. Chem.*, **1983**, *57* (Engl. Transl.).
40. Yadav, D.; Sharma, M. M. *Ind. Eng. Chem. Process Des. Dev.*, **1981**, *20*, 385.
41. Wang, L. K. *J. Am. Oil Chem. Soc.*, **1975**, *52*, 339.
42. Sirovsky, F. S.; Velichko, S. M.; Treger, Yu. A.; Chimishkyan, A. L.; Panova, M. V.; *Kinet. Catal.*, **1985**, *26* (Engl. Transl.).
43. Halpern, M.; Sasson, Y.; Rabinowitz M. *J. Org. Chem.*, 1983, **48**, 1022.
44. Starks, C. M.; Liotta, C. L.; Halpern, M. *Phase-Transfer Catalysis. Fundamentals, Applications and Industrial Perspectives;* Chapman & Hall: New York, London, 1994.
45. Halpern, M. *PTC Communication,* **1995**, *1*, 1.
46. Halpern, M. *This Book*, p. 000.
47. Mulyashov, S. A.; Sirovsky, F. S.; Grechishkina, O. S. *Kinet. Catal.*, **1989**, *30* (Engl. Transl.).
48. Pat. 157932 (CSSR), *R. Zh. Khim.* **1977**, 9Í156Í.
49. Dakka, J.; Zoran A.; Sasson, Y. Eur. Pat. Appl. EP 318399 (1989) [CA 111:134924].
50. Penniston, J. T.; Beckett, L.; Bentley, D. L.; Hansch, C. *Mol. Pharmacol.*, **1969**, *5*, 333.
51. Dearden, J. C.; Townsend M. S. *Proceedings of Symposium on Chemical Structure—Biological Activity Relationships, Quantitative Approaches.*, Suhl, G.D.R., October 1976.
52. McFarland, J. W. *J. Med. Chem.*, **1975**, *13*, 1192.

Chapter 7

A New Interfacial Mechanism for Asymmetric Alkylation by Phase-Transfer Catalysis

Irina A. Esikova[1], Theresa Stines Nahreini, and Martin J. O'Donnell[2]

Indiana University–Purdue University at Indianapolis,
Indianapolis, IN 46202

The kinetics of alkylation of the Schiff base esters of amino acids in the presence of *cinchona* salts demonstrate features similar to those of enzyme-promoted reactions: variable orders, substrate saturation, catalyst inhibition and non-linear Arrhenius-type plots. Interfacial reactions result in the formation of two catalyst-substrate complexes. The mechanism involves deprotonation of the Schiff base substrate by hydroxide ion. A tight coordination of the substrate by electrostatic interaction with the quaternary nitrogen atom of the *cinchona* salt provides a favorable chiral environment for the asymmetric alkylation.

Asymmetric synthesis, the "artificial" preparation of optically active substances, is one of the major challenges facing organic chemistry. Phase transfer catalysis (PTC) is an attractive method for catalytic asymmetric synthesis, especially if the chiral phase transfer catalysts can be derived directly from inexpensive precursors. Asymmetric phase transfer reactions have been recently reviewed by one of us (1).

In previous publications (2-6) it was shown that the PTC alkylation of Schiff base esters of amino acids with alkyl halides in the presence of *cinchona*-derived salts as catalyst (Scheme 1) typically proceeds with an enantioselectivity (% ee) of 40-65%, depending on the conditions and type of catalyst used. Furthermore, it was demonstrated that under the reaction conditions the OH-group of the *cinchona*-derived catalyst is converted *in-situ* to the O-alkylated form (5,6). Two important properties of the benzophenone Schiff base esters, such as **1** or **2**, have been exploited for the synthesis of α-amino acids: selective monoalkylation and the resistance of the monoalkyl derivatives to racemization (7).

We report here further studies concerning optimization, the kinetics and the mechanism of this important reaction (8).

Process Optimization

Under standard reaction conditions (9), the starting Schiff base (**1**, Z_2CH_2) was converted to product (**2**, Z_2CHR) in 30-60 minutes. The chemical yield of the major products [(R)-**2** and (S)-**2**] is 83%, and approximately 17% of benzophenone is produced. It was established that the non-catalytic alkylation is 17 times slower than alkylation in the presence of catalyst.

[1]Current address: Chiron Corporation, 4560 Horton Street, Emeryville, CA 94608
[2]Corresponding author

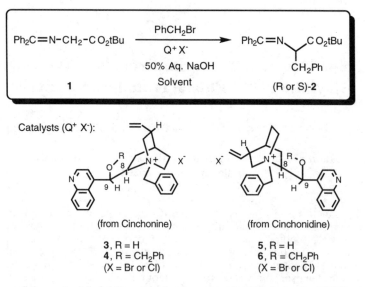

Scheme 1. Model Reaction for Asymmetric Alkylation by PTC.

The rate of alkylation of **1** in the presence of 50% aqueous NaOH and catalyst **6** is strongly dependent on the speed of stirring of the reaction mixture, a dependence described by an S-shaped curve. It was observed that the ratio of volumes of the aqueous to the organic phase is very important. Increasing this ratio from 0.1 to 0.25 doubles the rate of the reaction. When this ratio is more than 0.25, the rate of alkylation is practically independent of the ratio. Increasing either the speed of stirring or the relative amount of alkali presumably increases the contact surface area between the phases, which would cause an increased alkylation rate. This implies that the reaction consists of one or more interfacial steps.

Table I.   Influence of Different Bases and Solvent Systems

Base	Solvent (v/v)	Time (h)	% Chem. Yield	% ee
50% aq. NaOH	CH2Cl2	1	100	61
"	PhMe	2	100	69
"	70% PhMe/30% CH2Cl2	0.5	85	70
KOH	70% PhMe/30% CH2Cl2	4	75	77
KOH/K2CO3 (1:1)	"	4	72	76
K2CO3	"	4	0	--
CsF	PhMe	6	19	64

Reaction conditions: 25 °C; [Z2CH2] = [PhCH2Br] = 0.1 M, [**6**] = 0.01 M; organic solvent = 10 mL. Volume 50% aqueous NaOH / volume organic = 2.5 / 10. Amount solid bases: KOH, 8 g. pellets; K2CO3, 8 g. powder; KOH / K2CO3 (1:1), 4 g. each; CsF, 4 equivalents based on Z2CH2.

The influence of different solid bases (K_2CO_3, KOH or a 1:1 mixture of these two bases) was also studied (Table I). Using solid KOH, alone or mixed with K_2CO_3, resulted in a higher enantioselectivity than that obtained with 50% aqueous NaOH. Interesting results were obtained when solid CsF was used as base. No alkylation was observed with CH_2Cl_2 as solvent because the solvent reacted with the base. Several other solvents were tried, but, because of poor catalyst solubility, the rate of alkylation with CsF was very slow. Importantly, however, the enantioselectivity for the alkylation in the presence of CsF was practically the same as that obtained with 50% aqueous NaOH under similar conditions.

The reactivity and enantioselectivity of the different catalysts are dependent on their solubility in the organic phase and their stability under the PTC reaction conditions (Table II). It was found that decomposition of the *cinchona* salts in the

Table II. Catalysts: Solubility, Stability, Rate and Enantioselectivity

Catalyst	Solubility, M		Cat. Decomp.		$r_0 \times 10^4$	% ee*
	PhMe	CH_2Cl_2	PhMe	CH_2Cl_2	(Ms-1)*	
3 (X = Cl)	10^{-4}	10^{-3}	--	--	2.9	67% (R)
3 (X = Br)	10^{-4}	10^{-2}	<1%	10%	6.2	62% (R)
4 (X = Br)	10^{-3}	10^{-1}	0	0	10.2	71% (R)
5 (X = Cl)	--	--	--	--	22.0	49% (S)
5 (X = Br)	10^{-4}	10^{-1}	5%	33%	19.0	52% (S)
6 (X = Br)	10^{-3}	10^{-1}	3.5%	10%	9.3	67% (S)

Reaction conditions (see Table I). *Reactions in PhMe/CH_2Cl_2 = 70/30 (v/v).

presence of 50% aqueous NaOH in the non-polar solvent toluene is considerably slower than in CH_2Cl_2. Unfortunately, the use of pure PhMe is not practical because of the low solubility of the catalysts in this solvent. However, the situation can be improved by using a mixture of PhMe and CH_2Cl_2. The composition of the binary solvent influences both the rate and the enantioselectivity of the alkylation. Addition of a small amount of PhMe increases the rate of alkylation. However, when the amount of PhMe in the mixed solvent equals or exceeds 30%, the reaction rate decreases. In pure PhMe, the rate is four times slower than in the mixed solvent (30% PhMe/70% CH_2Cl_2). The enantioselectivity is also sensitive to the presence of PhMe in the mixed solvent. For example, the % ee is increased from 61% to 70% in going from pure CH_2Cl_2 to the optimal solvent mixture, 70% PhMe/30% CH_2Cl_2 (see Table I). Such an effect of the nonpolar aromatic component of the solvent is likely connected not only with the increased stability of the *cinchona* salt but also with the effect of the non-polar solvent on the transition state of the reaction.

Studies of different catalysts derived from the *cinchona* salts showed that asymmetric induction is possible in the presence of catalysts with either of a free -OH group in the position beta to the quaternary nitrogen or one in which this position is blocked by an alkoxy group (5,6) (Table II). In the free -OH catalyst series, catalyst **3**, from cinchonine, was less active than catalyst **5**, from cinchonidine. Replacement of the -OH group in the chiral catalysts by an -OBenzyl group increased the reactivity of the cinchonine-derived catalyst (r_0, **4** > **3**) and decreased the reactivity of the cinchonidine-derived catalyst (r_0, **6** < **5**). These catalysts (**3** or **4**) and (**5** or **6**) gave up to 71% ee and 67% ee for (R)-**2** and (S)-**2**, respectively. The O-alkylated forms of the catalyst are preferable to the free -OH forms because they give higher enantioselectivities and they are more stable to the reaction conditions.

Variation of the reaction temperature also affects the level of asymmetric induction. The alkylation in the mixed solvent (70% PhMe/30% CH_2Cl_2) showed that the chiral induction in the presence of catalyst **6** increased from 67% to 81% when the temperature was decreased from 25 °C to +5 °C (Scheme 2). This optimized result is comparable to the best result obtained in the Merck indanone alkylation using a cinchonidinium salt (10).

Scheme 2. Optimized Conditions for Catalytic Enantioselective PTC Alkylation.

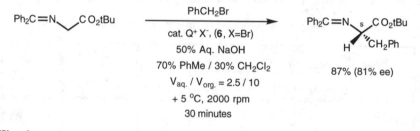

Kinetics

The kinetics of alkylation were studied in the presence of 50% aqueous NaOH in CH_2Cl_2 as solvent (Table III). Analysis of the dependence of the initial reaction rates

Table III. Dependence of the Reaction Rate on Reagent Concentrations

Z_2CH_2 (M)	RBr (M)	QX (M)	r_0 10^4 (Ms^{-1})	Z_2CH_2 (M)	RBr (M)	QX (M)	NaBr (M)	Z_2CHR (M)	r_0 10^4 (Ms^{-1})
0.02	0.10	0.01	2.3	0.10	0.41	0.01			10.0
0.05	0.10	0.01	5.4	0.10	0.47	0.01			9.9
0.10	0.10	0.01	8.3	0.10	0.49	0.01			10.1
0.20	0.10	0.01	9.0	0.10	0.10	0			0.62
0.30	0.10	0.01	10.2	0.10	0.10	0.002			5.4
0.40	0.10	0.01	10.3	0.10	0.10	0.005			6.6
0.43	0.10	0.01	11.2	0.10	0.10	0.01			8.3
0.47	0.10	0.01	9.8	0.10	0.10	0.02			10.1
0.10	0.01	0.01	2.4	0.10	0.10	0.023			10.0
0.10	0.04	0.01	4.8	0.10	0.10	0.034			12.9
0.10	0.06	0.01	5.8	0.10	0.10	0.045			13.0
0.10	0.11	0.01	8.3	0.10	0.10	0.01	0.03		5.9
0.10	0.22	0.01	8.7	0.10	0.10	0.01	0.10		4.1
0.10	0.32	0.01	11.0	0.10	0.10	0.01	0.20		2.9
0.10	0.36	0.01	9.6	0.10	0.10	0.01		0.10	8.9

Reaction conditions: 20 °C; catalyst **6**; $V_{aq.}/V_{org.}$ = 1.5/10; mechanical stirring at 5800 rpm.

(r_0) on the reagent concentrations results in variable orders with respect to the Schiff base substrate (Z_2CH_2), benzyl bromide (RX) and chiral catalyst (QX). These data are in agreement with a mechanism of alkylation that involves formation of an intermediate with the participation of both the Schiff base ester and benzyl bromide.

A working model for the mechanism of the PTC alkylation is based on formation of an ion pair adsorbed on the interface. Interaction of this adsorbed ion pair (**TC**, ternary complex) with alkyl halide gives a new adsorbed intermediate (**QC**,

quaternary complex), which then reacts to give alkylated product. This process is
depicted in Scheme 3.

Scheme 3. Working Model for Alkylation Mechanism.

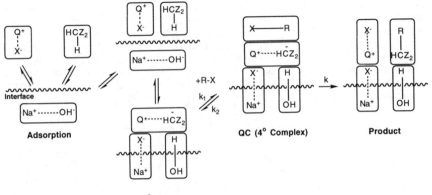

TC (3° Complex)

Formation of **TC** from the reagents Z_2CH_2, QX and NaOH is an equilibrium
process characterized by K_{eq}.

Assuming that the intermediates **TC** and **QC** are at steady state, the rate law for
this process includes the concentrations of the substrate (Z_2CH_2), alkyl bromide,
catalyst and base both in the numerator and in the denominator [Equation (1)] (11):

$$r_0 = \frac{kB[Z_2CH_2]_0[RBr]_0[QX]_0[NaOH]_0}{1 + B[Z_2CH_2]_0[RBr]_0\{[QX]_0 + [NaOH]_0\}} \qquad \text{where } B = K_{eq}\left(\frac{k_1}{k+k_2}\right) \quad (1)$$

$$
\begin{aligned}
k &= 0.77 \text{ sec}^{-1}\\
[NaOH] &= 1.2 \times 10^{-3} \text{ M}\\
B &= 2.3 \times 10^5 \text{ M}^3
\end{aligned}
$$

$$\frac{1}{r_0} = \frac{1}{kB[QX]_0[NaOH]_0[RBr]_0} \cdot \frac{1}{[Z_2CH_2]_0} + \frac{[QX]_0 + [NaOH]_0}{k[QX]_0[NaOH]_0} \qquad (2)$$

$$\frac{1}{r_0} = \frac{1}{kB[QX]_0[NaOH]_0[Z_2CH_2]_0} \cdot \frac{1}{[RBr]_0} + \frac{[QX]_0 + [NaOH]_0}{k[QX]_0[NaOH]_0} \qquad (3)$$

$$\frac{1}{r_0} = \left\{\frac{1}{k} + \frac{1}{kB[NaOH]_0[Z_2CH_2]_0[RBr]_0}\right\}\frac{1}{[QX]_0} + \frac{1}{k[NaOH]_0} \qquad (4)$$

The reaction rate is not changed under certain initial concentrations of Z_2CH_2, RBr,
QX and NaOH. By inverting both sides of Equation (1), Equations (2) - (4) are
obtained. Plots of $1/r_0$ vs. $1/[Z_2CH_2]_0$ and $1/r_0$ vs. $1/[RBr]_0$ are linear and do not
extrapolate to the origin (Figure 1), which implies substrate saturation typical of
enzyme-catalyzed reactions.

Different initial concentrations can be used to find the values of k and B shown
in Equation (1). Based on the value of the constant B, characterizing the steps leading

to **QC**, the adsorption step is favorable. Additionally, the reactivity of the **QC** is high ($k = 0.77$ s^{-1}).

The limiting factor in this reaction is a low concentration of NaOH molecules on the interface. The rate of reaction is proportional to the number of NaOH molecules at the interface and available for contact with the organic phase. It is suggested that the concentration of NaOH be determined as the ratio of the number of moles of NaOH on the interface to the volume of the organic phase. The numerical estimation of this concentration is difficult. It is assumed that a constant concentration of NaOH can be established at the interface and that this varies according to the volume of the 50% aqueous NaOH and/or the mixing speed. In the present case, by applying the numerical solution of Equations (2) - (4) based on the primary kinetic data, a value of 1.2×10^{-3} M was obtained as the concentration of NaOH at the interface. This value is comparable to the concentration of the catalyst in the reaction.

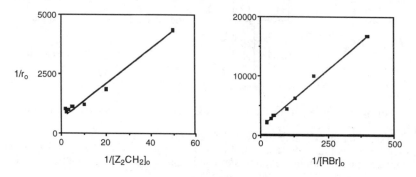

Figure 1. Plots of $1/r_0$ vs. $1/[Z_2CH_2]_0$ and $1/r_0$ vs. $1/[RBr]_0$.

Interfacial alkylation involving relatively low concentrations of NaOH at the interface might be inhibited by product (11). However, a noticeable effect on the reaction rate was not observed on addition of alkylated product (Z_2CHR) (Table III). But, added NaBr slowed the alkylation, and there is a linear dependence of $1/r_0$ vs. $[NaBr]_0$. This inhibition is likely due to the adsorption of NaBr on the surface of the 50% aqueous NaOH and a competition between the Schiff base ester and NaBr for the NaOH. These data provide further support for the adsorption mechanism.

The above kinetic data provides the first evidence for the formation of adsorption intermediates under liquid/liquid PTC conditions. A similar mechanism of catalysis for solid/liquid PTC was reported earlier (12).

Thermodynamic and Activation Parameters of Alkylation

Analysis of the temperature dependence of the reaction by using the Arrhenius equation gives a non-linear dependence (Figure 2). This could be due to two factors: the complex character of the observed rate constant: $k_{obs} = f(K_{eq}, k_1, k_2, k)$ which includes both equilibrium and rate constants and/or an exothermic equilibrium process (12).

Variation of the reaction temperature can have different effects on the constants for individual steps of the alkylation. Because of this, Equation (1) could change so that at low temperature the activation energy (8.95 kcal/mole) is that of the step involving destruction of **QC** while at high temperature (0.25 kcal/mole) it also involves the enthalpy of formation of **QC** (12). Using this approach, it is possible to estimate the activation parameters for the alkylation (Table IV) using the following equations: $\Delta H^{\ddagger} = E_a - RT$; $\Delta G^{\ddagger} = -RT \ln k = \Delta H^{\ddagger} - T\Delta S^{\ddagger}$.

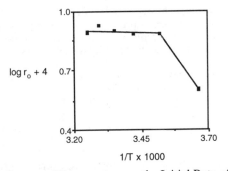

Figure 2. Influence of Temperature on the Initial Rate of Alkylation.

The estimation of the enthalpy of formation of **QC** was obtained from the observed energies of activation at low and high temperature [$\Delta H = E_{high} - E_{low}$] (12) and the entropy of formation was obtained from $\Delta G = \Delta H - T\Delta S$. The negative enthalpy and entropy for formation of **QC** are related to the adsorption step.

Table IV. Formation of **QC** (4° Complex) and Activation Parameters

Formation of QC		Activation Parameters for QC Destruction	
ΔH	-9.2 kcal/mole	E_a	8.9 kcal/mole
ΔG	-7.3 kcal/mole (298°)	$\Delta H^{\ddagger}$	8.4 kcal/mole
ΔS	-24.4 eu (298°)	$\Delta G^{\ddagger}$	0.15 kcal/mole (298°)
		$\Delta S^{\ddagger}$	27.7 eu (298°)

Mechanism of Alkylation

Many features of the kinetics of alkylation of the Schiff base esters discussed above (variable orders, substrate saturation, inhibition by NaBr, non-linear Arrhenius plots) are similar to those of enzyme-promoted reactions (13). Also, as in enzymatic reactions, electrostatics provide a major driving force.

Alkylation of the Schiff base ester of an amino acid involves cleavage and formation of several chemical bonds between various atoms of the substrate, alkyl halide, chiral catalyst and base. In such a reaction it is preferable to have a strong cooperativity between all reagents. The chiral catalyst adsorbed on the interface of 50% aqueous NaOH is able to react with the substrate in such a way that the starting materials are transformed to products because of their mutual interaction. Such structural conformity is a necessary condition for a fast and selective reaction.

Rearrangement on the interface results in the formation of the catalyst-substrate complex (**TC**), an adsorbed ion pair. Interaction of this complex with the second reagent (alkyl halide) gives a new adsorbed intermediate (**QC**), destruction of which yields the alkylated product. Thus, the mechanism of catalysis involves formation of two catalyst-substrate complexes on the interface.

The "enzyme mimic" in this case, which has been generated under PTC conditions, is a molecule of the *cinchona* salt adsorbed on the interface as **QX•NaOH**. This "enzyme mimic" has several functions during alkylation. First, hydroxide ion deprotonates the substrate (Z_2CH_2). Second, the quaternary nitrogen atom of the catalyst cation forms an ion pair with the substrate anion and assists in delocalizing the charges on the atoms of the **TC**. The third and most important property of this species is to provide a chiral environment for the anion of the substrate, resulting in a "penetrated ion pair" in which the asymmetric alkylation can occur (14,15).

In summary, the mechanism of the PTC alkylation demonstrates many of the properties inherent in heterogeneous processes such as adsorption on the interface, fluctuation of the reagents and complex formation. In addition, this reaction shares many common features with typical enzyme-catalyzed processes.

Acknowledgments. We gratefully acknowledge the National Institutes of Health (GM 28193) for support of this research. We wish to thank Professors C. A. Bunton, R. Larter, and S. Sen and Dr. J. Magrath for helpful discussions.

References

1. O'Donnell, M. J., "Asymmetric Phase-Transfer Reactions." In *Catalytic Asymmetric Synthesis*; I. Ojima, Ed.; VCH: NY, 1993, Chapter 8.
2. O'Donnell, M. J.; Bennett, W. D.; Wu, S. *J. Am. Chem. Soc.* **1989**, *111*, 2353-2355.
3. Lipkowitz, K. B.; Cavanaugh, M. W.; Baker, B.; O'Donnell, M. J. *J. Org. Chem.* **1991**, *56*, 5181-5192.
4. O'Donnell, M. J.; Wu, S. *Tetrahedron: Asymm.* **1992**, *3*, 591-594.
5. Esikova I. A.; Nahreini, T.; O'Donnell, M. J. 206th ACS National Meeting, Chicago, IL, August, 1993, ORGN 288.
6. O'Donnell, M. J.; Wu, S.; Huffman, J. C. *Tetrahedron* **1994**, *50*, 4507-4518.
7. O'Donnell, M. J.; Bennett, W. D.; Bruder, W. A.; Jacobsen, W. N.; Knuth, K.; Leclef, B.; Polt, R. L.; Bordwell, F. G.; Mrozack, S. R.; Cripe, T. A. *J. Am. Chem. Soc.* **1988**, *110*, 8520-8525.
8. For a related mechanistic paper and two recent reviews, see: (a) Lasek, W.; Makosza, M. *J. Phys. Org. Chem.* **1993**, *6*, 412-420; (b) Landini, D.; Maia, A. *Gazz. Chim. Ital.* **1993**, *123*, 19-24; (c) Atherton, J. H. *Res. Chem. Kinet.* **1994**, *2*, 193-259.
9. Preparation of starting materials and catalysts, analysis of the reaction mixture and the determination of enantiomeric excess have been described (2,4,6). All experiments were done in a three-necked flask, submerged in a water bath in which the temperature was controlled to ±0.5 °C by a Haake DI recirculator. Mechanical stirring was performed with a Barnant Model 700-5400 Series 10 Mixer at 5800 rpm. The initial rate was estimated using the method described by Algranati (16).
10. (a) Dolling, U. -H.; Davis, P.; Grabowski, E. J. J. *J. Am. Chem. Soc.* **1984**, *106*, 446-447; (b) Hughes, D. L.; Dolling, U. -H.; Ryan, K. M.; Schoenewaldt, E. F.; Grabowski, E. J. J. *J. Org. Chem.* **1987**, *52*, 4745-4752.
11. Equation (1) was written according to general rules in the following: (a) Hammett, L. P. *Physical Organic Chemistry. Reaction Rates, Equilibria and Mechanisms*, McGraw-Hill: N.Y., 1970; (b) Schmid, R.; Sapunov, V. N. *Non-Formal Kinetics*, Verlag Chemie: Weinheim, 1982; (c) Berezin, I. V.; Klesov, A. A. *Practical Course in Chemical and Enzyme Kinetics*, Moscow State University: Moscow, 1976.
12. (a) Esikova, I. A.; Yufit, S. S. *J. Phys. Org. Chem.* **1991**, *4*, 149-157; (b) Esikova, I. A.; Yufit, S. S. *J. Phys. Org. Chem.* **1991**, *4*, 341-345; (c) S. S. Yufit, S. S.; Esikova, I. A. *J. Phys. Org. Chem.* **1991**, *4*, 336-340.
13. (a) Jencks, W. P. *Catalysis in Chemistry and Enzymology*, McGraw-Hill: N. Y., 1969; (b) Bowden, E. C., *Principles of Enzyme Kinetics*, Butterworths: Boston, 1976.
14. Pochapsky, T. C.; Stone, P. M.; Pochapsky, S. S. *J. Am. Chem. Soc.* **1991**, *113*, 1460-1462.
15. Boche, G. *Angew. Chem., Int. Ed. Engl.* **1992**, *31*, 731-732.
16. Algranati, I. D. *Biochim. Biophys. Acta* **1963**, *73*, 152-155.

Chapter 8

Integrated Guideline for Choosing a Quaternary Ammonium Salt as a Phase-Transfer Catalyst To Enhance Reactivity and Separation

Marc E. Halpern[1]

PTC Technology, 1040 North Kings Highway, Suite 627, Cherry Hill, NJ 08034

An oversimplified, though useful, guideline is described for choosing a quaternary ammonium ("quat") salt as a phase-transfer catalyst during the screening stage in the evaluation of a new phase-transfer catalysis candidate application. Choosing quat structure to simultaneously enhance reactivity and catalyst separation can be guided by anticipating the identity of the rate determining step and by deciding ahead of time the desired mode of separation (extraction, distillation, etc.). A variety of process demands can be accommodated by the guideline.

Many factors must be taken into account when choosing a quaternary ammonium ("quat") salt as a phase-transfer catalyst. These factors include reactivity, separation of the catalyst from the product, ability to induce selectivity, catalyst stability, cost, availability, toxicity, environmental and others factors. This article will describe a useful, though somewhat oversimplified, guideline for choosing a quaternary ammonium salt as a phase-transfer catalyst which integrates theoretical and empirical aspects of reactivity and catalyst separation. The guideline presented may be used as an aid in screening phase-transfer catalysts at the outset of the evaluation stage of a project. The guideline is not intended to provide final optimal reactions conditions.

Background

Choosing a phase-transfer catalyst for a new application can be confusing. Makosza published scores of successful high yield PTC syntheses using NaOH and TEBA (triethyl benzyl ammonium chloride).[1,2] On the other hand, TEBA induces little or no reactivity at all in many other successful PTC/NaOH reactions.[3,4,5,6] In a classic paper, Landini convincingly showed[7] that "the effectiveness of a phase-transfer catalyst depends mainly on its organophilicity, with other structural factors much

---

[1]E-mail address: halpern@phasetransfer.com

less important." On the other hand, a study on the effect of the structure of 19 symmetrical and non-symmetrical quats on an alkylation reaction[8] showed that quats with shorter alkyl chains gave higher reactivity than quats with longer alkyl chains. Then again, in Chapter 6 of this book, Sirovski shows that quats containing a mixture of at least one long chain with several short chains, provide highly desirable results. In addition, some reactions which should be obvious PTC candidates do not seem to work with any catalyst. And finally, since many chemists just do not have the time to screen dozens of phase-transfer catalysts for every application, many of them choose TBAB (tetrabutyl ammonium bromide) out of convenience. This is not a bad choice since previous trial and error shows that TBAB usually works, more or less, and it can be washed out of the organic phase at the end of the reaction. One excellent kinetic study[9] explained the choice of catalyst as follows: "Unfortunately, a universal guideline is unavailable for selecting the proper phase-transfer catalyst to enhance the reaction rate. Thus, commonly commercial phase-transfer catalysts, such as tetra-n-butylammonium bromide, etc., are employed in the present study." Excellent discussions of the criteria for choosing a phase-transfer catalyst have been published by Starks and others.[10,11,12]

Who is right? They are all right! In other chapters in this book, Starks and Dehmlow point out that there cannot be a single catalyst or a single guideline that is optimal for all PTC applications since each application is different. They further point out that it would be erroneous to assume a single mechanism to be valid for all PTC reactions. It should therefore not be surprising that the choice of phase-transfer catalyst is not an off-handedly trivial decision. Nevertheless, in the real world, chemists *do* have to start projects and the choice of catalyst is usually the first, and possibly the psychologically most burdensome task. The purpose of the following discussion is to provide a guideline, which for the first time, integrates some of the key criteria for choosing a phase-transfer catalyst, into a single decision tree. Even though this guideline integrates only two factors (reactivity and separation), its value lies in using it as a starting point when initiating research for new applications of phase-transfer catalysis. This guideline should help reduce the development time during the screening stage by offering an alternative to screening every phase-transfer catalyst in the lab supply catalog. It should be noted that every PTC application will have its own individual set of performance requirements and this limited integrated guideline will not be sufficient to predict the fully optimized choice of catalyst. A further purpose of presenting this integrated guideline is to use its simplicity to entice chemists to screen PTC a lot more often than before, when confronted with their new synthetic or industrial process development challenge.

Discussion

Reactivity and catalyst separation are the two main factors which must be satisfied in order to perform a successful PTC reaction. Once the structural properties of the catalyst are defined which meet the desired reactivity and separation criteria, the other catalyst requirements (cost, availability, waste treatment, etc.) can usually be superimposed on the structure requirements. Special cases exist in which catalyst structure needs to meet special selectivity or thermal stability criteria. These special cases must be evaluated on a case-by-case basis and cannot be subjected to a single simple general guideline. However, when initiating many new PTC evaluation programs, a good place to start is to screen catalysts for the combination of reactivity and separation. The discussion will begin with separate treatments of reactivity and separation and will conclude with the integrated guideline.

Reactivity. As Starks pointed out in Chapter 2 of this book, conceptually, the rate determining step(s) of a PTC reaction may be considered to be "transfer," "intrinsic reaction," some combination of the two or even another reaction step for more complicated PTC systems, such as in most hydroxide reactions. The effect of quat structure on reactivity will be different for transfer rate limited reactions ("T-reactions") than for intrinsic reaction rate limited reactions ("I-reactions"). T-reactions usually represent reactions in which the overall rate is limited by the *transfer* of a reactant (usually an anion) from one phase to another (usually from an aqueous or solid phase into an organic or third phase). I-reactions usually represent reactions in which the overall rate is limited by a chemical reaction, in which *covalent bonds are actually formed and/or broken*, which takes place in the reaction phase, which is usually the organic or third phase. The essence and nature of the action of bond forming/breaking is quite different from the essence and nature of the action of transfer. Since (1) the phase-transfer catalyst actively participates in transfer and participates in or is in close proximity to bond forming/breaking and (2) the natures of transfer and bond forming/breaking are so different, then the structural attributes of a quat which are required to enhance the reaction rate of an I-reaction are very likely to be different from the structural attributes which would enhance the reaction rate of a T-reaction. In addition, transfer rate and intrinsic reaction rate are both affected by concentration. The interaction between concentration and quat structure is different for T-reactions and I-reactions. For all of these reasons, it is not surprising that quat structure affects the rate of I-reactions differently from the rate of T-reactions.

The quat structural factors which enhance the rate of I-reactions are better understood than for T-reactions. I-reaction rates are enhanced by *organophilicity* and by *anion activation*. Quat organophilicity increases as the number of carbons on the quat increases. As organophilicity increases, the equilibrium concentration of the quat-anion pair in the organic phase increases. Since the reaction rate of an I-reaction increases with increasing quat-anion pair concentration in the reaction phase (usually organic), enhanced organophilicity should result in higher reactivity in an I-reaction.

As Starks pointed out in Chapter 2, large bulky quats are considered to be "anion activating" due to the reduced electrostatic interaction between the quat and the anion. This anion activation can be manifested only if the activity of the anion is a factor in the rate determining step, i.e., an I-reaction. One must be careful not to equate the structural attribute of bulkiness, which leads to anion activation, with organophilicity. Bulkiness and anion activation are usually described in the literature as relating to symmetrical tetraalkyl ammonium quats, since symmetrical quats are commonly used and since it is easy to calculate ionic radii of such quats and their interaction energies with anions. However, some commonly used quats are quite organophilic and not symmetrical, e.g., methyl trioctyl ammonium. As depicted in Figure 1, both tetraoctyl ammonium and methyl trioctyl ammonium are organophilic. Both quats can quantitatively transfer anions to an organic phase. However, tetraoctyl ammonium will activate the anion more than methyl trioctyl ammonium due to a tighter electrostatic interaction with the more accessible positive charge on the nitrogen of the latter. If the rate determining step is the attack of the anion on a substrate, then tetraoctyl ammonium should enhance reactivity more than methyl trioctyl ammonium based on charge accessibility alone.

The effect of quat structure on the rate of T-reactions is harder to predict or rationalize. The ability of the quat structure to reduce the interfacial tension between a reservoir phase and a reaction phase should enhance the transfer process. The relationship between quat structure, interfacial tension and overall reaction rate was demonstrated in the alkylation of deoxybenzoin with ethyl bromide.[13] In this study, only symmetrical quats were examined. The methylation of deoxybenzoin with dimethyl sulfate was the subject of the most comprehensive systematic study of the effect of quat structure on reactivity using homologous series of symmetrical and non-symmetrical quats.[8] The non-symmetrical quats studied included all combinations of $R^1NR^2_3$, where R^1 and R^2 were methyl, ethyl, butyl and octyl. The results suggested that, empirically, as the accessibility of the positive charge on the nitrogen increased, so did the reactivity (until the quats became so extremely hydrophilic that their activity diminished, such as with tetramethyl ammonium). For example, orders of reactivity were $MeNBu_3 > EtNBu_3 > Bu_4N > OctNBu_3$; $MeNOct_3 > EtNOct_3 > BuNOct_3 > Oct_4N$; $RNEt_3 > RNBu_3 > RNOct_3$. It is possible that these orders of reactivity may be explained by the effect of quat structure on interfacial tension. Unfortunately, interfacial tensions were not measured for the quats studied. It can be speculated that since anions such as hydroxide which are hard to transfer from an aqueous phase to an organic phase, have a high charge density, they may "prefer" to associate (according to Hard Soft Bases Theory) with cations with an accessible cationic center. The accessibility of the cationic center increases as the length of the linear alkyl chains decreases, with a disproportionately large effect assigned to the shortest alkyl chain. Due to asymmetry, accessibility is not the opposite of bulkiness.

An empirical quantitative parameter has been suggested for characterizing quat accessibility, "q."[12] The q-value is calculated by adding the reciprocals of the number of carbons on each of the four alkyl chains (if linear). This calculation assigns proportionately higher values to quats with a methyl or several ethyl groups. For example, tetrahexyl ammonium, which has 24 carbons has a q-value of 0.67 (1/6 + 1/6 + 1/6 + 1/6) whereas methyl trioctyl ammonium, which has 25 carbons has a radically different q-value of 1.38 (1/1 + 1/8 + 1/8 +1/8). Ethyl trioctyl ammonium, which has 26 carbons, has a q-value of 0.88. Thus, it can be seen that small differences in the number of carbons can have a large effect on the

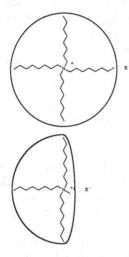

Oct$_4$N$^+$

Soluble in organic > 99%

Positive charge
not very accessible

C# = 32
q = 0.5

Oct$_3$NMe$^+$

Soluble in organic > 99%

Positive charge
somewhat accessible

C# = 25
q = 1.38

Figure 1. Anion Activation and Accessibility

q-value. Intuitive examination of these three examples of quat structure leads one to conclude that due to asymmetry, accessibility is not necessarily the opposite of bulkiness just as the q-value is not necessarily related to the total number of carbons on a quat (organophilicity). Many PTC reactions which involve high charge density anions, most notably hydroxide and fluoride, often achieve highest reactivity when the q-value is in the range 1.0 to 2.5, especially 1.5 to 2.0. Such reactions include the multitude of base-promoted reactions published by Makosza.[1] After examining a large number of published reactions, this author identified a rough empirical relationship (not absolute) between quat structure (as represented by q value and the total number of carbons on the quat) and reactivity based on the speculated rate determining step (Table I).[11,12,14]

Table I. Empirical and Semi-Theoretical Relationship Between Quat Structure and Reactivity Based on the Rate Determining Step

Quat Structural Parameter	Transfer Rate Limited	Intrinsic Reaction Rate Limited
q-value	1.0-2.5, preferably 1.5-2.0	important to be less than 1
total # of quat carbons	can vary; q more influential	16-32

Other quat structural factors which may affect reactivity of both T-reactions and I-reactions include those which affect quat decomposition and emulsion formation. Obviously, if the quat decomposes under the reaction conditions, reactivity will decrease (if the quat catalyzes the reaction). Quats with alkyl groups with two or more carbons (especially ethyl) may decompose readily in the presence of strong aqueous base at high temperature by Hoffmann degradation. Methyl and benzyl quats may decompose in PTC nucleophilic substitutions by methylating or benzylating the nucleophile. The degree of decomposition by Hoffmann degradation and by nucleophilic substitution depends greatly on a number of factors, most notably temperature and concentration. The majority of PTC applications successfully use ammonium quats which indicates that they are stable enough to provide desired results under a variety of useful the reaction conditions. Applications in which ammonium quats are not sufficiently stable can utilize complexants such as cyclic or open polyethers or other specialty phase-transfer catalysts which are more resilient to harsher PTC conditions. Another manner in which quat structure can affect reactivity is if the surfactant properties of the quat cause emulsions under the reaction conditions. If an emulsion is formed, the mode of action may be changed radically and reactivity is likely to be altered. Emulsions may actually enhance reactivity, but they are likely to result in separation problems. Emulsions can be minimized by minimizing the use of quats with one long chains and three methyl groups. Some PTC reactions can be performed with these quats without emulsion formation, as will be shown by Sirovski in Chapter 6.

Predicting the Likelihood of a Reaction Being an I-Reaction or a T-Reaction.
If empirical relationships are valid for relating the reaction rates of I-reactions and
T-reactions with such factors as bulkiness and accessibility, then it would be useful
to be able to predict the rate determining step of a given PTC reaction.
Unfortunately, almost any PTC reaction can be transfer rate limited, intrinsic
reaction rate limited or a combination of both, depending on the specific reaction
conditions. However, after examining hundreds of PTC reactions performed under
"convenient" reaction conditions, patterns have emerged which suggest that certain
reactions have a greater tendency to be either I-reactions or T-reactions. The
Halpern pKa guidelines[11,12,14] summarize the patterns observed. Similar to the
integrated guideline presented in this paper, **the pKa guidelines are not absolute
models**, rather they are guidelines for initiating synthesis or process development
projects. According to the pKa guidelines, one should examine the pKa of the
conjugate acid of the anion which is to be reacted (Table II). If the pKa of the
conjugate acid is below 16 (as for most of the common inorganic nucleophiles and
oxidants such as CN^-, N_3^-, Br^-, I^-, SCN^-, NO_2^-, SO_3^-, MnO_4^-, $Cr_2O_7^-$, ClO_4^-, etc. or
carboxylates, β-diketonates, phenoxides, mercaptides and other acidic organic
anions) then the reaction is likely to be intrinsic reaction rate limited. If the pKa of
the conjugate acid is in the range 16 to 23 (such as the anions derived from
phenylacetonitrile, phenylacetone, fluorene, indole and many heterocycles and
diactivated methylene groups) then the reaction rate is likely to be limited by
hydroxide transfer, since (1) many of these substrates are organic soluble and need
to be deprotonated by base prior to reaction and (2) hydroxide is a difficult anion to
transfer from an aqueous or solid phase to an organic phase. If the pKa of the
conjugate acid is in the range 24-38 (such as for the anion derived from
diphenylmethane) then the reaction is likely to be intrinsic reaction rate limited,
since the deprotonation becomes the difficult rate limiting step (even more difficult
than hydroxide transfer). Since deprotonation of substrates with a pKa higher than
38 is not yet known using PTC/NaOH conditions, reactions of the anions of these
substrates under known PTC conditions is considered unlikely at this time
(hopefully the pKa 38 barrier will be broken in the future).

Table II. pKa Guidelines (First Pass) for Anticipating Rate Determining Step

pKa of Conjugate Acid of Reacting Anion	Likely Rate Determining Step
< 16	Intrinsic Reaction
16 - 23	Transfer
23 - 38	Intrinsic Reaction

The only way to truly know if a reaction is an I-reaction, a T-reaction or a
combination is to perform a kinetic study. Of course, if there is time to perform a
comprehensive study in the lab, then there is probably time to properly screen a
variety of catalysts for the candidate PTC reaction. All of the guidelines presented

here (integrated, pKa and q-value/total C#) are to be used just as a first pass starting point for research.

Catalyst Separation. The four primary methods of separating a quat from the product or from the reaction mixture are (in order of decreasing use): (1) extraction, (2) distillation of the product, (3) recrystallization and (4) adsorption. In industrial processes, extraction, distillation and recrystallization are used. Adsorption is used only in laboratory synthesis and even then only rarely. Adsorption is performed by passing the reaction mixture through a column containing materials such as silica, which adsorb quats. Adsorption is expensive, even if the sorbant is regenerated.

Extraction is used most often to separate the quat from the product since (1) the products of most PTC reactions are soluble in the organic phase and not in the aqueous phase and (2) many quats are soluble in water. In order to separate the quat by extraction, the two (or multiple) phase reaction mixture is generally separated (decanted). This is usually necessary since the aqueous or solid phase has such a high ionic strength that the quat is salted out and partitions mostly or exclusively into the organic phase. Once the organic phase is separated (containing the product, the quat, an optional solvent, excess reactant and by products), it can then be washed, sometimes repeatedly, usually with water. The quat is extracted into the water washes and can then be treated, recovered and/or recycled. In order for aqueous extraction to be effective, the quat must be water soluble and the quat must furthermore be able to partition into the aqueous phase in the presence of the organic solvent. Most quats derived from three short chain tertiary amines such as trimethylamine, triethylamine and even tributylamine can be extracted into water from a variety of organic phases.

Extraction is used also for cases in which the product in not organic soluble, such as crosslinked polymers. In such cases the quat can be extracted from the polymer by an organic solvent.

Distillation and recrystallization are preferred in industry primarily for cases in which it is necessary or particularly easy to distill or recrystallize the product away from the reaction matrix. Since distillation and recrystallization separate the quat (and its decomposition products) from the product, it is usually desirable that the quat remains soluble in one phase (i.e., the organic phase), to avoid dealing with multiple waste streams (e.g., aqueous phase) containing the quat. Most quats derived from three long chain tertiary amines, such as trioctylamine, are suitable for PTC reactions which undergo distillation or recrystallization as the quat separation method.

Integrated Guideline: Choosing Quat for Reactivity and Separation. The discussion above showed that quat structure affects the rates of I-reactions and T-reactions and affects separation effectiveness. Since these effects are relatively independent of one another and as long as they are not mutually exclusive, then an integrated guideline can be composed for choosing quat structure for (1) reactivity, based on the rate determining step and (2) separation, based on choosing a phase for solubilizing the quat at the end of the reaction.

The four extreme cases are (1) T-reaction with the quat being extractable into the aqueous phase, T_{aq}, (2) T-reaction with the quat organophilic enough to remain nearly quantitatively in the organic phase, T_{org}, (3) I-reaction with the quat being extractable into the aqueous phase, I_{aq}, (4) I-reaction with the quat organophilic enough to remain nearly quantitatively in the organic phase, I_{org}.

The flow of thought during screening for new synthesis or process development would be as follows: (1) Estimate the pKa of the conjugate acid of the anion to be reacted. (2) Assume that the reaction is an I-reaction or a T-reaction according to the pKa guideline. (3) If the reaction is anticipated to be a T-reaction, consider quats with a q-value in the range of 1.0 to 2.5, preferably 1.5 to 2.0. If the reaction is anticipated to be an I-reaction, consider quats with 16 to 32 carbons. (4) Determine if your process is amenable to distillation or recrystallization or if extraction is required. If distillation or recrystallization is feasible, choose a quat (within the q-value or C# category determined in step 3) which is organophilic enough to remain in the organic phase. If extraction is required, choose a quat (within the q-value or C# category determined in step 3) which is soluble enough in the aqueous phase to be extractable. A summary of the guideline is shown in Table III, together with typical values for 4 linear alkyl groups of an ammonium quat.

Table III: Integrated Guideline for Choosing PTC Quat for Reactivity and Separation

quat = $R^1NR^2_3{}^+$	Water Soluble Separation by Extraction		Organic Soluble Separation by Distillation or Recrystallization	
Rate Limiting Step	R^1: # of Carbons	R^2: # of Carbons	R^1: # of Carbons	R^2: # of Carbons
Transfer Rate Limited	1*	up to 4*	1	8
Intrinsic Reaction Rate Limited	at least 4	4	at least 4	8
Fast Reactions	any water soluble quat		any organic-only soluble quat	
Slow Reactions	quat needs to be chosen carefully primarily based on reactivity; higher temperature may be needed than quat can withstand and complexants or specialty phase-transfer catalysts may be needed			

* can also be R^1 = any length, if R^2 = methyl or ethyl, as long as emulsions are not formed or emulsions can be broken easily

It should be noted that reactions can be fast, slow or not clear cut I- or T-reactions. Fast reactions have few constraints for choosing quat based on reactivity. Quats for fast reactions will usually be chosen based upon convenience, unless the quat is chosen to *slow down* an exothermic reaction. In contrast, slow reactions will usually need to be investigated in more depth than a quick and dirty evaluation can provide. Slow reactions as well as mixed I-/T-reactions may also benefit from the use of two catalysts, careful optimization of hydration level, solvent, agitation and other process parameters. In addition, high temperature PTC reactions may require special ammonium quats, phosphonium quats, crown ethers, polyethylene glycols or other complexants. Nonetheless, the majority of routine PTC applications are amenable to preliminary screening using the four cases, T_{aq}, T_{org}, I_{aq}, I_{org}, as the guide for choice of catalyst. In fact, a resource-efficient method to discover that an application is a "special" application (i.e., slow, high temperature, or mixed I/T) is to run the screening experiments according to the T_{aq}, T_{org}, I_{aq}, I_{org} methodology. Assigning "special" status to a reaction would occur after the reaction did not work under a standard set of screening conditions with the screening catalysts outlined in Table III.

Conclusion

The integrated guideline presented describes the basis for choosing a quaternary ammonium salt as a phase-transfer catalyst with the purpose of maximizing reactivity and catalyst separation during the screening stage in the evaluation of a new PTC candidate application. The integrated guideline uses previous assumptions and guidelines which include the pKa guideline and theoretical and empirical considerations relating to anion activation and accessibility as a function of quat structure. The integrated guideline should be effective for the development of many routine PTC applications. If the guideline does not provide rapid promising results, other parameters (e.g., hydration, solvent, specialty quats or complexants, agitation) should be modified in an expanded experimental program. Awareness of the integrated guideline should result in an increase in the number of new or retrofit processes which will be evaluated using PTC technology.

Literature Cited

1. Makosza, M. *Pure Appl. Chem.*, **1975**, *43*, 439 and extensive references cited therein
2. Makosza, M.; Fedorynski, M. *Adv. Catal.*, **1987**, *37*, 375 and extensive references cited therein
3. Herriott, A.; Picker, D. *J. Amer. Chem. Soc.*, **1975**, *97*, 2345
4. Halpern, M.; Sasson, Y.; Rabinovitz, M. *J. Org. Chem.*, **1983**, *48*, 1022
5. Dehmlow, E.; Lissel, M. *Tetrahedron*, **1981**, *37*, 1653
6. Halpern, M.; Lysenko, Z. *J. Org. Chem.*, **1989**, *54*, 1201
7. Landini, D.; Maia, A.; Montanari, F. *J. Amer. Chem. Soc.*, **1978**, *100*, 2796
8. Halpern, M.; Sasson, Y.; Rabinovitz, M. *Tetrahedron*, **1983**, *38*, 3183

9. Wang, M.; Wu, H. *J. Org. Chem.*, **1990**, *55*, 2344

10. Starks, C. *Chemtech*, **1980**, 110

11. Starks, C.; Liotta, C.; Halpern, M. *"Phase-Transfer Catalysis: Fundamentals, Applications and Industrial Perspectives,"* New York, **1994**, Chapman & Hall, Chapters 1, 4, and 6

12. Halpern, M. *Phase Trans. Catal. Comm.*, **1995**, *1*, 1

13. Mason, D.; Magdassi, S.; Sasson, Y. *J. Org. Chem.*, **1990**, *55*, 2714

14. Halpern, M. *Ph.D. Thesis*, Hebrew University of Jerusalem, **1983**

Chapter 9

How To Influence Reaction Paths by Phase-Transfer Catalyst Structure

Eckehard Volker Dehmlow

Fakultät für Chemie, Universität Bielefeld, D–33501 Bielefeld, Germany

Reaction branching under phase tranfer catalysis can be influenced by the nature of the catalyst. Several systems with competing reaction ramifications have been investigated. The catalysts can be classed according to their reaction path directing abilities. Whereas the "normal", typical onium catalysts (such as tetrabutylammonium salts, benzyltriethylammonium chloride) exhibit little selectivity, catalyst of groups (a) and (b) may have strong (and opposing) influences. Group (a) comprises quaternary ammonium cations with at least three methyl groups, benzo-crown-5 and oftentimes dibenzo-18-crown-6. In group (b) there are large, bulky onium cations (*e.g.* tetraphenylarsonium and branched-alkyl quaternary ammonium ions) and especially highly delocalized phospho-iminium salts.

Selection of the most suitable catalyst for a given reaction is basic to all PTC work, and an instructive general guideline to that has been given by Dr. M. Halpern in an accompanying paper of this volume. We shall be concerned here with catalyst influences on reaction branching. It is obvious that the "best" catalyst for one reaction path must not be the optimal one for a competing direction. On the other hand, interactions between catalyst cation and substrate anion are considered to be weak, so that no or very small effects might be expected. Actually, such internal ion pair interactions can be very substantial, for instance in enantioselective PT catalysis. There catalysts must be "tailored" to specific application. A slight variation either in catalyst or in substrate structure results often in a dramatic loss in enantioselectivity [surveys: see reff.(*1,2*)].

Frequentioselectivity

We shall turn for a moment to an aspect of PTC selectivity that has found limited attention: In certain cases it is desirable to differentiate between mono- and poly-reaction at similar or identical structural elements. We have dubbed this type of selectivity "frequentioselectivity" (*3*). An example is the C-benzylation of 4,4-dimethyl-3-pentanone. Even an excess of benzyl chloride and extended reaction periods do not force double alkylation with many different catalysts (NaH as a base, toluene, reflux). The attempted PTC alkylation of the mono-benzyl compound is similarly unsuccessful under the same conditions. Benzo-15-crown-5 (and to some extent other crowns), however, foster bisalkylation selectively. The effect seems to be due to selective solubilization of the bis-anion (*3*).

Mono-, bis-, and tris-additions of dichlorocarbene to 1,5,9-dodecatriene, for instance, are strongly affected by the nature of the catalyst under standardized PTC conditions: NMe_4Cl and $PhCH_2NMe_3Cl$ give 72 or 65 % of mono-adduct, TEBA leads to 81 % bis-adduct, whereas $C_{16}H_{33}NMe_3Br$ gives 81 % of tris-adduct (*4*). Although the catalysts exert a frequentioselectivity under the applied conditions, the general situation is different from the first example: Forcing conditions lead to polycyclopropanation with all catalysts. This is explained by the fact that the mono product favoring catalysts are the slowest acting ones.

Reaction Branching in Dihalocarbene Conversions

Addition *vs*. Hydolysis Competion. When dihalocarbenes are generated from 50% aqueous sodium hydroxide and haloform, a part of the reagent is hydrolyzed. It can be shown that hydrolysis (which is very slow in the absence of catalyst) is accelerated by the catalyst and partially suppressed by the presence of an alkene. The relative proportion of addition *vs*. hydrolysis depends both on the nucleophilicity of the olefin and the nature of the catalyst (*5,6*). Thus, under certain conditions, cyclohexene utilizes 78 % of the consumed bromoform (formation of dibromonorcarane; 22 % of the $HCBr_3$ is hydrolyzed) with NMe_4Cl as the catalyst and 85 % with $NOct_4Cl$. The respective percentages for dibromocarbene additions to n-hexene under the same circumstances are 59 and 71 % (*6*).

Halide Exchange in Dihalocarbenes. Mixed haloforms HCX_2Y undergo rapid halide exchange under most PTC conditions *via* the so-called halide exchange cascade.

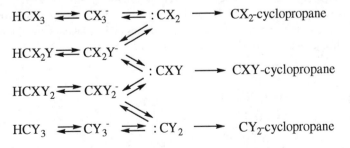

It is important to note that the more lipophilic halide can be introduced more easily, at least when equilibrating the HCX_3 haloform with inorganic halide under PTC/OH-conditions. Following up on earlier observations from Makosza's school (*7 ,8*) we have looked into such exchange reactions repeatedly (*6, 9-11*). When $HCBr_2Cl$ is the starting material one might expect exclusive loss of bromide. The experiment reveals, however, that with most catalysts large quantities of adducts are formed from intermediate $:CBr_2$ and even from $:CCl_2$ (although <u>only one chlorine</u> atom is present in the educt !). Very many catalysts were tested, and representative results for this conversion with cyclohexene are given below. It turned out that dibenzo-18-crown-6 and some of its derivatives as well as NMe_4Cl give $:CBrCl$ with almost total selectivity whereas other catalysts effect almost random scrambling of halides. Tetra-methylammonium - being very hydrophilic- is extremely slow. The rate can be improved considerably by using pinacol as cocatalyst (*10,12*). It must be also mentioned that the selectivity for $:CBrCl$ depends somewhat on the kind of alkene and the relative concentrations of olefin and catalyst (*9,10*). Take notice of the very poor performance of benzo-15-crown-5 in this reaction.

Catalyst	A	:	B	:	C	total yield (%)
NMe_4Cl	99		1		0	12
benzo-15-crown-5	70		23		7	73
dibenzo-18-crown-6	98		2		0	48
$AsPh_4Cl$	52		40		8	69
TEBA	61		32		7	45

Dihalocarbene Addition *vs*. Trihalomethylide Substitution. There is another interesting aspect in the haloform PTC chemistry. No matter which of the following methods is used to generate the dihalocarbene, a trihalomethylide anion is always the intermediate. This is a nucleophile, and it is in equilibrium an electrophile, the carbene, providing for a rather rare dichotomy.

(i) HCX_3 (base) $\rightarrow CX_3^- \rightarrow CX_2$ (*13*)   (ii) $NaO_2C\text{-}CX_3 \rightarrow CX_3^- \rightarrow CX_2$ (*14*)

(iii) $Me_3SiO\text{-}COCX_3 \rightarrow CX_3^- \rightarrow CX_2$ (*15*) (iv) $Me_3SiO\text{-}CX_3 \rightarrow CX_3^- \rightarrow CX_2$ (*16*).

R. R. Kostikov et al. (*17*) and M.S. Baird (*18*) directed attention to allyl bromide as an appropiate substrate which reacts with both species under Makosza conditions (i). We have applied a broad range of catalysts to this and related test systems (*19,20*).

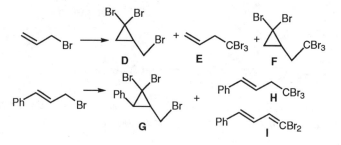

It can be demonstrated that product **F** is formed only from product **E**. With different catalysts, the carbene addition / substitution ratio (**D / E+F**) is between 7.5 and 0.05.NMe$_4$Cl and benzo-15-crown-5 give the highest proportion of carbene addition. Strongest preference for substitution is found with P[N=P(NMe$_2$)$_3$]$_4$Cl, Ph$_4$AsCl, and Ph$_3$P=N=PPh$_3$ Cl. An almost total switch of reaction channels is observed in the reaction of cinnamyl bromide. Cetrimid brings about a 92 : 1 ratio of **G** *vs.* **H+I**, whereas an 1 : 91 ratio is formed with Ph$_4$AsCl as a catalyst (*20*).

A related question concerns the reaction between α,β-unsaturated carbonyl compounds and esters with NaOH / HCX$_3$ / catalyst: Cyclopropanation might occur either by direct :CX$_2$ addition or *via* Michael type addition of CX $_3^-$ and subsequent loss of halide. M. Baird et al. (*21,22*) and our group (*23*) investigated various systems in this respect, including some with intramolecular conpetition for the trapping of the two species. There was a strong dependence of product ratios on the nature of the PT catalyst, but it was difficult to get a deeper insight.

It occurred to us that additions to cis and trans crotonic esters should permit a distinction: dihalocarbene would add stereospecifically whereas primary trihalomethylide Michael addition would lead to a non-stereospecific over-all addition. The experiment showed that **J** (X = Cl) is formed stereospecifically *only* with catalyst tetramethylammonium and in dichlorocarbene generation. Other catalysts for dichlorocarbene and all conversions with dibromocarbene give mixtures of **J** and **K**. Thus, most reactions occur probably to varying extents (depending on the type of alkene and of catalyst)*via* both pathways (*23*).

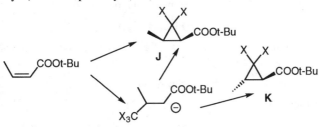

Summing up the various selectivity effects considered here, we can group the catalysts into three classes:
(a) Small, hard, sterically "available" ammonium ions and benzo-crown ethers . These favor selective :CXY formation and direct carbene additions.
(b) Large, sterically shielded, branched, or highly delocalized, "soft" ions. These favor reactions*via* trihalomethylide anions.
(c) Typical "normal" PT agents of medium size (for instance NBu$_4$X, TEBA) which do not have strong steering capabilities in either way.

This division is reinforced by the catalyst influence on the reaction between benzyl bromide and bromoform / NaOH. Only 1,1,1-tribromo-2-phenylethane and its dehydrobromination products are formed. The group (b) catalysts give high, the group (a) ones lowest yields under standardized conditions (20):

$$PhCH_2Br + HCBr_3 \ (NaOH \ /cat.) \ -> \ PhCH_2\text{-}CBr_3$$

Group (b) salts seem to stabilize trihalomethylide anions, but extensive experimentation did not allow to isolate such salts in substance. One might suspect further that reactions catalyzed by the (a) catalysts proceed *via* carbenoids rather than carbenes. This mechanistic possibility is ruled out, however, by careful selectivity comparisons of the usual type: Mixtures of alkenes (not carrying electron withdrawing groups) were reacted with bromoform or chloroform / concentrated sodium hydroxide in the presence of catalysts of all three groups. There was no selectivity difference within experimental error. Free dihalocarbenes must be involved in all of these cases. Thus, observed differences between catalysts in other reactions must the result of reaction branching with very differing individual rate constants (*24*).

Alkylation of Ambident Ions

Many aspects of desoxybenzoin alkylations have been studied (*25-28*). In fact, this is one of the best investigated PTC reactions. O / C Alkylation ratios (and rates) depend on very many parameters and the general trends are the expected ones for conversions of ambident ions. The O-alkylation is favored by these factors:

solvent	$\Rightarrow$ more polar ones
leaving group X	$\Rightarrow$ harder : $MeOSO_3^- > Cl^- > Br^- > I^-$
base concentration, hydratation of ion pair	$\Rightarrow$ solid KOH, NaOH > conc. aqueous > dilute
concentration of reagents	$\Rightarrow$ increasing
branching in alkylating agent	$\Rightarrow$ increasing
catalyst cation	$\Rightarrow$ "not accessible type" > "accessible"
alkali metal ion (non-PTC reaction)	$\Rightarrow Cs^+ > K^+ > Na^+ > Li^+$
original catalyst counter ion	$\Rightarrow HSO_4^- \approx Cl^- > Br^- > I^-$

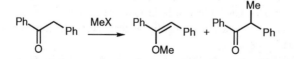

The effect of the anion brought in with the catalyst is rather prominent in alkylations with Me$_2$SO$_4$. Aparently iodide (and to a lesser extent bromide) transforms dimethyl sulfate partially into methyl iodide (or iodide), which brings about exclusive C-alkylation. Furthermore, there is a competing slow non-catalytic alkylation present here. The influence of a large range of catalyst cations was investigated in the same dimethyl sulfate reaction. O / C ratios were between 0.9 (benzo-15-crown-5) or 1.1 (NMe$_4$Br) and 29 ([P(-N=P(NMe$_2$)$_3$)$_4$] Cl), or 63 ([(Me$_2$N)$_3$P=N=P(NMe$_2$)$_3$] Cl) and 24 with MeN(isoBu)$_3$Cl . Very specific catalyst cation / substrate anion interactions must be operative which give the following O/C ratios with a series of symmetrical catalysts NR$_4$Br (R = CH$_3$ through C$_8$H$_{17}$) under identical conditions : 1.1, 3.8, 9.0, 3.8, 5.7, 2.8, 5.3, 2.4 (*28*).

Comparing catalyst actions in this field with the ones in the carbene reactions, we notice that benzo-15-crown-5 and the delocalized phosphonium ions are at the opposite ends of the scale, and the division of catalysts into groups (a) to (c) is also valid here. Unfortunately, these very interesting results cannot be generalized to simple ketones with smaller substituents. In such systems much smaller, often neglegible effects are found.

Early studies with ß-naphthol gave exclusive O-methylation even with methyl iodide using almost any catalyst. Tetramethylammonium alone lead to C-alkylation (benzo-15-crown-5 was not tested then). [Unpubl. work by Dr. V. Dryanska, Bulgaria, as guest in Bielefeld 1988]. Later our group started into this project again, noting that there was some general interest in the benzylation of substituted naphthols and real-

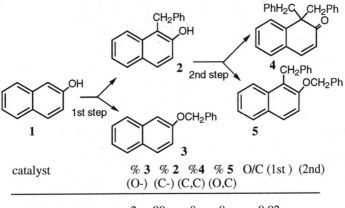

catalyst	% 3 (O-)	% 2 (C-)	%4 (C,C)	% 5 (O,C)	O/C (1st)	(2nd)
none	2	90	8	0	0.02	-
Me$_4$N Br	13	69	10	9	0.1	0.9
Ph(CH$_2$)$_2$NMe$_3$ Br	16	60	11	13	0.2	1.2
Ph-CH$_2$NMe$_3$ Br	27	45	10	18	0.4	1.8
Et$_4$N Br	34	35	9	22	0.5	2.4
Ph-CH$_2$NEt$_3$ Br	33	39	12	16	0.5	1.3
Bu$_4$N Br	40	28	10	22	0.7	2.3
Hept$_4$N Br	56	15	7	22	1.3	3.3
dibenzo-18-cr-6	31	40	13	17	0.3	1.2

izing further that there are two separate differentiation steps for O vs. C- alkylation in this system. Benzyl bromide was reacted with the sodium salt in toluene at 90°C. Take notice that there is a strong difference in O/C ratios between the various catalysts (more have been tested than given here) and that the ratios are different for the two reaction cycles (*29*)[*cf.* Scheme on preceding page]. Note changes brought about by going from NMe_4^+ over $PhCH_2CH_2NMe_3^+$, to $PhCH_2NMe_3^+$, then to NEt_4^+, TEBA and more hindered ones: The O/C ratio increases systematically. Notice also that the benzo-crown-ether stands out.

Similar observations on catalyst directing actions are made in the dimethyl sulfate alkylation of 1,1-dimethylindan-2-one. Again the O/C ratios are different in the two reaction steps. Intermediate **7** is not isolated. Extreme O/C values are found for the first reaction steps with these catalysts (many more were applied): Benzo-15-crown-5 0.85, 15-crown-5 1.0, 18-crown-6 1.65, NMe_4Br 1.1, cryptand [2.2.2] 21, Ph_4AsCl 10, [$(Me_2N)_3P=N=P(NMe_2)_3$] BF_4 21. (*30*)

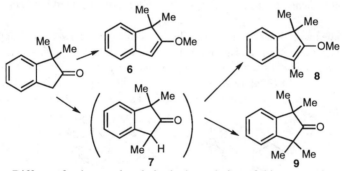

Different fundamentals rule in the benzylation of thiocyanate ions. Here benzyl thiocyanate is the kinetic, isothiocyanate the thermodynamic product. Depending on the reaction conditions, the catalysts can induce consecutive alkylation and isomerization. The more lipophilic and bulky ions favor the thermodynamic product. Both rate of conversion and rate of isomerization are higher with group (b) catalysts. Choice of catalyst for preparative purposes is not critical, however, because the reaction can be directed with temperature and solvent (solid/liquid or liquid/liquid PTC) (*31*).

$$PhCH_2Br + SCN^- \rightarrow Ph\text{-}S\text{-}CN + Ph\text{-}N=C=S$$

Attempts to exert directing influences on the nitroalkane / alkyl nitrite formation by use of different catalysts were not sucessful [unpublished work form the authors's laboratory].

$$R\text{-}X + NaNO_2 \rightarrow R\text{-}NO_2 \text{ (main product) } + R\text{-}O\text{-}NO \text{ (minor product)}$$

Two consecutive alkylations are also possible with many heterocycles. In 2-me-thyl-1-phenyl-5-pyrazolone, O-, N- and C-alkylation can compete. Typical results for two alkylations are given below. Effects of 7 tested catalysts are moderate only (*30*).

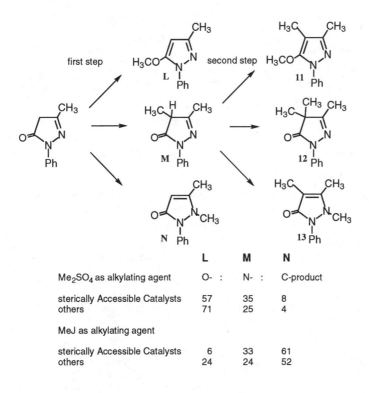

	L	**M**	**N**
Me₂SO₄ as alkylating agent	O- :	N- :	C-product
sterically Accessible Catalysts	57	35	8
others	71	25	4
MeJ as alkylating agent			
sterically Accessible Catalysts	6	33	61
others	24	24	52

4-Nitroimidazole and 5-nitrobenzimidazole anions carry also two non-equivalent sites for alkylation. Again we find small but distinct catalyst effects with the imidazole case: **O / P** ratios in the MeI alkylation, for instance, are in the range of 80-85 : 20-15 with several catalysts but 94 : 6 with benzo-15-crown-5. Charge density difference seems too small in 5-nitrobenzimidazole; 1 : 1 mixtures of benzylated and methylated products were obtained without any catalyst influences (*3*).

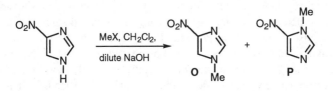

The more complicated [1]-benzothieno[2,3-d]triazole has even three sites for N-alkylation. Small but reproduceable catalyst directing effects can be found [unpublished result 1996, S.Schrader, E.V. Dehmlow]. Group (a) and group (b) catalysts behave oppositely.

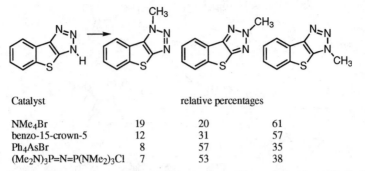

Catalyst		relative percentages	
NMe₄Br	19	20	61
benzo-15-crown-5	12	31	57
Ph₄AsBr	8	57	35
(Me₂N)₃P=N=P(NMe₂)₃Cl	7	53	38

Oxime anions are not exactly ambident anions. They can be O- or N-alkylated (to give nitrones). The PTC conversion of E- and Z-benzaldehyde oximes is known already from the literature (*32*) . We find **Q/R** ratios for benzophenone oxime methylation between 0.56 (catalyst: NMe₄Br) and 2.23 (NBu₄Br and NHex₄Br). Surprisingly "out of role" behaviour of Ph₄AsCl and benzo-15-crown-5 are observed: both give ratios of 1.17! [E.V.Dehmlow and C.Michalek, unpublished, 1996].

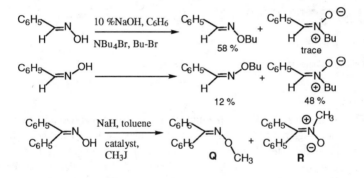

Other Reaction Branchings

Oxiranes and cyclopropanes are formed in two-step Darzens and Michael type reactions. Here a differentiation will probably occur in the initial addition, although this first step might also be reversible. Previous PTC workers on these two reactions include the Makosza-Jon´czyk groups from Poland (*33-35*) and C. Kimura and coworkers from Japan (*36*). In our study (*37*) different catalysts were used again. Z/E ratios differed relatively strongly depending on catalysts: 0.10 to 0.96 in reaction (1), 1.3 to 15.2 in reaction (2), and 0.6 to 1.4 in reaction (3). Here, however, the now more or less expected systematic trends were not present. Catalysts that behaved similarly in the conversions considered above above (and therefore grouped together) have opposing effects sometimes in the present reactions. It must be assumed that very specific interactions in the ion pairs of the intermediates are operative in the reactions considered here.

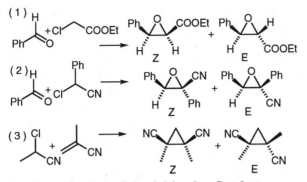

Synopsis: Reaction Path Selectivities by Catalysts

We have seen that there are a number of chemical reaction ramifications which can be influenced by the structure of the applied phase transfer catalyst. The division of catalysts into three classes of different steering capabilities was first observed in dihalocarbene related conversions. It could be extended to other reactions. Sizable effects were found especially with those ambident ions which contain relative rigid, aromatic ring systems. It became also apparent that unexpected "run-away" cases are not uncommon, and generalizations are not yet possible.

Group (a) and group (b) catalysts have opposing directing properties. The first mentioned ones favor generation of and reactions (including hydrolysis) of carbenes. Furthermore, reactions on the less electronegative site of ambident systems are promoted.

Group (a) catalysts comprehend ammonium ions $R\text{-}NMe_3^+$ and (most of the time) benzo-15-crown-5 and / or dibenzo-18-crown-6.

Group (b) catalysts are large, sterically shielded ammonium ions, especially ones with branched substituents, tetraphenyl arsonium and highly delocalized phospho-iminium ions.

Finally, group (c) catalysts are the common medium sized onium salts. They exhibit no marked selectivity preferences.

Initial experiments towards steering of untested reactions should be made with two catalysts each of group (a) [benzo-15-crown-5 and $PhCH_2CH_2NMe_3$ Cl for instance] and two of group (b) [commercially available Ph_4AsCl and $Ph_3P=N=PPh_3$ Cl for instance]. The results will probably show whether it is worthwhile to optimize further.

Tentative Explanations

What concepts do we have to explain these results? Two qualitative ideas come into mind: The hard-soft dichotomy and the accessibility of the quat. The accessibility q of an ammonium ion (a "Quat") has recently been defined as a semi-quantitative measure, the sum of reciprocals of the C atom numbers of the 4 chains on the central N atom: $q = 1/C\#1 + 1/C\#2 + 1/C\#3 + 1/C\#4$. [M. Halpern, *cf.* ref.(*38*)]

Quaternary cations with a q value > 1 are considered accessible. These must be the harder species. Typical catalysts have these q values: NBu_4^+ 1, Bu_3NMe^+ 1.75, NMe_4^+ 4, and Aliquat 336 1.38. It is open to question whether these numbers offer more than a rough estimate of properties. Let us apply these concepts nevertheless. We may assume that the hard / soft combination of the trichloromethylide ion pair with "available" NMe_4^+ is less stable and is decomposed faster than the soft / soft pair with tetraphenylarsonium. In the ß-naphtholate alkylation the hard / soft combination would give a tighter ion pair than the association of a sterically more bulky cation with the same delocalized anion. Still it is difficult to understand, why the amount of C-alkylation is decreased in the latter case. Although one could argue that the harder, more available cation will be closer to the oxygen of the ambident ion and the softer cation will be near the carbon site, these two positions are vicinal and differential shielding will not be effective when only steric bulk is considered.

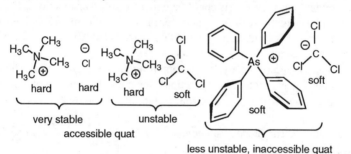

A concept to explain the predominance of C-alkylation of ß-naphthol in non-polar solvents was advanced by Kornblum in 1963 (*39*). He argued that the stabilization of the leaving group is a major factor in this type of reaction, and here C-alkylation will more favorable because of better charge distribution, as shown.

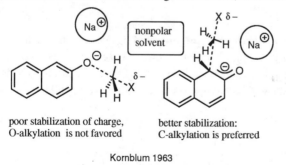

Kornblum 1963

Let us try to extend this concept tentatively to PTC. One important aspect (with which we shall deal with in a more general way below) must be mentioned first: Calculations have shown that the positive charge in an onium cation rests primarily in the α-CH$_2$-groups (*40*). Thus, hydrogen bonding by α-methylene groups is really the important factor here. An "available" quat in the shown position can be hydrogen bonded to both the oxygen and the leaving group X as indicated: the catalyst behaves much like an alkali metal cation. A larger cation will block the C-site of the naphtholate <u>both</u> by size and hydrophobic attractive interaction of two of its sidearms with the aromatic system. One or two other side arms might interact with the leaving group by α-CH$_2$ or ß-CH$_2$ hydrogen bonding.

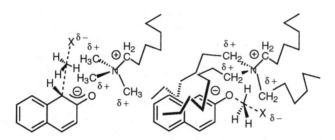

H-bonding between α-CH$_2$ s of quat
with phenolate and leaving group

We assume that very distinctive arrangements of cation and anion in the ion pairs govern the differences in chemical behavior of these species. How can we get actual knowledge about the relative positions of catalyst cation and substrate anion in solution? If we follow the leads given by our biochemistry colleagues, crystal structures will give a good idea about solution structures inspite of the fact that special lattice effects might interfere. Our group's attempts to characterize representative salts of ambient anions with several cations by X-ray crystal studies have not yet been sucessful. We continue work on this problem including experiments with homochiral quats.

Meanwhile M. Reetz et al. have published a series of crystal structures of tetrabutylammonium salts (group (c)) of such anions as substituted dialkyl malonates (*41*), cyclopentadiene (*42*), carbazole (*43*), and other C-H acidic compounds (*44*).These workers use the metal free ion pairs to trigger anionic polymerizations of methacrylates. All structures exhibit strong hydrogen bonding between cations and anions. Sometimes there are discrete ion pairs with relatively strong H-bonding (indicated by shorter-than-normal interatomic distances). In other cases ion pair dimers (ion quadruplets) were found which are held together by weaker H-bonds both in the crystal and in benzene solution. Intensive shielding of the carbanionic sites is frequently visible at least in the "frozen" X-ray structures. The structure of tetrabutylammonium ethylcyanoacetate (*41*) represents an especially remarkable example as interpreted by the Reetz group: In its dimeric species there are strong H-bonds between oxygen and an α-H of the first quat, a weaker one to a ß-H of the second quat, and finally a weak interaction of the nitrile N with a γ-H of the first quat!

These very subtle interactions make it understandable why seemingly very similar, related quaternary salts have quite different catalytic performances and/or steering effects oftentimes. One of the open question is the kind of shielding one would expect from crown ethers, especially as the results presented earlier in this text imply that similar crowns may give very different relative O/C ratios. It is not easy to comprehend these effects or the sizeable performance differences of the large delocalized phosphoiminium cations conceptually.

Some insight might be found in studying X-ray structures published recently by H. Bock and coworkers (45). They prepared salts of the disodium perylene dianion with three different, but very similar ligands for the sodium: monoglyme, diglyme and tetraglyme, which we may consider as open crown ethers. An incredibly fine tuning of structures was recorded: In the first case the counter ions were above and below the flat aromatic core, in the second instance they were arranged sidewise as "solvent seperated triple ions", and in the third case they stood inclined somewhat sidewise above and below the aromatic anion. The first and last arrangements were interpreted as solvent shared triple ions. Thus, as complexing of the sodium becomes more intensive the cations are more and more withdrawn from the anion.

Conclusions

We find:
(1) Certain catalyst discriminate between reaction channels. Strongest effects are observed in the trihalomethylide / dihalocarbene reaction branching.
(2) In the field of ambient anions, steering by catalysts can be strong in rigid planar delocalized systems, but might be small with simple enolates.
(3) There must be many more reactions that have not been investigated in which catalyst steering actions could be exploited.
(4) Tetramethylammonium often brings about the most extreme swing in one direction. In addition to this rather hydrophilic and slow reacting catalyst, its close congeners with one long carbon chain might do well. Benzo-15-crown-5 (and sometimes dibenzo-18-crown-6) exhibit relatively reliably strong influences in the same direction as NMe_4.
(5) The two just mentioned types of catalyst are the most likely candidates for directing conversions to the "soft" end of ambient ions.
(6) An opposite steering effect is observed with $[R_3P=N=PR_3]^+$ and Ph_4P^+ cations. Sometimes large Quats with four long chains do as good, as might ammonium ions with branched chains.

Outlook

Cation-anion interactions must be investigated further in detail. Although NMR and other spectroscopic techniques have been and probably will be of some help, X-ray crystallography might be the method of choice. Only a better understanding of such interactions might help to get away from the present stage of trial and error. This is especially important for enantioselective PTC. Investigation of catalyst directing effects may be extended to other reaction branchings, not necessarily of ambident ion type.

Acknowledgments

Our own results presented here have been obtained by dedicated coworkers whoes names appear in the literature citations. The author is indebted to Deutsche Forschungs-gemeinschaft and Fonds der Chemischen Industrie for financial support.

Literature Cited

1. Dehmlow, E. V.; Dehmlow, S. S. Phase Transfer Catalysis, 3rd rev. enl. Edit., VCH-Publishers, Weinheim (Germany), New York, N.Y., 1993.
2. O'Donnell, M. J. In Catalytic Asymmetric Synthesis; Ojima, I., Edit.; VCH-Publishers, New York, N.Y., Weinheim (Germany), 1993; chapter 8; pp 389-411.
3. Dehmlow, E. V.; Richter, R., Zhivich, A. B. *J. Chem. Res.* (S) **1993**, 504.
4. Dehmlow, E. V.; Prashad, M. *J. Chem. Res.* (S) **1982**, 354.
5. Dehmlow, E. V.; Lissel, M., Heider, J. *Tetrahedron* **1977**, *33* , 363.
6. Dehmlow, E. V.; Wilkenloh, J. *J. Chem. Res.* (S) **1984**, 396; (M) **1984**, 3744.
7. Fedoryn´ski, M. *Synthesis* **1977**, 783.
8. Fedoryn´ski, M.; Poplawska, M.; Nitschke, K.; Kowalski, W.; Makosza, M. *Synth. Commun.* **1977**, *7*, 287.
9. Dehmlow, E. V.; Slopianka, M. *Liebigs Ann. Chem.* **1979**, 1465.
10. Dehmlow, E. V.; Stütten, J. *Liebigs Ann. Chem.* **1989**, 187.
11. Dehmlow, E. V.; Broda, W. *Chem. Ber.* **1982**, *115* , 3894.
12. Dehmlow, E. V.; Thieser, R.; Sasson, Y.; Neumann, R.*Tetrahedron* **1986**, *42* , 3569.
13. Makosza, M.; Wawrzyniewicz, W. *Tetrahedron Lett.* **1969**, 4659.
14. Dehmlow, E. V.; Remmler, T. *J. Chem. Res.* (S) **1977**, 72; (M) **1977**, 766.
15. Dehmlow, E. V.; Leffers, W. *J. Organomet. Chem.* **1985**, *288*, C41.
16. Cunico, R. F.; Chou, B. B. *J. Organomet. Chem.* **1978**, *154*, C45.
17. Labeish, N. N.; Kharicheva, E. M.; Mandelshtam, T. V.; Kostikov, R. R. *Zh. Org. Khim.* **1978** *14*, 878 [Engl. transl. 815]
18. Baird, M. S.; Baxter, A. G. W.; Devlin, B. R. G.; Searle, R. J. G. *J.C.S., Chem. Commun.* **1979**, 210.
19. Dehmlow, E. V.; Wilkenloh, J. *Tetrahedron Lett.* **1987**, *28*, 5489.
20. Dehmlow, E. V.; Wilkenloh, J. *Liebigs Ann. Chem.* **1990**, 125.
21. Baird, M.S.; Gerrard, M. E. *Tetrahedron Lett.* **1985**, *26* , 6353.
22. Baird, M.S.; Gerrard, M. E. *J. Chem. Res.* (S) **1986**, 114.
23. Dehmlow, E. V.; Wilkenloh, J. *Chem. Ber.* **1990**, *123*, 583.
24. Dehmlow, E. V.; Fastabend, U. *J.C.S., Chem. Commun.* **1993**, 1241.
25. Halpern, M.; Sasson, Y.; Willner, I.; Rabinovitz, M. *Tetrahedron Lett.* **1981**,*22* , 1719.

26. Halpern, M.; Sasson, Y.; Rabinovitz, M. *Tetrahedron* **1982**, *38* , 3183.
27. Mason, D.; Magdassi, S.; Sasson, Y. *J. Org. Chem.* **1990**, *55*, 2714.
28. Dehmlow, E. V.; Schrader, S. *Z. Naturforsch.* **1990**, *45b* , 409.
29. Dehmlow, E. V.; Klauck, R. *J. Chem. Res.* (S) **1994**, 448.
30. Dehmlow, E. V.; Richter, R. *Chem. Ber.* **1993**, *126*, 2765.
31. Dehmlow, E. V.; Torossian, G. O. *Z. Naturforsch.* **1990**, *45b* , 1091.
32. Shinozaki, H.; Yoshida, N.; Tajima, M. *Chem. Lett.* **1980**, 869.
33. Fedoryn'ski, M.; Wojciechowski, K.; Matacz, Z.; Makosza, M. *J. Org. Chem.* **1978**, *43*, 4682.
34. Jon'czyk, A.; Kwast, A.; Makosza, M. *J.C.S., Chem. Commun.* **1977**, 902.
35. Jon'czyk, A.; Makosza, M. *Synthesis* **1976**, 387.
36. Kimura, C.; Kashiwaya, K.; Murai, K.; Katadu, H. *ind. Eng. Chem. Prod. Res. Dev.* **1983**, *22*, 118.
37. Dehmlow, E. V.; Kinnius, J. *J. Prakt.Chem.* **1995**, *337* , 153.
38. Starks, C. M.; Liotta, C. L.; Halpern, M. *Phase transfer Catalysis: Fundamentals, Applications, and Industrial Perspectives*; **1994**, Chapman and Hall, New York.
39. Kornblum, N.; Seltzer, R.; Haberfield, P. *J. Amer. Chem. Soc.* **1963**, *85*, 1148.
40. Reetz, M. T.; *Angew. Chem.* **1988**, *100* , 1026; *Angew. Chem., Int.Ed. Engl.* **1988**, *27*, 994: reference [10] therein.
41. Reetz, M. T.; Hütte, S.; Goddard, R. *J. Amer. Chem. Soc.* **1993**, *115*, 9339.
42. Reetz, M. T.; Hütte, S.; Goddard, R. *Z. Naturforsch.* **1995**, *50b* , 415.
43. Reetz, M. T.; Hütte, S.; Goddard, R.; Minet, U. *J. C. S., Chem. Commun.* **1995**, 275.
44. Reetz, M. T.; Hütte, S.; Goddard, R. *J. Phys. Org.Chem.* **1995**, *8*, 231.
45. Bock, H.; Näther, C.; Havlas, Z. *J. Amer. Chem. Soc.* **1995**, *117*, 3869.

Phase-Transfer Catalysis
in Organic and Polymer Synthesis

Chapter 10

Amino Acid and Peptide Synthesis Using Phase-Transfer Catalysis

**Martin J. O'Donnell, Irina A. Esikova[1], Aiqiao Mi[2],
Daniel F. Shullenberger, and Shengde Wu**

**Indiana University–Purdue University at Indianapolis,
Indianapolis, IN 46202**

The racemic and enantioselective synthesis of amino acids and the racemic synthesis of dipeptides by phase transfer catalysis are discussed. The various levels of selectivity in this process are outlined. New and more effective catalysts are reported.

Phase transfer catalysis (PTC) has been used for the preparation of numerous classes of compounds over the past 25 years (1,2). It is a synthetic method that often provides advantages over more conventional procedures: easy reaction procedures, mild conditions, safe and inexpensive solvents and reagents, and the ability to conduct reactions on large scale.

Since 1978, we have been engaged in a program to develop new methods for the preparation of amino acids (3), the "building blocks of life" (Scheme 1), using

Scheme 1. Chemistry of Schiff Base Protected Amino Acid Derivatives.

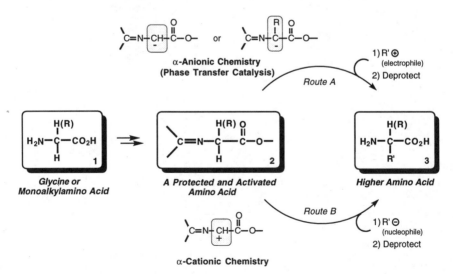

[1]Current address: Chiron Corporation, 4560 Horton Street, Emeryville, CA 94608
[2]Current address: Chengdu Institute of Organic Chemistry, Chinese Academy of Sciences, Chengdu 610041, People's Republic of China

carbon-carbon bond formation (4-6). The two major types of reactions used are the reaction of α-anionic glycine or higher amino acid equivalents with electrophiles (Route A) (4,5) or the complementary reaction of the α-cationic derivatives with nucleophiles (Route B) (6). These reactions can be accomplished from a single class of protected and activated amino acids, the Schiff base esters 2. More recently this methodology has been extended to the selective modification of peptides (7). Key features of the anionic chemistry (Scheme 1, Route A), which can conveniently be carried out under typical PTC conditions, will be discussed in this paper.

Issues of selectivity (8) are of crucial importance in the alkylation of the Schiff base esters 2 because of the multifunctional nature of both starting substrates and products. It is necessary to selectively deprotonate at the α-carbon with base but avoid reaction of the base as a nucleophile with either the imine or ester (*chemoselectivity*). Once the anion has been formed, reaction with the electrophile must be controlled to give the desired α-alkylated product without γ-alkylation (*regioselectivity*). With the simplest amino acid glycine, where an active methylene group is present in the substrate, one must be able to stop the reaction at the monoalkylation stage without subsequent dialkylation (*frequentioselectivity*) (9). In the case of glycine a prochiral substrate must be converted into a chiral, non-racemic product (*stereoselectivity*) under conditions that will not racemize the resulting product. In the related case with a higher amino acid a chiral, racemic or non-racemic substrate is converted to a chiral non-racemic product. In this latter case, product racemization is not a concern since the newly formed carbon-carbon bond is part of an α-quaternary center.

Schiff Base Derivatives in Amino Acid Chemistry: Racemic Products

Early research from our laboratory showed that the doubly-protected and doubly-activated Schiff base esters of amino acids could be conveniently alkylated under various PTC conditions (Scheme 2) (4). An important observation was that α-monoalkylated products (5) are readily prepared from the corresponding benzophenone Schiff base ester of glycine (4, a ketimine) (4a-c,e-h) while the α,α-dialkylated derivatives (8) are typically prepared from an aldimine (7) of a higher amino ester (4d-g,4j,10). The alkylation products, which in these cases are racemic since there is no chiral control element, could then be deprotected to yield the amino acids.

Scheme 2. Amino Acid Synthesis by Phase Transfer Catalysis (PTC).

Monoalkylation of a Glycine Anion Equivalent:

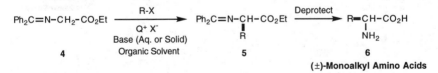

Monoalkylation of a Higher Amino Acid Anion Equivalent:

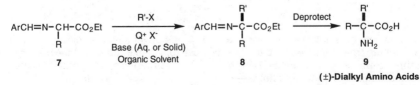

The resistance of the monoalkylated benzophenone imine product 5 to further alkylation is key to both the frequentioselectivity (avoiding dialkylation of the active methylene substrate) and, ultimately, to developing an enantioselective synthesis based on PTC. Thus, following a possible stereoselective alkylation of the anion of **4**, it is crucial that the monoalkylated product not be deprotonated under the basic reaction conditions, thus leading to racemization upon reprotonation.

There is a large difference in acidity (~10^4) in the benzophenone Schiff base series between the starting substrate **10** and the monoalkylation product **11** (Scheme 3), whereas in the aldimine series related compounds are of similar acidities (pKa in DMSO of 4-ClC$_6$H$_4$CH=NCH(R)CO$_2$Et: R=H, 18.8; R=Me, 19.2) (11). The large acid-weakening effect in **11** is likely due to the potential for A$_{1,3}$ strain in the benzophenone series. The origin of this effect can be seen by examining the crystal structures of compounds **10** and **11** (Scheme 3) (12). With starting substrate **10**, the two π systems flanking the α-carbon are almost coplanar and, to the extent that there is not extensive reconformation upon deprotonation, the anion of **10** will be stabilized by delocalization over a planar five atom system. In contrast, the two π systems in **11** are nearly orthogonal, a result of minimization of the strain caused by proximity of the imine phenyl (cis to the α-carbon) and the alkyl group on the α-carbon. Here deprotonation without reconformation would result in two orthogonal anionic π systems, either an enolate ion or a carbanion α to the nitrogen of an imine. Thus, the anion of **11** would be expected to be less stable (more basic) than the anion of **10**, which in turn implies that **11** should be considerably less acidic than **10**.

Scheme 3. Crystal Structures of Starting Material and Monalkylated Product.

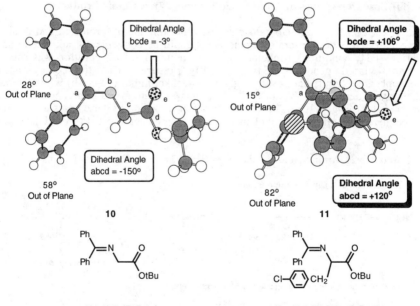

pKa 18.7 (Et Ester)   pKa 23.2 (PhCH$_2$-, Et Ester)

The benzophenone Schiff base esters (**14**) are conveniently prepared by a room-temperature transimination reaction between benzophenone imine and the appropriate amino ester salt (Scheme 4) (13). The benzophenone Schiff bases of aminoacetonitrile, optically pure higher amino esters as well as dipeptide esters can also be prepared by the transimination procedure. This procedure replaces the earlier tedious preparation of these derivatives, which involves refluxing a xylene solution of

Scheme 4. Starting Material Preparation.

$$Ph_2C=NH + H_2N-CH_2-CO_2R \cdot HCl \xrightarrow[RT]{CH_2Cl_2} Ph_2C=N-CH_2-CO_2R$$

$$12 \qquad\qquad 13 \qquad\qquad\qquad\qquad 14$$

benzophenone and the free amino ester in the presence of boron trifluoride etherate followed by a high-temperature vacuum distillation and then recrystallization (4a)! The benzophenone imine is an attractive protecting group for a primary amine for a variety of reasons: easy introduction from stable and readily available precursors, stability to routine basic conditions, strong chromophore for UV detection in TLC or HPLC systems, stability to flash chromatography and, finally, easy removal either with mild acid or by hydrogenolysis. The aldimine Schiff base esters (**7**) are prepared according to the method of Stork and coworkers (14) by reacting the aldehyde with the amino ester salt in the presence of a mild base such as triethylamine.

The mode of addition of reagents in PTC alkylation reactions is very important. Typically, in related anhydrous alkylations of similar systems (4a), the anion of the substrate is first generated by addition of a strong base under anhydrous conditions at low temperature (e.g. LDA, THF, -78 ºC.) followed by addition of the alkyl halide in a second, independent step. In contrast, in the PTC alkylation reaction, *it is important to have the alkyl halide present when the base* (aqueous NaOH or solid KOH, K$_2$CO$_3$ or mixed KOH/K$_2$CO$_3$) *is added* to avoid side reactions of the starting substrate. In fact, the entire course of the reaction can be changed if the base is added first (Scheme 5) in conjunction with a sterically non-hindered methyl ester and a full equivalent of the quaternary ammonium phase transfer reagent. In this case, the ester is saponified to give a quaternary ammonium carboxylate (**16**) which can be O-alkylated to give new esters (**17**) (15). Alternatively, **16** can be converted into a mixed anhydride and then treated with nucleophiles, such as amines or amino esters, as a novel route to amides or dipeptides **18** (16). Other applications of Schiff base protected and related amino acid derivatives include N-alkylation chemistry (17), selective mono- or diphenylation by nucleophilic aromatic substitution chemistry (18,19), Wittig reactions of an α-keto amino acid equivalent (20), and preparation and reactions of an α-aminophosphonate cation equivalent (21).

Scheme 5.   Carbonyl Chemistry of Schiff Base Protected Amino Acid Derivatives.

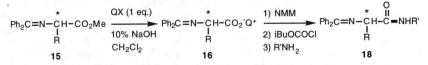

"Transesterification":

Amide and Peptide Synthesis via Mixed Anhydrides:

Recently, we have extended the utility of the PTC alkylation methodology to dipeptides (7). It is possible to selectively alkylate the N-terminal residue of dipeptides by initial conversion into the appropriate imine [benzophenone imine in the case of a N-terminal glycine (**19**) or aldimine for N-terminal higher amino acids (**20**)] followed by PTC alkylation (Scheme 6). The scale of this reaction has been reduced to 50 mg of substrate and the alkylation occurs without racemization of the adjacent C-terminal α-carbon. As expected, the stereoselectivity of alkylation at the N-terminal α-carbon is low.

Scheme 6. Selective PTC Alkylation of Dipeptide Derivatives.

N-Terminal Glycine Alkylation by PTC:

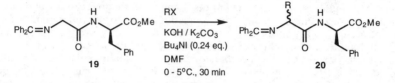

N-Terminal Higher Amino Acid Alkylation by PTC:

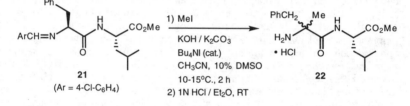

Catalytic Enantioselective Synthesis of Amino Acid Derivatives by PTC

The previous sections have described development of PTC systems for the preparation of amino acid derivatives in which stereochemistry at the newly created stereogenic center is not controlled. While it is normally possible to resolve such racemic products or separate mixtures of diastereomers, it is desirable to control the stereochemistry of these reactions so as to obtain optically pure products.

Two major techniques for controlling stereochemistry in product amino acids differ by location of the chiral control. Use of a chiral phase transfer catalyst is a catalytic enantioselective reaction since the control is used in *less than a stoichiometric amount*. This "holy grail" of PTC chemistry will be the focus of the remainder of this chapter (2). The second method for controlling stereochemistry uses the chiral control in a *stoichiometric amount* by introducing a chiral auxiliary in the substrate. Both methods have their advantages and disadvantages. The catalytic route uses only a small amount of the usually costly control but, since the products are enantiomers, normally involves a difficult separation of enantiomers if the asymmetric induction is not complete. On the other hand, the stoichiometric use of chiral auxiliaries can be costly, especially on large scale, but the resulting products are diastereomers, which are normally easy to separate.

Our program in the catalytic enantioselective PTC synthesis of amino acids started following a 1984 report from the Merck group (22a). This pioneering research (22) involved the highly enantioselective alkylation of an active methine substrate (indanone **23**) under PTC conditions using a quaternary ammonium salt catalyst derived from the *cinchona* alkaloids (see structures **27** and **28** in Scheme 8 below). Molecular models of one of our substrates, the benzophenone imine of glycine benzyl ester (**24**), showed that it might be possible to achieve the same types of π-π stacking interactions between the anion of **24** and the chiral, quaternary ammonium catalyst that were so effective in the Merck system (anion of **23** and the

Scheme 7. Merck Substrate and Initial Starting Substrate for this Research.

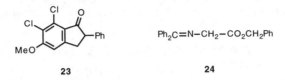

23   **24**

same catalyst) (Scheme 7) (23). Two complicating factors in our system are the acyclic nature of substrates such as **24** and, as discussed earlier, the fact that, since they are active methylene compounds, monoalkylation needs to be controlled to avoid both product racemization and the formation of undesired dialkylated product.

Alkylation of **24** using conditions similar to those used in the Merck study gave disappointingly poor results: 5% ee (52.5% one enantiomer and 47.5% other enantiomer). A detailed study to optimize reaction conditions and the nature of the ester protecting group led to considerable improvement in this process (5a); i.e., alkylation of the benzophenone imine of glycine t-butyl ester (**10**) in methylene chloride with the pseudoenantiomeric catalysts (**27** and **28**) derived from cinchonine and cinchonidine, respectively, provides a simple preparation of either enantiomer of the desired monoalkylated glycine derivative in up to 66% and 64% ee (Scheme 8) (5a, 24).

Scheme 8. Pseudoenantiomeric Catalysts in Enantioselective PTC Alkylations.

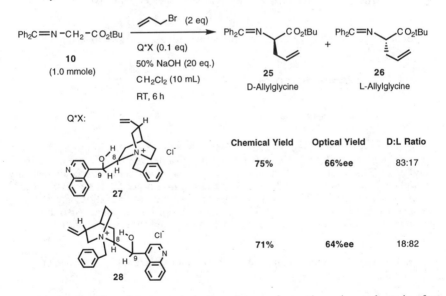

	Chemical Yield	Optical Yield	D:L Ratio
27	75%	66%ee	83:17
28	71%	64%ee	18:82

As noted earlier, the separation of a mixture of enantiomeric products is often problematic. Although not a general solution, in some cases it is possible to accomplish separation by recrystallization (25). This was accomplished in the case of the benzophenone Schiff base of 4-chlorophenylalanine t-butyl ester derivative (Scheme 9). A single recrystallization of the crude product removed nearly racemic crystals, leaving the highly enriched filtrate (>99% ee). Removal of solvent and deprotection then gave optically pure 4-chlorophenylalanine (**29**) (5a). This technique has been used by others to obtain optically pure amino acid products from **10** (26). Other useful techniques include racemic (27) or enantioselective (28) PTC alkylation followed by enzymatic hydrolysis, and enantioselective PTC-type alkylation followed by coupling with a homochiral amino acid derivative and subsequent chromatographic separation of the resulting diastereomeric dipeptides (29).

Scheme 9. "Large Scale" Preparation of Optically Pure *p*-Chloro-D-Phenylalanine.

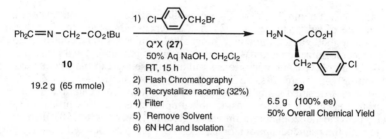

Two other variables were briefly explored in this early research (30). A study of the steric and electronic effects on the Schiff base led to a small improvement (69% ee for enantioselective PTC allylation with starting substrate **30** compared with 66% for the same reaction using **10**). Introduction of a chiral ester in combination with the enantioselective PTC alkylation (double asymmetric induction) showed more promise: a 78% ee was obtained with the (-)-menthol ester Schiff base **31**.

Scheme 10. Modified Schiff Base Substrates for PTC Alkylations.

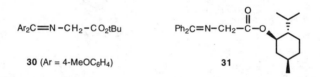

Extension of this chemistry to preparation of α,α-dialkylamino acids is also possible (Scheme 11) (5b). Using the 4-chlorobenzaldehyde Schiff base of alanine t-butyl ester (**32**) and the cinchonine-derived catalyst, a variety of protected α-methyl phenylalanine derivatives (**33**) were prepared in up to 50% ee (R:S = 3:1). Interestingly, substrate **32** from L-alanine (S-absolute configuration) gave slightly higher enantioselectivity than did **32** from either D- or D,L-alanine. Also, use of the pseudoenantiomeric catalyst **28**, from cinchonidine, led to less efficient induction in this series.

Scheme 11. α,α-Dialkylamino Acids by Enantioselective PTC: Active Alkyl Halides.

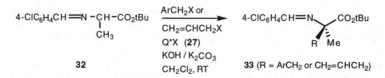

As noted earlier, the mechanism of the PTC alkylation of active methylene compounds is complicated on various levels. It involves a heterogeneous reaction system (either liquid-liquid or solid-liquid PTC) with substrates, catalysts and products, which are capable of undergoing a variety of undesirable reactions. Additionally, the asymmetric induction involves the diastereoselective alkylation of ion-pairs (**34**) formed during the reaction. Scheme 12 depicts our mechanistic model for the non-stereoselective portions of the reaction. Studies based on this model led to the proposal of a new active catalyst species (**35**), which is formed by an *in situ* deprotonation of the alcohol in the starting catalyst followed by O-alkylation (5c).

Scheme 12. Working Model for PTC Alkylation of Active Methylene Compounds.

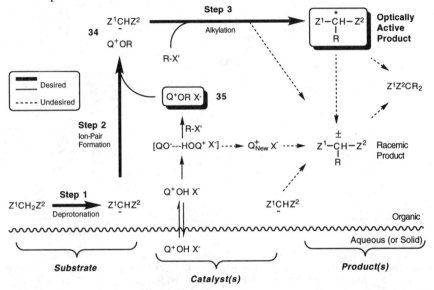

An example of catalyst-type **35**, N,O-dibenzylcinchonidinium bromide (**36**), was prepared independently from cinchonidine and X-ray crystallography verified the proposed structure (Scheme 13) (5c). In chemical tests, catalyst **36** gave within experimental error identical enantioselectivity in the PTC benzylation of **10** as did either the free alkaloid (cinchonidine) or the free -OH catalyst, N-benzylcinchonidinium bromide. These results support the proposal that the active catalyst in these alkylations is the N,O-dialkylcinchona salt.

Scheme 13. Structure of New Catalyst **36**.

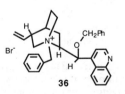

36

It is sometimes advantageous in the case of liquid alkyl halides to conduct the PTC reaction in the absence of solvent (Scheme 14) (30). Thus, product of 58% ee was obtained by isobutylation of **32** using catalyst **27** and no solvent while reaction with the same catalyst in CH_2Cl_2 gave product **37** in only 36% ee. However, caution is warranted because the same reaction with 4-fluorobenzyl bromide in the absence of solvent led to a decreased enantioselectivity: 30% ee with catalyst **27** and no solvent and 50% ee with the same catalyst in CH_2Cl_2.

A further study of different types of N-alkylated-O-alkylated cinchonine-derived salts was carried out in the alkylation of the alanine Schiff base ester (**32**) with isobutyl bromide (Scheme 14) (30). In this system, results with the O-benzyl salt **40** are only slightly better than with the free -OH catalyst **27**. However, a substantial improvement is realized by using the O-allyl-N-benzyl derivative (**42**). The effect of the counterion in the catalyst, while small, is significant: chloride is typically better than bromide.

Scheme 14. PTC Alkylations with Catalysts Derived from Cinchonine.

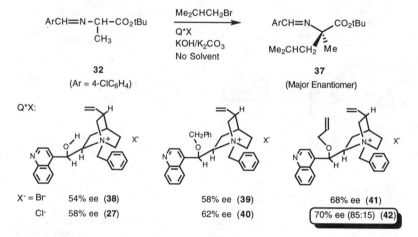

A detailed mechanistic study was carried out for the monobenzylation of the benzophenone Schiff base of glycine t-butyl ester (**10**) with cinchonidine-derived catalyst **36**. The optimal results from this study, which gave product **43** in 81% ee and 87% chemical yield are summarized in Scheme 15. The HPLC trace of the crude reaction mixture shows the minor enantiomeric product followed by the major enantiomer (**43**). This study is described in detail in a separate chapter of this monograph (31).

Scheme 15. α-Amino Acids by Enantioselective PTC: Optimal Results.

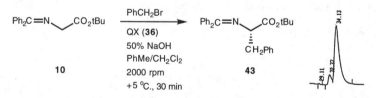

Finally, a catalytic enantioselective PTC synthesis of the protected derivative **44** of the important α,α-dialkylamino acid, α-methyl tryptophan (4j,32), was developed using the optimal catalyst, reagents and conditions derived from our earlier studies (33). Using a solid-liquid PTC system of the mixed base system, KOH/K₂CO₃, in the solvent mixture PhMe/CH₂Cl₂ (70/30, v/v) product **44** was prepared in 75% ee and >85% chemical yield in four hours at room temperature. The HPLC trace shows the mixture of enantiomers present in the crude reaction product (Scheme 16).

Scheme 16. α,α-Dialkylamino Acids by Enantioselective PTC.

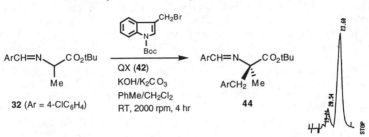

32 (Ar = 4-ClC$_6$H$_4$)

QX (42)
KOH/K$_2$CO$_3$
PhMe/CH$_2$Cl$_2$
RT, 2000 rpm, 4 hr

In summary, the use of Schiff base derivatives of amino acids and peptides in conjunction with phase transfer catalysis provides a simple yet powerful route to a variety of products. Typically, these reactions can be realized on either large or small scale under mild conditions and with considerable selectivity. The various mechanistic and product studies described here emphasize the importance of careful optimization of the numerous reaction variables in these systems.

Acknowledgments. One of us (MJO) would like to acknowledge the considerable effort and skill displayed by his students and research associates throughout the years. The names of many of these people are listed in the accompanying references. We also gratefully acknowledge the National Institutes of Health (GM 28193) for major financial support of this research.

References

1. Monographs concerning phase transfer reactions: (a) Brändström, A. *Preparative Ion Pair Extraction*; Apotakarsocieten/Hässle: Läkemedel, Sweden, 1974; (b) Weber, W. P.; Gokel, G. W. *Phase Transfer Catalysis in Organic Synthesis*; Springer-Verlag: Berlin, 1977; (c) Caubère, P. *Le transfert de phase et son utilisation en chimie organique*; Masson: Paris, 1982; (d) Keller, W. E. *Phase-Transfer Reactions. Fluka Compendium*; Georg Thieme Verlag: Stuttgart, 1986; Vol. 1; (e) *Phase Transfer Catalysis: New Chemistry, Catalysts and Applications*, Starks, C. M., Ed.; American Chemical Society: Washington, D.C., 1987; (f) Keller, W. E. *Phase-Transfer Reactions. Fluka Compendium*; Georg Thieme Verlag: Stuttgart, 1987; Vol. 2; (g) Keller, W. E. *Phase-Transfer Reactions. Fluka Compendium*; Georg Thieme Verlag: Stuttgart, 1990; Vol. 3; (h) Goldberg, Y. *Phase Transfer Catalysis. Selected Problems and Applications*; Gordon and Breach: London, 1992; (i) Dehmlow, E. V.; Dehmlow, S. S. *Phase Transfer Catalysis;* 3rd Ed.; VCH: Weinheim, 1993; (j) Starks, C. M.; Liotta, C. L.; Halpern, M. *Phase-Transfer Catalysis: Fundamentals, Applications, and Industrial Perspectives*; Chapman & Hall, N.Y., 1994.
2. O'Donnell, M. J., "Asymmetric Phase-Transfer Reactions." In *Catalytic Asymmetric Synthesis*; Ojima, I., Ed.; VCH: NY, 1993, Chapter 8.
3. (a) "α-Amino Acid Synthesis," *Tetrahedron*, Symposium-in-Print, M.J. O'Donnell, Ed.; Pergamon Press: London, 1988, Vol. 44, Issue 17; (b) Williams, R. M. *Synthesis of Optically Active α-Amino Acids*; Pergamon: Oxford, 1989; (c) Duthaler, R. O. *Tetrahedron* **1994,** *50*, 1539-1650.
4. (a) O'Donnell, M. J.; Boniece, J. M.; Earp, S. E. *Tetrahedron Lett.* **1978,** 2641-2644; (b) O'Donnell, M. J.; Eckrich, T. M. *Tetrahedron Lett.* **1978,** 4625-4628; (c) Ghosez, L.; Antoine, J. -P.; Deffense, E.; Navarro, M.; Libert, V.; O'Donnell, M. J.; Bruder, W. A.; Willey, K.; Wojciechowski, K. *Tetrahedron Lett.* **1982,** *23*, 4255-4258; (d) O'Donnell, M. J.; LeClef, B.; Rusterholz, D. B.; Ghosez, L.;

Antoine, J. -P.; Navarro, M. *Tetrahedron Lett.* **1982,** *23,* 4259-4262; (e) O'Donnell, M. J.; Bruder, W.; Wojciechowski, K.; Ghosez, L.; Navarro, M.; Sainte, F.; Antoine, J. -P. *Pept.: Struct. Funct., Proc. 8th Am. Pept. Symp.* **1983,** 151-154; (f) O'Donnell, M. J.; Bruder, W. A.; Eckrich, T. M.; Shullenberger, D. F.; Staten, G. S. *Synthesis* **1984,** 127-128; (g) O'Donnell, M. J.; Wojciechowski, K.; Ghosez, L.; Navarro, M.; Sainte, F.; Antoine, J.-P. *Synthesis* **1984,** 313-315; (h) O'Donnell, M. J.; Barney, C. L.; McCarthy, J. R. *Tetrahedron Lett.* **1985,** *26,* 3067-3070; (i) McCarthy, J. R.; Barney, C. L.; O'Donnell, M. J.; Huffman, J. C. *Chem. Commun.* **1987,** 469-470; (j) O'Donnell, M. J.; Rusterholz, D. B. *Synth. Commun.* **1989,** *19,* 1157-1165.

5. (a) O'Donnell, M. J.; Bennett, W. D.; Wu, S. *J. Am. Chem. Soc.* **1989,** *111,* 2353-2355; (b) O'Donnell, M. J.; Wu, S. D. *Tetrahedron: Asymmetry* **1992,** *3,* 591-594; (c) O'Donnell, M. J.; Wu, S.; Huffman, J. C. *Tetrahedron* **1994,** *50,* 4507-4518.

6. (a) O'Donnell, M. J.; Bennett, W. D.; Polt, R. L. *Tetrahedron Lett.* **1985,** *26,* 695-698; (b) O'Donnell, M. J.; Falmagne, J.-B. *Tetrahedron Lett.* **1985,** *26,* 699-702, (c) O'Donnell, M. J.; Falmagne, J.-B. *Chem. Commun.* **1985,** 1168-1169; (d) O'Donnell, M. J.; Bennett, W. D. *Tetrahedron* **1988,** *44,* 5389-5401; (e) O'Donnell, M. J.; Yang, X.; Li, M. *Tetrahedron Lett.* **1990,** *31,* 5135-5138; (f) O'Donnell, M. J.; Li, M.; Bennett, W. D.; Grote, T. *Tetrahedron Lett.* **1994,** *35,* 9383-9386; (g) O'Donnell, M. J.; Zhou, C.; Mi, A.; Chen, N.; Kyle, J. A.; Andersson, P. G. *Tetrahedron Lett.* **1995,** *36,* 4205-4208; (h) O'Donnell, M. J.; Zhou, C.; Chen, N. *Tetrahedron: Asymmetry* **1996,** *7,* 621-624.

7. O'Donnell, M. J.; Burkholder, T. P.; Khau, V. V.; Roeske, R. W.; Tian, Z. *Polish J. Chem.* **1994,** *68,* 2477-2488.

8. Warren, S. *Organic Synthesis: The Disconnection Approach,* John Wiley & Sons: Chichester, England, 1982, Chap. 5 (chemoselectivity), Chap. 14 (regioselectivity), Chap. 12 and 38 (stereoselectivity).

9. Dehmlow, E. V.; Richter, R.; Zhivich, A. B. *J. Chem. Res. (S)* **1993,** 504-505.

10. Typically the undesirable second alkylation of benzophenone imines of glycine derivatives can be readily avoided by the choice of mild conditions and reagents. However, see reference 4f for a practical synthesis of 1-aminocycloalkane-1-carboxylic acids by a dialkylation involving initial intermolecular monoalkylation followed by a second, intramolecular alkylation. For a comparison of the selectivity of mono- vs. dialkylation under ion-pair extraction (IPE) and PTC conditions as well as rate constants for the mono- and dialkylation of benzophenone imines under anhydrous conditions, see reference 11. Recently, Ezquerra, Moreno-Mañas and coworkers have reported preparation of dialkylated products by reaction of the benzophenone imines of amino esters with active alkyl halides under solid-liquid PTC conditions, see: Ezquerra, J.; Pedregal, C., Moreno-Mañas, M., Pleixats, R.; Roglans, A. *Tetrahedron Lett.* **1993,** *34,* 8535-8538.

11. O'Donnell, M. J.; Bennett, W. D.; Bruder, W. A.; Jacobsen, W. N.; Knuth, K.; LeClef, B.; Polt, R. L.; Bordwell, F. G.; Mrozack, S. R.; Cripe, T. A. *J. Am. Chem. Soc.* **1988,** *110,* 8520-8525.

12. O'Donnell, M. J.; Huffman, J. C.; Wu, S.; unpublished results.

13. O'Donnell, M. J.; Polt, R. L. *J. Org. Chem.* **1982,** *47,* 2663-2666.

14. Stork, G.; Leong, A. Y. W.; Touzin, A. M. *J. Org. Chem.* **1976,** *41,* 3491-3493.

15. O'Donnell, M. J.; Cook, G. K.; Rusterholz, D. B. *Synthesis* **1991,** 989-993.

16. O'Donnell, M. J.; Shullenberger, D. F.; unpublished results.

17. O'Donnell, M. J.; Bruder, W. A.; Daugherty, B. W.; Liu, D.; Wojciechowski, K. *Tetrahedron Lett.* **1984,** *25,* 3651-3654.

18. O'Donnell, M. J.; Bennett, W. D.; Jacobsen, W. N.; Ma, Y. -a.; Huffman, J. C. *Tetrahedron Lett.* **1989,** *30,* 3909-3912.

19. O'Donnell, M. J.; Bennett, W. D.; Jacobsen, W. N.; Ma, Y. -a. *Tetrahedron Lett.* **1989,** *30,* 3913-3914.

20. O'Donnell, M. J.; Arasappan, A.; Hornback, W. J.; Huffman, J. C. *Tetrahedron Lett.* **1990,** *31,* 157-160.

21. O'Donnell, M. J.; Lawley, L. K.; Pushpavanam, P. B.; Burger, A.; Bordwell, F. G.; Zhang, X. -M. *Tetrahedron Lett.* **1994**, *35*, 6421-6224.
22. (a) Dolling, U.-H.; Davis, P.; Grabowski, E. J. J. *J. Am. Chem. Soc.* **1984**, *106*, 446-447; (b) Bhattacharya, A.; Dolling, U.-H.; Grabowski, E. J. J.; Karady, S.; Ryan, K. M.; Weinstock, L. M. *Angew. Chem., Int. Ed. Engl.* **1986**, *25*, 476-477; (c) Conn, R. S. E.; Lovell, A. V.; Karady, S.; Weinstock, L. M. *J. Org. Chem.* **1986**, *51*, 4710-4711; (d) Dolling, U.-H.; Hughes, D. L.; Bhattacharya, A.; Ryan, K. M.; Karady, S.; Weinstock, L. M.; Grabowski, E. J. J. In *Phase Transfer Catalysis (ACS Symposium Series: 326)*; Starks, C. M. Ed.; American Chemical Society: Washington, D.C., 1987; Vol. 326; pp. 67-81; (e) Dolling, U.-H.; Hughes, D. L.; Bhattacharya, A.; Ryan, K. M.; Karady, S.; Weinstock, L. M.; Grenda, V. J.; Grabowski, E. J. J. In *Catalysis of Organic Reactions (Chem. Ind.: 33)*; Rylander, P. N.; Greenfield, H.; Augustine, R. L. Eds.; Dekker, 1988; pp. 65-86; (f) Hughes, D. L.; Dolling, U.-H.; Ryan, K. M.; Schoenewaldt, E. F.; Grabowski, E. J. J. *J. Org. Chem.* **1987**, *52*, 4745-4752.
23. For a computational study of the Merck PTC system and our early enantioselective PTC alkylation chemistry, see: Lipkowitz, K. B.; Cavanaugh, M. W.; Baker, B.; O'Donnell, M. J. *J. Org. Chem.* **1991**, *56*, 5181-5192.
24. For other studies of the catalytic enantioselective PTC alkylation of Schiff base **10**, see: (a) Dehmlow, E. V.; Knufinke, V. *Liebigs Ann. Chem.* **1992**, 283-285; (b) Dehmlow, E. V.; Nachstedt, I. *J. Prakt. Chem.* **1993**, *335*, 371-374; (c) Dehmlow, E. V.; Schrader, S. *Polish J. Chem.* **1994**, *68*, 2199-2208.
25. It is often useful to first prepare the racemate of a particular product using an achiral phase transfer catalyst (e.g. Bu₄NBr) and determine if the racemate is crystalline. However, it is noted again that this method is not general. For example, in one case studied, the first recrystallization of a non-racemic sample gave racemic crystals and enriched filtrate. A second recrystallization from the filtrate then gave nearly optically pure crystalline product.
26. (a) Tohdo, K.; Hamada, Y.; Shioiri, T. *Pept. Chem. 1991*; Suzuki, A., Ed. **1992**, 7-12; (b) Imperiali, B.; Fisher, S. L. *J. Org. Chem.* **1992**, *57*, 757-759; (c) Tohdo, K.; Hamada, Y.; Shioiri, T. *SYNLETT.* **1994**, 247-249; (d) Imperiali, B.; Roy, R. S. *J. Am. Chem. Soc.* **1994**, *116*, 12083-12084; (e) De Lombaert, S.; Blanchard, L.; Tan, J.; Sakane, Y.; Berry, C.; Ghai, R. D. *Bioorg. Med. Chem. Lett.* **1995**, *5*, 145-150; (f) Imperiali, B.; Roy, R. S. *J. Org. Chem.* **1995**, *60*, 1891-1894.
27. For the use of enzymatic resolution following racemic PTC alkylation with Schiff base esters, see: (a) Chenault, H. K.; Dahmer, J.; Whitesides, G. M. *J. Am. Chem. Soc.* **1989**, *111*, 6354-6364; (b) Knittel, J. J.; He, X. Q. *Pept. Res.* **1990**, *3*, 176-181; (c) Bjurling, P.; Långström, B. *J. Label. Cmpd. Radiopharm.* **1990**, *28*, 427-432; (d) Pirrung, M. C.; Krishnamurthy, N. *J. Org. Chem.* **1993**, *58*, 954-956; (e) Pirrung, M. C.; Krishnamurthy, N. *J. Org. Chem.* **1993**, *58*, 957-958.
28. Imperiali, B.; Prins, T. J.; Fisher, S. L. *J. Org. Chem.* **1993**, *58*, 1613-1616.
29. Kim, M. H.; Lai, J. H.; Hangauer, D. G. *Int. J. Pept. Protein Res.* **1994**, *44*, 457-465.
30. O'Donnell, M. J.; Wu, S.; unpublished results.
31. Esikova, I. A.; Nahreini, T.; O'Donnell, M. J., "A New Interfacial Mechanism for Asymmetric Alkylation by Phase Transfer Catalysis," chapter in this monograph.
32. Horwell, D. C.; Hunter, J. C.; Kneen, C. O.; Pritchard, M. C. *Bioorg. Med. Chem. Lett.* **1995**, *5*, 2501-2506.
33. O'Donnell, M. J.; Mi, A.; unpublished results.

Chapter 11

Use of Chiral Quaternary Salts in Asymmetric Synthesis

Takayuki Shioiri, Akira Ando, Moriyasu Masui, Toshio Miura, Toshiaki Tatematsu, Akemi Bohsako, Masayo Higashiyama, and Chiharu Asakura

Faculty of Pharmaceutical Sciences, Nagoya City University, Tanabe-dori, Mizuho-ku, Nagoya 467, Japan

Preparative methods and synthetic use of chiral quaternary salts have been developed for an enantioselective α-hydroxylation of cyclic ketones and asymmetric silyl-mediated aldol reactions. The former proceeds in good chemical yields with up to 79% ee while the latter proceeds by use of various quaternary ammonium fluorides with up to 70% ee. Tetraphenylphosphonium hydrogendifluoride proved to be a solid substitute for tetra-n-butylammonium fluoride for aldol reactions and Grignard type addition reactions. Use of chiral phosphonium hydrogendifluorides was attempted for asymmetric aldol reactions.

Phase transfer catalysis has now become a very important and general method in synthetic organic reactions. In contrast, asymmetric reactions utilizing chiral phase transfer catalysts are, in general, still in their infancy and remain to be investigated further (1, 2). We have had an interest in the design of chiral phase transfer catalysts and their application to asymmetric synthesis, on which our recent studies have been described in this article.

1. Enantioselective α-Hydroxylation of Cyclic Ketones

We have developed an enantioselective α-hydroxylation of cyclic ketones **2** with molecular oxygen catalyzed by chiral ammonium salts (3), shown in Scheme 1. We prepared and used a series of phase transfer catalysts derived from cinchona alkaloids, ephedrine, or (R,R)-cyclohexanediamine. Among them, ammonium bromide **1** derived from cinchonine (4) was found to be most efficient and the method of choice. One of the advantages of this asymmetric hydroxylation is that the reaction proceeds without resorting to impractically low temperatures. The model **4** for the structure of the transition state would account for the stereochemical course.

One of the most notable examples of the α-hydroxylation procedure is the α-hydroxylation of 7-chloro-5-methoxy-2-methyl-1-tetralone (**2a**), giving the corresponding α-hydroxyketone **3a**. The reaction smoothly proceeded in 95% yield with 79% ee using catalyst **1** in 50 % aqueous NaOH-toluene in the presence of triethylphosphite. Furthermore, (E)-2-ethylidene-1-tetralone (**5**) was oxidized to the α-hydroxy ketone **6** under the same reaction conditions using catalyst **1** in 73% yield with 55% ee.

Scheme 1

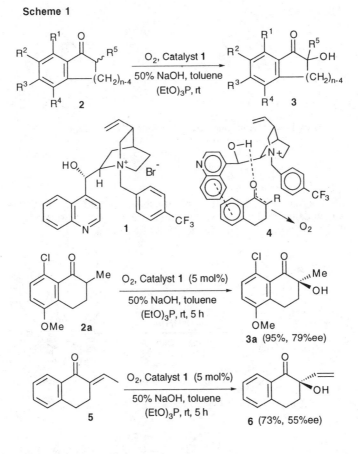

2. Asymmetric Silyl-Mediated Aldol Reactions

Utilization of the chemical affinity of the silyl groups for the fluoride anion has recently proved to be a versatile tool in organic synthesis. However, application of this tool to asymmetric synthesis has rarely been investigated. In this context, we thought that chiral ammonium fluorides would be useful for asymmetric aldol reactions between enol silyl ethers and carbonyl compounds (5). Thus, the methods for the preparation of chiral ammonium fluorides and their application to silyl-mediated aldol reactions were investigated.

2.1 Preparation of Chiral Ammonium Fluorides

We employed cinchonine as a chiral source because its derivative 1 afforded the best result in the α-hydroxylation of cyclic ketones, as described in Chapter 1. The reaction conditions to convert the cinchoninium bromide 7 to the corresponding fluoride 8a were investigated as summarized in Scheme 2 (6). Among four methods we tried, method C proved to be most easily and conveniently carried out and the method of choice. The fluoride 8a thus obtained was dried over phosphorus pentoxide and used for aldol reactions. The ^{19}F NMR spectrum of 8a showed a peak of the fluoride anion at about -124 ppm.

Scheme 2

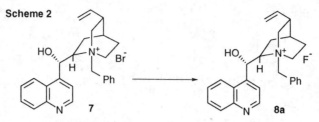

Method A 1) Amberlite IRA-410 F⁻ Form 2) Evaporation
       B 1) Amberlyst A-26 F⁻ form 2) Evaporation
       C 1) Amberlyst A-26 OH⁻ form 2) 1N HF 3) Evaporation
       D 1) AgF 2) Filtration 3) Evaporation

    Various other ammonium fluorides **8b-8m** shown in Table 1 were prepared from cinchona alkaloids, as outlined in Scheme 3 (7). All of the fluorides thus obtained showed ¹⁹F-signals at ca. -124 ppm in the ¹⁹F NMR spectra.

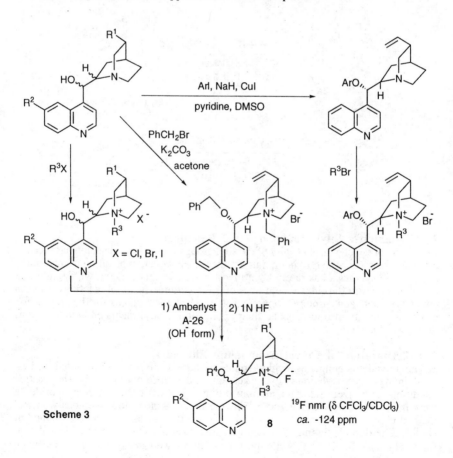

Scheme 3

Table 1. Fluoride Catalysts **8** Derived from Cinchona Alkaloids

catalyst	R^1	R^2	R^3	R^4	C_8-H	C_9-OR^4	origin
8a	$CH_2=CH$	H	$PhCH_2$	H	R	S	cinchonine
8b	$CH_2=CH$	MeO	$PhCH_2$	H	R	S	quinidine
8c	CH_3CH_2	MeO	$PhCH_2$	H	R	S	quinidine
8d	$CH_2=CH$	H	$PhCH_2$	Ph	R	S	cinchonine
8e	$CH_2=CH$	H	$PhCH_2$	1-naphthyl	R	S	cinchonine
8f	$CH_2=CH$	H	$PhCH_2$	$PhCH_2$	R	S	cinchonine
8g	$CH_2=CH$	H	$4-CF_3C_6H_4CH_2$	Ph	R	S	cinchonine
8h	$CH_2=CH$	H	$4-CF_3C_6H_4CH_2$	1-naphthyl	R	S	cinchonine
8i	$CH_2=CH$	H	$4-MeOC_6H_4CH_2$	H	R	S	cinchonine
8j	$CH_2=CH$	H	$PhCH_2$	H	S	R	cinchonidine
8k	$CH_2=CH$	MeO	$PhCH2$	H	S	R	quinine
8l	$CH_2=CH$	H	$PhCH_2$	H	R	R	9-epi-cinchonine
8m	$CH_2=CH$	H	$PhCH_2$	H	S	S	9-epi-cinchonidine

2.2 Asymmetric Aldol Reaction Catalyzed by Chiral Ammonium Fluorides Derived from Cinchona Alkaloids

With various chiral ammonium fluorides **8** in hand, we investigated the catalytic aldol reaction of the enol silyl ether **9** of 2-methyl-1-tetralone with benzaldehyde (*6, 7*). The crude aldol was treated with hydrochloric acid-methanol to remove the trimethylsilyl function, giving a mixture of the erythro and threo aldols **10** as shown in Scheme 4.

Scheme 4

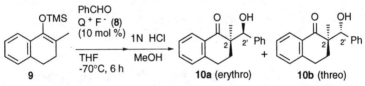

Q^+F^- (**8**): quaternary ammonium fluorides derived from cinchona alkaloids

Table 2 shows the results of the aldol reaction. The relative and absolute stereochemistries of the products **10** were assigned by the analysis of ^1H NMR spectra of the isolated diastereomers and those of the corresponding MTPA esters. The results in Table 2 suggest that the stereochemical course of the aldol reaction mainly depends on the stereochemistry of the hydroxymethyl-quinuclidine fragment, indicating that the hydroxymethyl-quinuclidine fragment is most responsible for binding with the enolate generated from the enol silyl ether **8**. Although the C-9 hydroxyl groups of the catalyst were anticipated to provide a directional handle for the ionic interaction via a hydrogen bond to the enolate anion, it is not essential in this reaction because protection of the hydroxyl group has little influence on the stereochemical outcome. Furthermore, the benzylic group and quinoline ring which might cause the π-π interaction proved to be of lesser importance. So far, the catalyst **8a** derived from cinchonine afforded a superior result: the ratio of the main erythro isomer **10a** to the minor threo isomer **10b** was 70:30 and the enantioselectivity (ee) of the erythro isomer **10a** was 70% while that of the threo isomer **10b** was 20%.

Table 2. Silyl-Mediated Aldol Reactions Catalyzed by Chiral Quaternary Salts **8** - 1

run	catalyst	isolated yield(%)	aldol **10** erythro/threo[a]	% ee erythro (conf.)[b]	threo (conf.)[b]
1	**8a**	74	70/30	70 (R,S)	20 (R,R)
2	**8b**	57	64/36	48 (R,S)	30 (R,R)
3	**8c**	46	61/39	41 (R,S)	23 (R,R)
4	**8d**	90	54/46	55 (R,S)	25 (R,R)
5	**8e**	80	49/51	51 (R,S)	27 (R,R)
6	**8f**	78	69/31	56 (R,S)	4 (R,R)
7	**8g**	37	61/39	14 (R,S)	13 (R,R)
8	**8h**	54	54/46	35 (R,S)	16 (R,R)
9	**8i**	44	76/24	64 (R,S)	17 (R,R)
10	**8j**	35	37/63	5 (R,S)	15 (S,S)
11	**8k**	75	33/67	12 (R,S)	47 (S,S)
12	**8l**	0[c]	-	-	-
13	**8m**	4.5	56/44	1.1 (R,S)	0.05 (R,R)

a) Determined by HPLC (SUMICHIRAL OA-4100) analysis. b) Configuration at C-2 and C-2' positions of the major product. c) The reaction does not proceed at all because the catalyst **8l** does not dissolve.

Incidentally, the enol silyl ethers **11** from acetophenones and pinacolone were subjected to the fluoride ion-catalyzed asymmetric aldol reaction using **8a** and **8j**, as shown in Table 3. Again, the stereochemistry of the products depends on the configuration of the hydroxymethyl-quinuclidine fragments. Bulkiness of the tert-butyl group favors the asymmetric efficiency in the aldol reaction.

Table 3. Silyl-Mediated Aldol Reactions Catalyzed by Chiral Quaternary Salts **8** - 2

a : Ph   b : 4-ClC$_6$H$_4$ c : 4-MeOC$_6$H$_4$ d : 2,4-(MeO)$_2$C$_6$H$_3$ e : Me$_3$C

run	catalyst	R	product	isolated yield of **12** (%)	% ee	config. of major product
1	**8a**	Ph	**12a**	76	39.5[a]	S[d]
2	**8j**	Ph	**12a**	46	35.5[a]	R[d]
3	**8a**	4-ClC$_6$H$_4$	**12b**	55	42[a]	S[d]
4	**8a**	4-MeOC$_6$H$_4$	**12c**	70	36.5[a]	S[d]
5	**8a**	2,4-(MeO)$_2$C$_6$H$_3$	**12d**	73	25[b]	S[d]
6	**8a**	Me$_3$C	**12e**	62	62[c]	S[e]
7	**8j**	Me$_3$C	**12e**	55	39[c]	R[e]

a) Determined by ^{1}H NMR analysis of the corresponding acetyl ester in the presence of Eu(hfc)$_3$. b) Determined by ^{1}H NMR analysis of the corresponding (R)-MTPA ester. c) Determined by HPLC (SUMICHIRAL OA-4100) analysis. d) Determined by the specific rotation value (S. H. Mashuragui and R. M. Kellog, *J. Org. Chem.*, **1984**, *49*, 2513). e) Determined by the comparison of retention times of HPLC (SUMICHIRAL OA-4100).

2.3 Asymmetric Aldol Reaction Catalyzed by Chiral Ammonium Fluorides Derived from (R)-1-Phenylethylamine

To improve the asymmetric efficiency in the aldol reaction shown in Scheme 4, we investigated the utilization of the chiral ammonium fluorides **14a** and **14b**, which were prepared from (R)-1-phenylethylamine (**13**) as shown in Scheme 5.

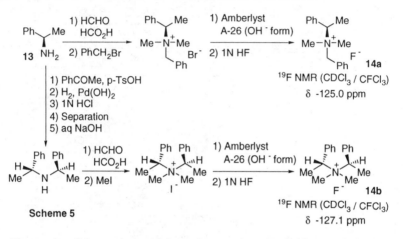

Scheme 5

The results of the catalytic aldol reaction are shown in Table 4 (8). The reaction analogously proceeded with 10 mol % of the fluoride **14a** or **14b** in tetrahydrofuran at -70°C, but both the chemical yield and stereoselectivity could not be improved. These fluoride catalysts were not readily soluble in tetrahydrofuran; therefore, a co-solvent such as acetonitrile, dimethylformamide, or dimethoxyethane was used. Addition of acetonitrile or dimethylformamide was found to increase the yield and diastereomeric excess while the enantioselectivity was not improved at all. Increase of the chemical yield could be explained by two factors: (1) increasing solubility by solvation of the ammonium cation in the dipolar aprotic solvents, and (2) increasing reactivity of the fluoride anion which suffered less solvation.

Table 4. Silyl-Mediated Aldol Reactions Catalyzed by Chiral Quaternary Salts **14**

run	catalyst (mol%)	reaction solvent	conditions	yield (%) erythro/threo[a]	%ee[a] erythro (R,S)[b]	threo (R,R)[b]
1	14a (10)	THF	-70°C, 6h	58 (56/44)	2.1	3.4
2	14a (20)	THF	-70°C, 6h	76 (28/72)	1.5	7.8
3	14a (10)	THF-MeCN(7/3)	-70°C, 6h	71 (85/15)	0.5	4.5
4	14b (10)	THF	-70°C, 6h	25 (30/70)	0.3	5.7
5	14b (10)	THF-MeCN(7/3)	-70°C, 6h	94 (89/11)	1.8	0.2
6	14b (5)	THF-MeCN(7/3)	-70°C, 6h	76 (88/12)	0.6	0.2
7	14b (10)	THF-DMF(2/8)	-70°C, 0.5h	97 (92/8)	0.2	0.7
8	14b (10)	THF-DME(2/8)	-70°C, 6h	21 (46/54)	0.2	0.8

a) Determined by HPLC (SUMICHIRAL OA-4100) analysis.   b) Configuration at C-2 and C-2' positions of the major product.

The reaction would proceed under kinetic control via acyclic transition states **15a** and **15b**, in which the former would be favored because of the lack of steric interaction between the phenyl group of benzaldehyde and the cyclohexadienyl ring of the enolate, as shown in Scheme 6.

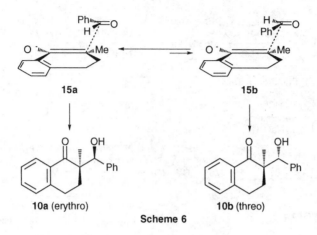

15a   **15b**

10a (erythro)   **10b** (threo)

Scheme 6

2.4 Asymmetric Aldol Reaction Catalyzed by Chiral Ammonium Fluorides Having Axial Asymmetry

We then turned our attention to utilizing 1,1'-binaphthyl compounds having axial asymmetry as a fluoride catalyst for the aldol reaction. The binaphthyl ammonium fluorides **17** were prepared from the known amine **16** (9, 10), as shown in Scheme 7 (8, 11).

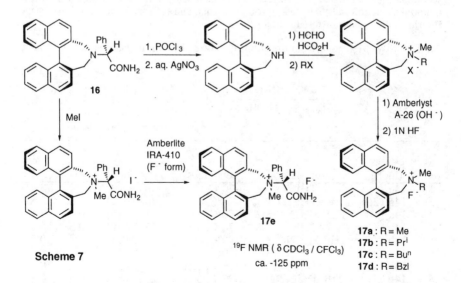

16

1. POCl₃
2. aq. AgNO₃

1) HCHO
   HCO₂H
2) RX

1) Amberlyst
   A-26 (OH⁻)
2) 1N HF

MeI

Amberlite
IRA-410
(F⁻ form)

17e

¹⁹F NMR (δ CDCl₃ / CFCl₃)
ca. -125 ppm

17a : R = Me
17b : R = Prⁱ
17c : R = Buⁿ
17d : R = Bzl

Scheme 7

Utilizing these ammonium fluorides **17**, we investigated the same asymmetric aldol reaction, as shown in Table 5 (*8*). The yields of the aldol products were moderate to excellent. The ratio of the erythro and threo isomers was dependent on the catalyst used, and the threo isomer was generally predominant. The enantiomeric efficiency was not superior. Use of a mixture of tetrahydrofuran and acetonitrile in the case of **17d** increased the chemical yield, but both the diastereo- and enantioselectivity of the threo isomer decreased.

Table 5. Silyl-Mediated Aldol Reactions Catalyzed by Chiral Quaternary Salts **17**

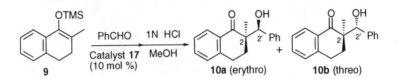

run	catalyst	config. of catalyst	reaction solvent	conditions	yield (%) erythro/ threo[a]	%ee[a,b] erythro	threo
1	**17a**	S	THF	-70°C, 6h	78 (41/59)	7.6 (R,S)	1.3 (R,R)
2	**17b**	S	THF	-70°C, 6h	83 (69/31)	1.9 (S,R)	1.7 (R,R)
3	**17c**	S	THF	-70°C, 6h	94 (50/50)	5.2 (R,S)	1.3 (R,R)
4	**17d**	S	THF	-70°C, 3h	80 (28/72)	1.7 (S,R)	3.8 (S,S)
5	**17e**	SR	THF	-70°C, 6h	30 (36/64)	8.5 (R,S)	8.0 (S,S)
6	**17e**	SR	THF-MeCN[c]	-70°C, 6h	61 (42/58)	7.9 (R,S)	0.6 (S,S)

a) Determined by HPLC (SUMICHIRAL OA-4100) analysis. b) Configuration at C-2 and C-2' positions of the major product. c) Ratio: 7:3.

2.5 Tetraphenylphosphonium Hydrogendifluoride as a Fluoride Anion Source

Although phosphonium salts are well-known as phase transfer catalysts, it is rather curious that no asymmetric synthesis using chiral phosphonium salts has been reported. Because tetraphenylphosphonium hydrogendifluoride, $Ph_4P^+HF_2^-$, proved to be a surprisingly powerful source of fluoride ion (*12*), we then explored the potential of $Ph_4P^+HF_2^-$ as a catalyst for aldol reactions, which might promise the application of chiral phosphonium fluorides to asymmetric aldol reaction. Thus, $Ph_4P^+HF_2^-$ was prepared from the corresponding bromide by our standard methodology (treatment with Amberlyst A-26 (OH- form) and then 1N hydrofluoric acid).

As shown in Table 6, various enol silyl ethers **18** rapidly and smoothly reacted with aldehydes **19** in dimethylformamide to give the aldol products **20** in good to excellent yields (*13*). When the formation of two diastereoisomeric products was possible, erythro isomers were predominantly formed. Dimethylformamide proved to be better than tetrahydrofuran, acetonitrile, dichloromethane, or diethyl ether as a reaction solvent.

Furthermore, extension of the $Ph_4P^+HF_2^-$ catalyzed reaction to the Grignard type addition of phenyltrimethylsilylacetylene (**21**) to carbonyl compounds **22** also afforded the alkynols **23** under analogous reaction conditions to those of the aldol reaction, as shown in Table 7 (*13*).

Table 6. $Ph_4P^+HF_2^-$ Catalyzed Aldol Reactions

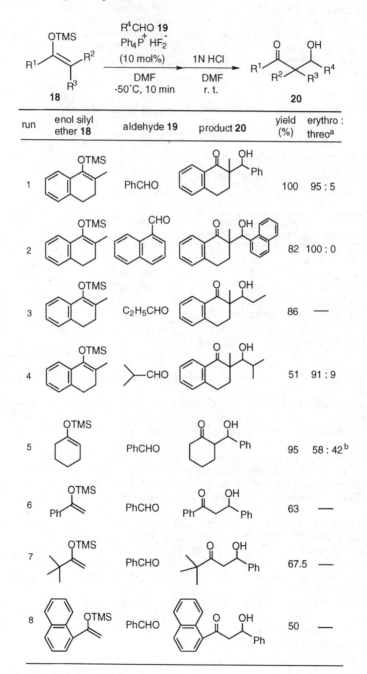

run	enol silyl ether **18**	aldehyde **19**	product **20**	yield (%)	erythro : threo[a]
1		PhCHO		100	95 : 5
2				82	100 : 0
3		C_2H_5CHO		86	—
4		CHO		51	91 : 9
5		PhCHO		95	58 : 42[b]
6		PhCHO		63	—
7		PhCHO		67.5	—
8		PhCHO		50	—

a) Determined by HPLC analysis using SUMICHIRAL OA-4100. b) Determined by the isolated yield.

Table 7. Addition of 1-Phenyl-2-trimethylsilylacetylene to Carbonyl Compounds

run	R¹COR² **22**	catalyst (mol%)	product **23**	isolated yield (%)
1	PhCHO	3		64
2	Ph₂CO	8		54
3		5		46
4	PhCCH=CHPh (trans)	3		49

2.6 Asymmetric Aldol Reaction Catalyzed by Chiral Phosphonium Hydrogendifluorides

In place of achiral phosphonium salts such as $Ph_4P^+HF_2^-$, we designed some binaphthylphosphonium hydrogendifluorides as a chiral catalyst for asymmetric aldol reactions. The known chiral phosphines **24** (*14*) were quaternized with alkyl halides, followed by treatment with Amberlyst A-26 (OH⁻ form) and then 1N hydrofluoric acid which afforded the required chiral phosphonium hydrogendifluorides **25**, as shown in Scheme 8 (*8*).

Table 8 shows the results of the utilization of these chiral phosphonium salts **25** in the aldol reaction between the enol silyl ether **9** of 2-methyltetralone and benzaldehyde (*8*). Application of similar reaction conditions used for the $Ph_4P^+HF_2^-$ catalyzed reactions in dimethylformamide revealed that the diastereoisomeric efficiency was good to yield the erythro isomer **10a** as the major product while very poor enantiomeric efficiency was observed. One of the reasons for poor enantiomeric efficiency might be due to the possible solvation of the phosphonium cation with dimethylformamide. Replacement of dimethylformamide with tetrahydrofuran or diethyl ether, however, resulted in a poor chemical yield as well as lower diastereoselectivity, though slightly better enantioselectivity of the erythro isomer **10a** was observed. Addition of acetonitrile to tetrahydrofuran raised the yield and diastereoselectivity but with poor enantioselectivity.

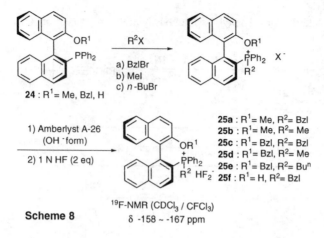

24 : R^1= Me, Bzl, H

R^{2}X
a) BzlBr
b) MeI
c) *n*-BuBr

1) Amberlyst A-26
   (OH$^-$ form)

2) 1 N HF (2 eq)

25a : R^1= Me, R^2= Bzl
25b : R^1= Me, R^2= Me
25c : R^1= Bzl, R^2= Bzl
25d : R^1= Bzl, R^2= Me
25e : R^1= Bzl, R^2= Bu^n
25f : R^1= H, R^2= Bzl

Scheme 8

^{19}F-NMR (CDCl$_3$ / CFCl$_3$)
δ -158 ~ -167 ppm

Table 8. Silyl-Mediated Aldol Reactions Catalyzed by Chiral Quaternary Salts **25**

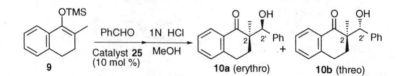

9

PhCHO   1N HCl

Catalyst **25**   MeOH
(10 mol %)

10a (erythro)   +   **10b** (threo)

run	catalyst	mol%	solvent	reaction conditions	yield (%) erythro/threo[a]	%ee[a] erythro	threo
1	25a	10	DMF	-50°C, 6h	50 (89/11)	0.2	1.2
2	25a	10	THF-MeCN(7/3)	-70°C, 6h	69 (82/18)	4.3	0.4
3	25b	20	DMF	-50°C, 10 min	85 (87/13)	0.2	1.9
4	25c	10	DMF	-50°C, 6h	60 (91/9)	0.1	0.2
5	25c	10	THF	-70°C, 6h	25 (54/46)	8.8	1.3
6	25c	10	Et$_2$O	-70°C→-5°C, 8h	37 (45/55)	0.4	0.5
7	25c	10	THF-MeCN(7/3)	-70°C, 6h	45 (81/19)	5.0	1.7
8	25d	10	THF-MeCN(7/3)	-70°C, 6h	94 (97/3)	2.1	0.8
9	25e	5	THF-MeCN(7/3)	-70°C, 6h	28 (54/45)	0.4	0.5
10	25f	10	THF	-70°C, 6h	3 (53/47)	0.6	1.0
11	25f	10	THF-MeCN(7/3)	-70°C, 6h	8 (59/41)	0.1	2.4

a) Determined by HPLC (SUMICHIRAL OA-4100) analysis.

So far we have investigated the α-hydroxylation of cyclic ketones and the silyl-mediated aldol reaction; the chiral quaternary ammonium salts derived from cinchonine have given the best results. We are now further exploring much superior efficient chiral quaternary salts.

Acknowledgment We thank the Ministry of Education, Science, Sports and Culture, Japan for Grants-in-Aid for Scientific Research for partial financial support of this work.

References and Notes
1. O'Donnell, M.J. Asymmetric Phase Transfer Reactions, In *Catalytic Asymmetric Synthesis*; Ojima, I., Ed.; VCH Publishers, Inc.: New York, 1993, Chapter 8, pp 389-411.
2. Shioiri, T. Chiral Phase Transfer Catalysis, In Handbook of Phase Transfer Catalysis; Sasson, Y., Neumann, R., Ed.; Blackie Academic & Professional: Glasgow, in press.
3. Masui, M.; Ando, A.; Shioiri, T. *Tetrahedron Lett.* **1988**, *29*, 2835-2838.
4. Hughes, D.L.; Dolling, U.-H.; Ryan, K.M.; Schoenewaldt, E.F.; Grabowski, E.J.J. *J. Org. Chem.* **1987**, *52*, 4745-4752 and references therein.
5. Nakamura, E.; Shimizu, M.; Kuwajima, I.; Sakata, J.; Yokoyama, K.; Noyori, R. *J. Org. Chem.* **1983**, *48*, 932-945.
6. Ando, A.; Miura, T.; Tatematsu, T.; Shioiri, T. *Tetrahedron Lett.* **1993**, *34*, 1507-1510.
7. Shioiri, T.; Bohsako, A.; Ando, A. *Heterocycles* **1996**, *42*, 93-97.
8. Shioiri, T.; Higashiyama, M. Unpublished results.
9. Maigrot. N.; Mazaleyrat, J.P.; Welvart, Z. *J. Org. Chem.* **1985**, *50*, 3916-3918.
10. Stará, I.G.; Stary, I.; Závada, J. *Tetrahedron: Asymmetry* **1992**, *3*, 1365-1368.
11. The quaternary ammonium nitrogen of **17** is also a chiral center, and a mixture of diastereomers might be formed by quaternization. However, all of the chiral ammonium salts seem to be stereochemically pure because the N-methyl signal of each compound shows a singlet peak in each ^{1}H NMR spectrum, though the absolute stereochemistry of the chiral nitrogen center is not clear.
12. Brown, S.J.; Clark, J.H.; Macquarrie, D.J. *J. Chem. Soc. Dalton Trans.* **1988**, 277-280 and references therein.
13. Bohsako, A.; Asakura, C.; Shioiri, T. *Synlett* **1995**, 1033-1034.
14. Uozumi, Y.; Tanahashi, A.; Lee, S.-Y.; Hayashi, T. *J. Org. Chem.* **1993**, *58*, 1945-1948.

Chapter 12

Fluoride Anion as a Base and a Nucleophile in Phase-Transfer Catalysis of Uncharged Species

Yoel Sasson, Noam Mushkin, Eli Abu, Samuel Negussie,
Salman Dermeik, and Ami Zoran

Casali Institute of Applied Chemistry, Hebrew University of Jerusalem,
Jerusalem 91904, Israel

Fluoride anion in apolar environment is highly receptive to protic compounds. Non stoichiometric hydrogen bonded adducts, of the general formula $R_4N^+F^-(HY)_n$ are readily formed, in aprotic solvents, between quaternary ammonium (quat) fluoride salts and molecules such as water, hydrogen halides, alcohols, amines and even carbon acids. Collateral interactions exist between the hydrogen bond donor HY and the acceptor F^-. While the nucleophilicity of the latter is diminished, induced nucleophilicity is conferred upon HY. In this study the above principle was practiced in several phase transfer reactions of neutral HY molecules. Selective nucleophilic aromatic substitutions and β–dehydrohalogenations were perused under mild conditions. Synthesis and some properties and other related applications of quat fluorides are also presented.

Phase Transfer Catalysis of Neutral Molecules

Introduction. Lipophilic quaternary ammonium salts (quats) are, by far, the most prevalent and frequently used phase transfer catalysts. The well authenticated phase transfer mechanism perceive the ammonium catalyst as a liquid anion exchanger which both extract and activate a base or a nucleophile. According to this concept quats are "single site" catalysts with capacity limited to negatively charged reactants or reagents. Some recent studies suggest that protic neutral molecules interact with quats in organic apolar media via hydrogen bond formation which result in extending the scope of phase transfer catalysis to the domain of *neutral* reactants. Numerous uncharged molecules such as hydrogen halides, water, hydrogen peroxide, hypochlorous acid, hydrogen cyanide, ammonia, metal salts and colloids, hydrophilic alcohols carboxylic acids were transferred and reacted in organic phase.

H-Bonded Quat Adducts. In an apolar environment the anion paired with a large lipophilic quaternary onium cation tend to interact with protic molecules to form stable solvates with the general formula $Q^+X^-(HY)_n$.*(1)* The adduct is non stoichiometric and the magnitude of n depends primarily on the nature, mainly the electronegativity and hardness of the anion X and of the acid HY, the dimension and geometry of Q and the characteristics of the solvent. The solvating compound does not necessarily have to be protic. Polarizable, soft acid, molecules, such as halogens, notably bromine, also form stable adducts with quats. The latter with the general formula $Q^+X^-(Br)_n$ or $Q^+Br_n^-$.

Particularly stable adducts are formed with various Bronsted acids when Q is a large cation (and consequently a soft acid) and X^- is a small and highly electronegative anion such as fluoride or hydroxide. Such "naked anions" are characterized as strong bases and strong nucleophiles and are prone to complex formation via hydrogen bonding. Very strong interactions are thus observed between e.g. quat fluorides and hydrogen fluoride or between quat hydroxides with water.

Stability of Quats and Quat Adducts. The hygroscopic nature of quats, particularly those paired with a hydrophilic anion, is well known. The unique chemical, electrical and thermal properties of quat hydrates render them useful materials for various applications.

When the aptitude of fluoride or hydroxide ion pairs towards hydrogen bond complex formation cannot be satisfied, e.g. in a highly anhydrous environment, intramolecular interactions are eminent . This frequently results in self destruction of the quaternary ammonium cation *via* Hoffman elimination or *via* nucleophilic reverse Menschutkin substitution. Quats which carry C-H bond β to the ammonium cation are particularly susceptible to Hoffman elimination. This is shown in equation 1:

$$2R_4N^+F^- \longrightarrow R_4N^+HF_2^- + R_3N + R'CH=CH_2 \tag{1}$$

Consequently, tetraethylammonium fluoride (TEAF) is by far less stable (decomposition at 80º) than the tetramethylammonium (TMAF) counterpart (decomposition at 180º). Steric hindrance which restrains the access to the sensitive β position also assists in improving stability. The decomposition temperature of symmetrical quats with alkyl chains from C3 to C8 is gradually increasing. Unsymmetrical quat fluorides with one methyl group such as tricaprylmethylammonium (aliquat 336-F) and trihexadecylmethylammonium, are particularly stable with the latter compound decomposing only above 240º.(2)

Thus, hydration, and solvation in general, reduces the nucleophilicity of the anion in phase transfer systems but at the same time contribute to the stability of the catalyst.

Induced Nucleophilicity. The interaction in the hydrogen bonded complex $Q^+F^-(HY)_n$ is mutual. Thus, as the nucleophilicity of X^- is decreasing, sometimes dramatically,(3) upon solvation, the induced nucleophilicity of HY is increasing accordingly. Maximum impact is induced in HY when 1:1 stoichiometry exists (n=1). Thus the basicity and nucleophilicity of one molecule of water H-bonded to quat fluoride is higher than that of a dihydrate and so on with the trihydrate and the tetrahydrate.With certain number of solvating water, alcohol or carboxylic acid molecules, quat fluorides loose their potential as fluorinating agents and transmute to a hydroxide, alkoxide or carboxylate source. Even carbon acids with pKa as high as 27 such as acetylenes, halomethanes or aliphatic nitriles can be ionized to some extent through H-bonding to fluoride ion pairs in non-polar milieu.

Quantitative measure of the charge distribution in various fluoride H-bonded adducts can be found in the 19F NMR shifts of the fluoride anion.(4)

Quat Bifluoride Adducts. Exceptionally stable hydrogen bonded complexes are formed between ammonium compounds with hydrogen fluoride. The extraction of hydrogen fluoride from water into organic phase has been critically reviewed by Eyal.(5,6) Cousseau and Albert have prepared tetrabutylammonium dihydrogentrifluoride by contacting tetrabutylammonium fluoride in dichloroethane with aqueous mixture of KF and HF or of KHF_2 and HF. (equation 2) (7)

(2)

$$Bu_4N^+ F^-_{(org)} + HF\text{-}KHF_{2(aq)} \longrightarrow Bu_4N^+ H_2F_3^-_{(org)}$$

This trifluoride compound as well as its polymeric derivative were found to be far more stable than the corresponding K, Cs, Rb and even tetramethylammonium counterparts (note, however, report by Tamura et al.(8) who have utilized potassium dihydrogen trifluoride for the ring opening of epoxides). $Bu_4NH_2F_3$ was utilized for the hydrofluorination of activated acetylenes under very mild conditions and for the regio- and stereoselective conversion of epoxides to fluorohydrines.(9) Tetrabutylammonium bifluoride was prepared by Landini (10) and coworkers and by Bosch (11) and colleagues. It was utilized in both groups for aliphatic and aromatic nucleophilic substitutions.(12)

Quaternary ammonium fluorides are preprepared, or advantageously, made up *in situ* by an anion exchange process, customarily with potassium fluoride as the primary fluoride source. A typical example is the Michael reaction which readily takes place in the presence of potassium fluoride under phase transfer conditions.(13). The acidic Michael donor was presumably activated *via* hydrogen bonding with the fluoride ion-pair.

Fluoride reagents are relatively mild and non-hydrolytic bases. Consequently they are anticipated to function as more selective reagents in base mediated reactions ordinarily effectuated by hydroxylic bases. In addition, fluoride bases can potentially be recycled by simple thermal treatment. An authoritative survey of the synthetic applications of fluoride salts as bases was published by Clark.(14)

Preparation of Quaternary Ammonium Fluoride Salts

Of the many routes leading to quat fluorides the most economical and straight forward procedure is the direct halide exchange between potassium fluoride and quats chlorides or bromides in methanol at 25° (equation 3): (1)

(3)

$$Q^+X^-_{(org)} + KF_{(aq)} \rightleftharpoons Q^+F^-_{(org)} + KX_{(aq)}$$

This exchange reaction attains useful equilibrium composition only when performed in polar, preferably protic solvents, in the presence of strictly controlled amount of water. Over 97% conversion was measured when phenol/toluene solvent mixture was applied with the potassium fluoride originally containing 4% (w/w) water. For practical reasons, methanol, that gave 95% conversion, was advocated as the solvent of choice. The role of water in this exchange reaction is particularly interesting and relates to the significance of water in phase transfer systems in general. Thus, both in the total absence of water and with a large excess of water no exchange is observed.

It was inferred that interchange can take place only with hydrated fluoride anion under conditions where the leaving anion (chloride or bromide) remained unhydrous and thus inactive.(15) This resolved the inactivity of sodium fluoride in this exchange reaction. The solubility ratio of KF/KCl in water at 25° is 2.7 while the corresponding figure for NaF/NaCl is 0.12. Evidently, it can be deduced that in the sodium system, when minute amount of water is present, chloride, and not fluoride, would be preferentially hydrated thus inhibiting the exchange.

The exchange process in methanol is best carried out in two consecutive stages with intermediate filtration to yield an almost pure methanol solution of the quat fluoride. This solution is carefully evaporated under vacuum at no higher than room temperature to avoid decomposition. Almost anhydrous quat fluoride is obtained. This could be stabilized by controlled addition of water to form hydrates with a known composition.

A flowchart for the synthesis of quaternary ammonium fluorides in overall isolated yield of 90% is shown in Figure 1.

Quaternary Ammonium Fluoride Salts as Analytical reagents

Total Anion Determination. A remarkable characteristic of long-chain quaternary ammonium salts is their high solubility in water. Aqueous solutions of up to 0.3M of tetraoctylammonium fluoride (TOAF) could be prepared. This solution is a useful universal analytical reagent for titration of practically any anion or anion mixture. Any dissolved anion, with the exception of hydroxide, instantly forms an insoluble ion pair upon addition of TOAF. Thus, even nitrate anion would precipitate as oily TOA salt according to the following (equation 4):

$$(C_8H_{17})_4N^+F^-_{(aq)} + NO_3^-_{(aq)} \longrightarrow (C_8H_{17})_4N^+NO_3^-_{(org)}\downarrow + F^-_{(aq)}$$

(4)

This reaction and its end-point are readily monitored by coductometry or by potentiometry. Figure 2 presents the conductometric titration curve of hydrogen sulfate anion using TOAF as titrant. In a similar fashion various anion mixtures could be assayed with good reproducibility and precision.(*16*)

Detection of Water in Organic Solvents. The interaction of quat fluorides with various indicators receptive to hydrogen bonding often results in color change. Thus 4-nitroaniline, which is yellow in anhydrous aprotic organic solvents, turns red upon the addition of TOAF or other quat fluoride. When, however if some water, or another protic compound, is added to the medium the color instantly turns yellow again (equation 5).

(5)

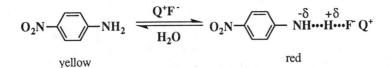

yellow   red

Due to the non-stoichiometric nature of the hydrogen bonded complexes (both of the indicator and the protic additives) this phenomenon could be applied only as a qualitative method for detection of presence of e.g. water or alcohols in various aprotic solvents.

β- Eliminations Induced by Quat Fluoride:

Dehydrochlorination of PVC. A useful fluoride catalyzed β -elimination reaction was introduced by Dermeik (*17*) who applied quaternary ammonium fluoride salts (mainly Aliquat 336® in its fluoride form) on polyvinyl chloride (Mn= 46000) in tetrahydrofuran or in *ortho* dichlorobenzene at 60º (equation 6).

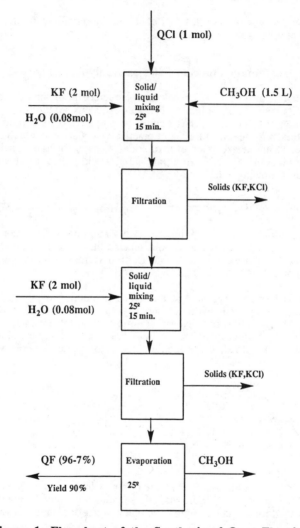

Figure 1: Flowchart of the Synthesis of Quat Fluorides (based on 1 mol QCl)

(6)

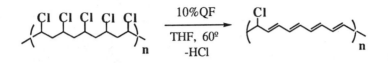

Polyene sequence with 8-16 conjugated double bonds (as determined by UV spectroscopy) were obtained with minimal consecutive cross-linking. In a second series of experiments benzene solution of quat fluorides was used for surface treatment of solid PVC films to yield 3-4 mμ thick exterior polyacetylene layer. The treated films exhibited measurable electrical conductivity which was lost after short exposure to air due to oxidation. This system was found to be milder and more selective then the corresponding phase transfer PVC elimination using hydroxide bases.(*18,19*)

Catalytic Dehydrobromination of 2-Phenylethyl Bromide. The outstanding activity of quat fluorides in promoting stoichiometric elimination reactions had been recognized many years ago.(*20*) TEAF was found to enhance dehydrohalogenations 800 times faster than potassium *tert*- butoxide. Quats are known to act as catalysts in these eliminations also in the absence of additional base.(*21*) We have now examined the elimination reaction of 2-phenylethyl bromide in benzonitrile at 145º to yield styrene and HBr, and compared the reaction rate in four different systems as follows: [1] With one equivalent of TBAB. [2] With two equivalents of potassium carbonate and 10 mol% of TBAB. [3] With one equivalent of TBAF. [4] With two equivalents of KF and 10 mol% of TBAB. The last experiment can be formulated as follows (equation 7):

(7)

$$PhCH_2CH_2Br + 2KF \xrightarrow[\text{145º, 30 min}]{\text{PhCN, 10\% TBAB}} PhCH=CH_2 + KBr + KHF_2$$
$$85\%$$

The following yields of styrene (mol%) were measured after 30 min: [1] 20 [2] 11 [3] 100 [4] 85. These results clearly suggest that quat fluorides prepared *in-situ* are a feasible alternative to the preprepared reagents. Unfortunately, however, our anticipation that the inorganic products would consist of recyclable KF.HBr adducts did not prove feasible.

Role of water. As could be expected for a solid/liquid phase transfer system the water content of the system had a marked effect on the observed rate of reaction 6. Figure 3 presents the first order rate constant as function of the water weight fraction in the potassium fluoride. A maximum is obtained at 12% of water. This result, which is notably different than the maximum at 4% observed in the exchange reaction 3, evidently suggest that the rate determining step in the dehydrobromination reaction 6 is not the fluoride/bromide exchange (despite the different nature of the solvent system). Since the mechanism apparently involves formation of a QF.HBr adduct in the bulk of the organic phase in a step that could only be retarded by the presence of water and therefore cannot possibly be the RDS. The only possible RDS is the subsequent neutralization of QF.HF by potassium fluoride at the solid/liquid interface which is seemingly at optimal conditions with the observed 12% of water.

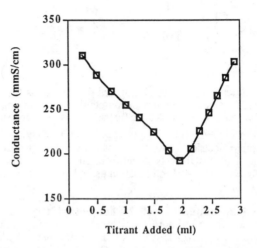

Experimental Conditions: 100 ml of 2.6×10^{-4}N KHSO$_4$ titrated with 0.03N TOAF.
Reprinted from reference 16: "Fatty Ammonium Fluoride Salts as Universal Analytical
Reagents for Total Anion Determination" (Copyright 1990) with kind permission of
Elsevier Science-NL Sara Burgerhartstraat 25, 1055 KV Amsterdam, The Netherlands.

Figure 2: Conductometric Titration of Hydrogen Sulfate with TOAF
(Reproduced with permission from reference 16. Copyright 1990 Elsevier Science.)

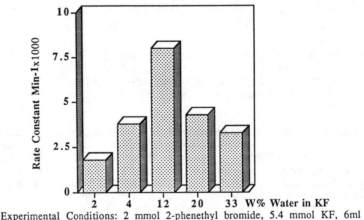

Experimental Conditions: 2 mmol 2-phenethyl bromide, 5.4 mmol KF, 6ml
benzonitrile, 10% TBAB, 145°

**Figure 3: The Effect of Water on the Rate of Fluoride Promoted
Dehydrobromination (equation 7)**

Aromatization of Hexachlorocyclohexane. Applying KF/TBAB catalytic system we were able to demonstrate the effectual and selective triple elimination reaction of hexachlorocyclohexane (γ isomer) to yield 1,2,4- trichlorobenzene (equation 8).

(8)

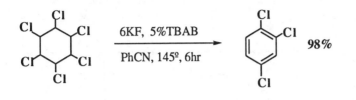

$$\text{6KF, 5\%TBAB} \quad \text{PhCN, } 145^\circ, 6\text{hr}$$

98%

Proton-Halonium and Deuterium Exchange in Carbon Acids Catalyzed by Quat Fluorides

Chlorinating of Malonate Esters. If no neutralization of an acid is required in a given reaction, quat fluorides can function as a true chemical catalysts. A unique example of this class of reactions is the equilibrium of halonium-proton exchange between perhaloalkanes and carbon acids. Thus dialkylalkylmalonates readily react with carbon tetrachloride in the presence of 1 mol% TBAF trihydrate according to the following: (equation 9)

(9)

$$\text{EtCH(COOEt)}_2 + \text{CCl}_4 \;\underset{25^\circ}{\overset{1\% \text{ TBAF}}{\rightleftharpoons}}\; \text{EtCCl(COOEt)}_2 + \text{CHCl}_3$$

The equilibrium reaction proceeds to a conversion of 99% which is achieved after 1 hour at 25°. As shown in Figure 4, the rate of reaction 9 follows a second order kinetics with a rate constant k= 1.49 lit/mol.min (neglecting the slow reverse reaction). The catalyst suffers from activity loss due to various side reactions and partial protection is achieved with addition of small amount of solid potassium carbonate (that had no direct effect on the catalytic cycle).

Chlorination of Other Substrates. Other acidic substrates that could be chlorinated by CCl4/TBAF system were acetylenes, fluorenes, nitriles and other polyhalomethanes. Similar to PTC/CCl4/NaOH (22) systems, selective chlorination is observed only with monovalent carbon acids. With unsubstituted malonate esters, dimers are the major products (equation 10) but conversion is very low as the catalyst is rapidly deactivated due to the release of HCl.

(10)

$$2\,\text{CH}_2(\text{COOEt})_2 \;\xrightarrow[25^\circ \quad -\text{HCl}]{\text{TBAF/CCl4}}\; (\text{EtOOC})_2\text{C=C(COOEt)}_2 \quad \text{low conversion}$$

A related exchange is the complete scrambling of halogens in the reaction of carbon tetrachloride and bromoform: (equation 11)

(11)

$$\text{CHBr}_3 + \text{CCl}_4 \;\underset{25^\circ}{\rightleftharpoons}\; \begin{array}{l} \text{CCrCl}_3 + \text{CBr}_2\text{Cl}_2 + \text{CBr}_3\text{Cl} + \text{CBr}_4 + \\ \text{CHCl}_3 + \text{CHBrCl}_2 + \text{CHBr}_2\text{Cl} + \text{CHBr}_3 \end{array}$$

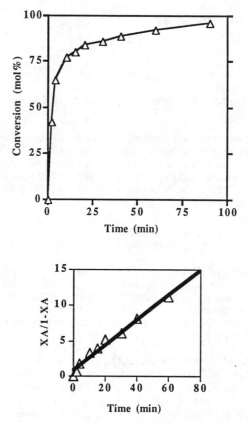

Experimental Conditions: 10 mmol CCl_4, 10 ml Diethyl Ethylmalonate, 0.2 mmol TBAF, 1 mmol K_2CO_3, 25º.

Figure 4: A Kinetic Profile and Second Order Plot of Reaction of Diethyl Ethylmalonate with CCl_4 Catalyzed by TBAF (equation 9)

Similar deuterium exchange of acidic compounds with $CDCl_3$ took place rapidly. Thus at 25° in presence of 1% TBAF complete H/D exchange was observed between $CDCl_3$ and malonate esters, phenyl acetate esters, fluorene, acetylene, ketones, aldehydes, nitroalkanes and other polyhalomethanes. Obviously no exchange could be detected with protic deuterium donors such as D_2O or CD_3OD that neutralize the catalyst by excessive hydration.

These reactions apparently proceed via a hydrogen bond complex formation between the carbon acid and the catalyst followed by concertic bond rearrangement (equation 12).

(12)

$$F^{\cdot-} - - - - H \longrightarrow R$$

$$Cl_3C \longrightarrow Cl$$

Arylation Reactions via Trifunctional Catalysis of Fluoride Anion

Fluoride is a strong attacking nucleophile, particularly under phase transfer conditions, and at the same time an agile leaving group in nucleophilic aromatic substitutions reactions that comply with SNAr mechanism in which the first step is rate determining. Integrating these facts with the aptitude of fluoride to induce nucleophilicity upon protic compounds convey to an interesting multifunctional role of fluoride in arylation reactions.

Halex Reaction and Fluoride catalyzed Hydrolysis of Activated Aryl Chlorides. We examined the solid/liquid Halex reaction of 3,4-dichloronitrobenzene (DCNB) with potassium fluoride in the presence of 10% TMAC catalyst in DMSO at 120° to yield 3-chloro-4-fluoronitrobenzene (FCNB).(23) First order rate constants measured were 0.167 1/hr in the absence of the catalyst and 1.436 1/hr in its presence. (Remarkably, TMAC performed better than any other quaternary ammonium catalyst examined). Traces of water had a dramatic effect on the conduct of the reaction, both the rate and the selectivity significantly curtailed with increased amount of water in the reaction mixture. Thus the selectivity to FCNB tumbled from 100% to 40% (at similar conversion) when KF containing 0.2% w/w water was replaced with KF with 10% water content. The reaction rate dropped accordingly by 50% and by 80% when the water content was 5% and 10% respectively. GC/MS analysis of the reaction mixtures revealed that in the presence of water hydrolysis reactions are taking place to yield phenols and ethers. (equation 13) These experiments are summarized in Figure 5.

(13)

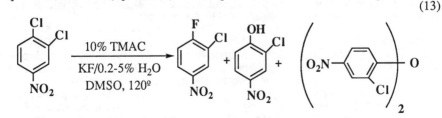

Under controlled condition the phenol and the ether became the major products. It was clarified that the hydrolysis reaction is consecutive to the fluorination step and not parallel to it. In other words, DCNB could not be directly hydrolyzed under the above conditions and FCNB was the intermediate in this process. When FCNB was exposed to the above conditions smooth conversion to hydrolysis products was observed. Similar findings were reported by Gallo and his colleagues.(*24*)

Fluoride Mediated Etherification of DCNB. Following these observations we attempted the direct aromatic etherification by applying alcohols and KF on DCNB and FCNB in presence of TMAC. Thus both substrates were reacted with phenol, methanol, isopropanol and *tert* butanol in KF/DMSO/TMAC system (equations 14,15).

$$
\text{ArF} + \text{ROH} + \text{KF} \quad \xrightarrow[\text{DMSO } 100\text{-}120^{\circ}]{10\% \text{ TMAC}} \quad \text{ArOR} + \text{KHF}_2 \tag{14}
$$

$$
\text{ArCl} + \text{ROH} + 2\text{KF} \quad \xrightarrow[\text{DMSO } 100\text{-}120^{\circ}]{10\% \text{ TMAC}} \quad \text{ArOR} + \text{KCl} + \text{KHF}_2 \tag{15}
$$

These reactions come about readily with very high selectivity and without any side nitro group reductions (e.g. formation of azoxy derivatives) which are typical to aromatic alkoxylations in presence of caustic bases.(*25,26*) The measured rate was approximately three times higher in reaction 14 relative to reaction 15, under otherwise identical conditions. Quantitative yields could be obtained, in both reactions at 120° using phenol or methanol as substrates. 90% conversion was measured in reaction 14 after 0.5 h with R=Ph and after 3.0 h with R=Me. In reaction 15 the corresponding figures were 3.5 h and 11 h respectively. Isopropanol and *tert* butanol reacted sluggishly yielding 27% and 25% respectively in reaction 14 and 8 and 5% in reaction 15 after 12h.

The Role of Soluble Fluoride. Both reactions progressed well also in the absence of the phase transfer catalyst. However, the rates were significantly improved upon introduction of more soluble fluoride. Figure 6 presents three profiles of reaction 14, with phenol as a substrate, in the absence of a catalyst, with 10% of TMAC and with stoichiometric quantity of tetramethylammonium fluoride. Second order rate laws were perceived in both reactions at a given catalyst concentration.

Arylation of Other Substrates. Under similar conditions (KF, DMSO, 10% TMAC, 120°, 12 h) we were able to prepare, in 60-90% yields, N-aryl and S-aryl derivatives by reacting phthalimide, morpholine, imidazole and thiophenol with FCNB and DCNB. Even C-arylation turned out to be feasible under F⁻ catalysis. Thus phenyl acetonitrile, diethylmalonate, ethyl phenylacetate and *para*- nitrotoluene were C-arylated under the above conditions, though in inferior yields (20-50%). The catalytic path in these reactions is probably partially inhibited by the product.(*27*)

Fluoride Catalyzed Oxidations. When the above reaction of FCNB with phenyl acetonitrile was carried out under air atmosphere, a rapid consecutive oxidation reaction took place to yield a benzophenone as the major product. The intermediate diphenylacetonitrile could be detected only in minute quantity. (equation 16)
This reaction probably proceed via the fluoride H-bonded complex of the intermediate which respond similarly to an ordinary carbanion upon exposure to oxygen.(*28*) The resulting cyanohydrine instantly decompose to the parent benzophenone. Oxidation of carbon acids (fluorene) by silica supported TBAF was reported by Clark.(*29*)

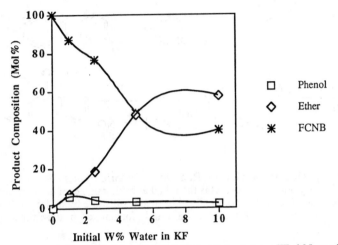

Experimental Conditions: 31.25 mmol DCNB, 62.5 mmol anhydrous KF, 6.25 mmol TMAC, 20 ml DMSO, 120º.

Figure 6: Arylation of Phenol by FCNB in the Presence of Various Fluoride Sources

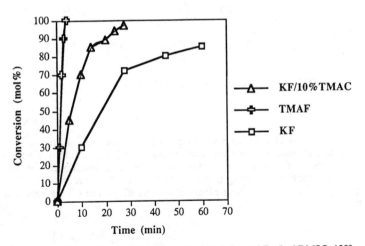

Reaction Conditions: 5 mmol FCNB, 5 mmol phenol, 5 mmol F⁻, 5 ml DMSO, 120º.

Figure 5: Effect of Water on The Product Composition Of TMAC Catalyzed Halex Reaction of DCNB

(16)

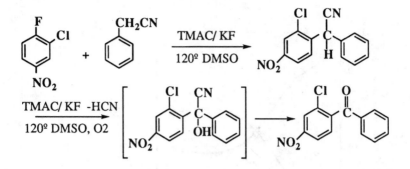

Role of Fluoride in Arylation Reactions. Fluoride anion has a triple role in these DCNB arylation reactions. It is the initial attacking nucleophile, the leaving group and the activating base that both induce the acidic substrate by hydrogen bonding and absorb the HCl released in the reaction. this is formulated in the following scheme (equation 17):

(17)

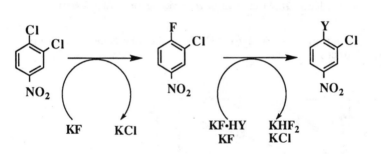

$$HY = ROH, ArOH, R_2NH, RSH, RH$$

Effect of Excess of the Substrate. In presence of the HY acid (equation 17), particularly when added in stoichiometric excess, the fluoride anion nucleophilicity is consequently impaired and the overall process is slowed down accordingly. This effect was studied in reaction 14 using different phenol/KF molar ratios. The results are shown in Figure 7. The inhibiting effect of excess of phenol on the reaction rate is apparent.

Summary

Numerous selective base mediated reactions can be performed using KF as a hydrogen bond acceptor and a mild non-hydroxide base(12) Phase transfer catalysis enhances and expands the scope of this concept and allows the combined simultaneous utilization of fluoride as a nucleophile and a base.

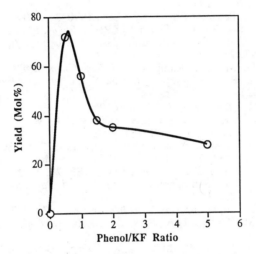

Reaction Conditions: 10 mmol DCNB, 20 mmol KF, 1 mmol TMAC, 10 ml DMSO, 80°. 6 hours.

Figure 7: Effect of Phenol Concentration on Yield of Arylation with DCNB Catalyzed by TMAC (equation 15)

162 PHASE-TRANSFER CATALYSIS

References

1   Dehmlow, E.V.; Dehmlow, S.S.; *Phase Transfer Catalysis 3rd Ed.* **1993**, VCH Weinheim.
2   Dermeik, S.; Sasson, Y. *J. Org. Chem.*, **1989**, 54, 4827.
3   Landini, D.; Maia, A.; Rampoldi, A. *J. Org. Chem.*, **1988**, 54, 328.
4   Christe, K.O.; Wilson, W.W.; Wilson, R.D.; Bau, R.; Fang, J. *J. Am. Chem. Soc.*, **1991**, 112, 7619.
5   Eyal, A. M. *Solvent Extr. Ion Exch.*, **1989**, 7, 951.
6   Eyal, A. M. *Solvent Extr. Ion Exch.*, **1989**, 7, 971.
7   Cousseau, J.; Albert, P. *Bull. Soc. Chim. Fr.*, **1986**, 910.
8   Tamura, M.; Shibakami,M.; Arimura,T.; Kurosawa,S.; Sekiya,S. *J. Fluorine Chem.*, **1995**, 70, 1.
9   Landini, D.; Penso,M.*Tetrahedron Lett.*, **1990**, 31, 7209.
10   Landini, D.; Molinari,H.; Penso, M.; Rampoldi, A. *Synthesis*, **1988**, 953.
11   Bosch, P.; Camps,F.; Chamorro,E.; Gasol,V.; Guerrero, A. *Tetrahedron Lett.*, **1987**, 28, 4733.
12   Landini, D.; Maia, A.; Rampoldi,A. *J. Org. Chem.*, **1989**, 54, 328.
13   Annunziata, R.; Cinquini, M.; Colonna, S. *J. Chem. Soc. Perkin Trans. I*, **1980**, 2422.
14   Clark, J.H. *Chem. Rev.*, **1980**, 80, 429.
15   Sasson, Y.; Arrad, O.; Dermeik, S.; Zahalka, H.A.; Weiss, M.; Wiener, H. *Mol. Cryst. Liq. Cryst.*, **1988**, 161, 495.
16   Dermeik, S.; Sasson, Y. *Anal. Chim. Acta* , **1990**, 238 ,389.
17   Dermeik, S. **1985**, Ph.D. Thesis, Hebrew University of Jerusalem.
18   Kise, H. *J. Polym. Sci., Polym. Chem.*, **1982**, 20, 3189.
19   Leplyanin, G.V.; Salimgareeva, V.N.*Izv. Akad. Nauk. Ser. Khim.*, **1995** 1886. CA124:177176.
20   Hayami,J.; Ono, N.; Kaji, A. *Tetrahedron Lett.* **1970**, 2727.
21   Halpern, M.; Zahalka, H.A.; Sasson, Y.; Rabinovitz, M. *J. Org. Chem.* **1985**, 50, 5088.
22   Jonczyk, A.; Kwast, A.; Makosza, M. *J. Org. Chem.*, **1979**, 44, 1192.
22   Rieux, C.; Langlois, B.; Gallo, R. *C.R. Acad. Sci. Paris*, **1990**, 310, Serie II, 25.
23   Sasson, Y.; Negussie, S.; Royz, M.; Mushkin, N. *J. Chem. Soc. Chem. Commun.*, **1996**, 297.
24   Rieux, C.; Langlois, B.; Gallo, R. *C.R. Acad. Sci. Paris*, **1990**, 310, Serie II, 25.
25   De la Zerda, J.;Cohen, S.; Sasson,Y. *J. Chem. Soc. Perkin Trans.II*, **1990**,1.
26   Bassani. A.; Prato, M.; Ramapazzo, P.; Quintily, U.; Scorrano, G. *J. Org. Chem.*, **1980**, 45, 2263.
27   Makosza, M.; Tomashewskij, A.A., *J. Org. Chem.*, **1995**, 60, 5425.
28   Hermann, C.K.F.; Sachdeva, Y.P.; Wolfe, J.F. *J. Heterocyclic Chem.* **1987**, 24, 1061.
29   Clark, J.H. *J. Chem. Soc. Chem. Commun.*, **1978**, 789.

Chapter 13

Anomeric Group Transformations Under Phase-Transfer Catalysis

R. Roy, F. D. Tropper[1], S. Cao, and J. M. Kim

Department of Chemistry, University of Ottawa, Ottawa, Ontario K1N 6N5, Canada

Liquid-liquid phase transfer catalysis (PTC) has been used for the synthesis of a wide range of anomeric glycosyl derivatives including C-, N-, O-, S-, and Se-glycosides. Moreover, glycosyl azides, enol ethers, esters, halides, hydroxysuccinimides, imides, phosphates, thiocyanates, and xanthates have also been stereospecifically obtained from glycosyl halides. PTC reactions have been successfully applied to mono- and disaccharides, sialic acid, 2-deoxy-2-acetamido sugars, and to D-xylose. All the reactions were efficiently accomplished at room temperature under mildly basic conditions. Ethyl acetate or in some cases dichloromethane appear to be the organic solvents of choice for fast and high yielding reactions. Examination of a number of PTC catalysts has indicated that one molar equivalent of tetrabutylammonium hydrogen sulfate provided optimum results. The reactions proceeded with good to excellent yields and were essentially complete within 3 hours. Preliminary mechanistic studies together with the scope and applications will be discussed.

Over the years, phase transfer catalysis (PTC) has emerged as a very useful and far reaching methodology for many common synthetic transformations. First devised to bring reagents and relatively simple organic substrates having incompatible solubilities together so that a chemical reaction could take place, PTC has evolved in a method where more complex substrates and chemical transformations can be adapted with great success. Included in this is the use of PTC methods in natural product synthesis. As a contribution to the use of PTC in natural product chemistry, we present here some of our synthetic work on important phase transfer catalyzed syntheses of carbohydrate derivatives. This chapter will deal with practical methods and the mechanistic aspects surrounding glycosylation reactions when performed under PTC conditions.

[1]Current address: Glycodesign, Inc., 480 University Avenue, Suite 900, Toronto, Ontario, Canada

Of the many types of natural products, carbohydrates are particularly challenging to the synthetic chemist. Carbohydrates are often delicate molecules that possess a variety of functional groups, ring sizes, branch points and stereocenters. As such they require special attention during synthesis. Traditional synthetic pathways are often long and tedious as a result of the need for elaborate chemo- and regioselective protection and deprotection sequences. These often require very reactive promoters and reagents that can be very noxious, costly and difficult to handle. There is often also the need for very polar and high boiling solvents that can prove tedious and time consuming to dry, purify and remove once the reaction is complete. Alternatively, PTC methodology is well suited to carbohydrate synthesis as reaction conditions can be quite mild and large scale preparations can be easily accommodated. Furthermore, the ability to use non-anhydrous conditions with relatively low boiling, non-anhydrous technical or reagent grade solvents can alleviate much of the technical work. In addition and of notable importance to the carbohydrate chemists is the fact that glycosides are prone to hydrolysis under acidic conditions. Thus PTC conditions are well suited to carbohydrates as reactions are carried out in near neutral or alkaline conditions that do not affect the glycosidic linkages.

Carbohydrate chemists have used PTC to perform important synthetic transformations but the practice is not common. A recent review describing PTC work surrounding carbohydrates has been published (1). Known examples of carbohydrate transformations by PTC methods include complete or regioselective protection of hydroxyl groups by acetalation (2), acylation (3), benzylation (4) or silylation (5). Chain elongation by Wittig reaction (6) as well as oxidation (7) and reduction (8) reactions have also been performed on carbohydrate substrates under PTC conditions. Although such traditional chemical transformations are equally important to the carbohydrate chemists, no other reaction is more closely identified with the carbohydrate chemists than glycosylation reactions. The anomeric center and the ability to control the stereochemistry of synthetic transformations at this position can be quite challenging and is perhaps the most important to the carbohydrate chemists.

So far, PTC has been used by other groups to provide only a limited number of glycosides (9-20). We present here our results demonstrating PTC as a simple and effective general methodology to perform a variety of stereoselective glycosylation reactions. We have shown that by using relatively labile glycosyl halides in an aqueous/organic two phase system, a variety of O-, -ON-, S-, N-, Se- and C-glycosides can be prepared in a mild, simple and efficient manner when soft nucleophiles in combination with catalysts such as tetraalkylammonium salts are used.

Using a PTC approach, we have stereoselectively synthesized a wide range of important glycosides in the form of prodrugs, glycoprobes such as lectin and enzyme substrates, precursors to neoglycoconjugates (protein and polymer), glycosyl donors for new and existing block oligosaccharide synthesis strategies, drug metabolites, glycopeptide and glycopeptoid precursors, as well as unusual glycosides that can be found in complex natural products. This chapter will describe our work in the area of PTC glycosylations and provide an insight to the mechanistic aspects surrounding this important synthetic transformation. Further, it will give an overview of some of the important glycosides that have been efficiently prepared using a PTC methodology and how they relate to the study fundamental biological phenomena.

Mechanistic Considerations

Ideally, anomeric nucleophilic substitutions under PTC conditions should occur stereospecifically. That is, α-glycosyl halides should provide exclusively the newly formed glycosyl derivatives with β-anomeric configurations. Alternatively, β-glycosyl halides should only furnish α-anomers (Scheme 1). For optimum predictability, this situation should be valid for all sugars irrespective of the halides, the nature of the protecting groups, and the conditions used.

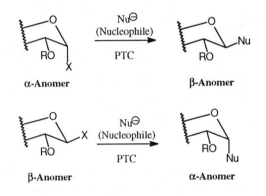

α-Anomer   β-Anomer

β-Anomer   α-Anomer

Scheme 1. Idealized anomeric nucleophilic substitutions under PTC conditions.

However, in principle, anomeric group transformations of glycosyl halides can trigger a cascade of events which may lead to complex reaction mixtures. For anomeric nucleophilic substitutions of glycosyl halides, the outcome of the reactions will ultimately depend on the structure of the carbohydrate itself, the leaving group, the nucleophile, the solvent, and the presence or absence of neighboring protecting groups. In general, when ethers are used as neighboring protecting groups, there is no possibility of anchimeric group participation. This feature, can provide easier mechanistic interpretations. Alternatively, when esters are used, their electron withdrawing properties tend to destabilize oxocarbenium ion intermediates or transition states depending on whether the reactions will undergo mono- or bi-molecular substitutions. If the process follows a pure S_N2 pathway, anomeric nucleophilic substitutions will inevitably provide glycosyl derivatives with inverted anomeric configurations (Scheme 2). Under these conditions, the nature of the protecting groups, if present, should not affect the stereochemical outcome of the reactions and the reactions will be stereospecific. However, if S_N1 conditions prevail, oxocarbenium ions will result and scrambling of the anomeric configuration will ensue. The reactions will then give rise to an anomeric mixture of the desired glycosyl derivatives. Moreover, under liquid-liquid PTC conditions, the released halide anions may compete with the nucleophiles for the phase transfer catalysts and may return in the organic phase, thus giving rise to glycosyl halides with inverted anomeric configurations. For this reason, phase transfer catalyst containing halide anions should be strictly avoided.

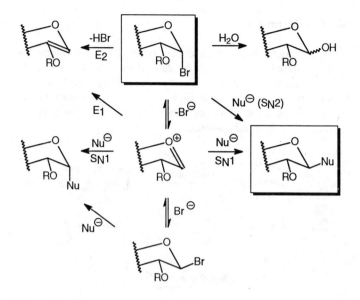

Scheme 2. Possible side-reactions for anomeric nucleophilic substitutions under liquid-liquid two phases PTC conditions.

Since it is generally easier to synthesize anomerically pure glycosyl halides having ester functionality as protecting groups, peracetylated glycosyl halides have been the most common starting materials for PTC studies. However, the hydrolysis of the ester function may occur with highly basic aqueous phase. Therefore, highly concentrated solutions of NaOH should similarly be avoided as a secondary liquid phase. Moreover, as most nucleophiles of interest are somewhat polar, the organic phase of choice should also contain solvents of relatively high polarity. Prior to our work, most PTC transformations were effected in refluxing benzene with 40% aqueous NaOH (1). We found that either EtOAc or CH_2Cl_2 and 1M NaHCO$_3$ or 1M Na$_2$CO$_3$ at room temperature gave most satisfactory results (21-23). In all the PTC reactions, elimination through E1 or E2 mechanisms together with hydrolysis accounted for some of the by-products obtained.

As mentioned above, the nature of the neighboring protecting groups may interfere with the anomeric substitutions. This is particularly true for esters such as acetates and benzoates. For instance, with 1,2-*cis* peracetylated glycosyl bromides, the lost of bromide anions followed by anchimeric group stabilization of the oxocarbenium cation by the neighboring 2-O-acetoxy group will provoke formation of an acyloxonium ion which will then block the α-side (bottom) of the molecule (Scheme 3). The approaching nucleophiles will then attack from the top (β-face) of the molecule with the resulting formation of a β-glycosyl derivative starting from an α-glycosyl halide. The neat result however, still appears as an S_N2-like reaction occurring by inversion of configuration. With harder nucleophiles, the reaction may even take the course of acyloxonium ion trapping with the resulting formation of an "orthoester" like products. For instance, 1,2-*cis* glycosyl halides **1,2**, and **5** provided exclusively thioglycosides **3** (11), **4** (18), and **6** (24) with inverted configuration (Scheme 4). The same anomeric inversions were obtained from **1** and **2** irrespective of the nature of the protecting groups on O-2.

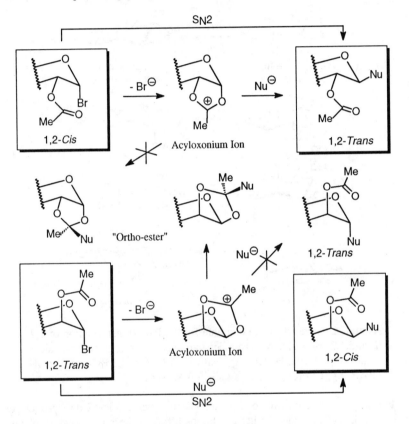

Scheme 3. Hypothesized effects of 2-acetoxy groups on the nucleophilic substitutions 1,2-*cis* and 1,2-*trans* peracetylated glycopyranosyl bromides under PTC conditions.

The most striking evidence accumulated against the "hypothesis" of anchimeric group participation in anomeric nucleophilic substitutions of glycosyl halides under PTC conditions came from experiments with 1,2-*trans* peracetylated glycosyl halides **7** and **9** (Scheme 3). In these cases, if 2-acetoxy group participation would have prevailed, glycosyl derivatives resulting from retention of configuration would have resulted. This has not been the situation in any example studied so far in our laboratory. For instance, peracetylated α-D-mannopyranosyl bromide (**7**) and α-L-rhamnopyranosyl bromide (**9**) with 1,2-*trans* diaxial configuration provided exclusively the phenyl 1-thio-β-glycosyl derivatives **8** or **10** with soft a nucleophile such as thiophenol (Scheme 4). One report however has described the exclusive formation of orthothioester when peracetylated β-D-glucopyranosyl chloride having 1,2-*trans* diequatorial configuration was treated with thiophenol in benzene and NaOH (18).

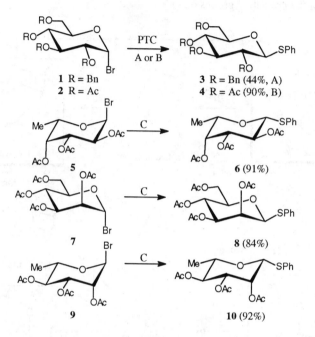

Scheme 4. Stereochemical outcome for the PTC transformation of 1,2-*cis* and 1,2-*trans* glycosyl bromides into phenyl 1-thio-glycopyranosides. Conditions, A: PhSH, CH₂Cl₂, NaOH, TEBAC, r.t.; B: PhH, NaOH, TBAB, r.t.; C: EtOAc, 1M Na₂CO₃, TBAHS, r.t.

Furthermore, treatment of peracetylated 2-acetamido-2-deoxy-α-D-gluco-pyranosyl chloride **11** under similar conditions afforded the corresponding phenyl 1-thio-glucopyranoside **12** with inverted configuration in essentially quantitative yield (25). This results and numerous others indicated that an acetamide protecting group in the C-2 position is compatible with the PTC conditions. Interestingly, the *a priori* anticipated product resulting from anchimeric group participation, that is, the oxazoline {2-methyl-(3,4,6-tri-*O*-acetyl-1,2-dideoxy-α-D-glucopyrano)-[2,1-d]-oxazoline} (see **16** below) did not constitute a major side reaction pathway for this particular reaction. Additionally, treatment of 3-deoxy ketoside **13** (sialic acid), having no protecting group suitably positioned for anchimeric group participation, provided thioglycoside **14** with inversion of configuration at the anomeric center, even though it contains a tertiary halide (Scheme 5). The fascinating result obtained for this particular carbohydrate derivative has been fully exploited for the syntheses of a wide number of other sialic acid analogs (26-32).

As anticipated, products resulting from hydrolysis, elimination, and anchimeric group participation have all been observed, although products resulting from these side-reactions were only observed in minor quantities. "Ortho-ester-like" products have only been occasionally encountered with peracetylated 1,2-*trans* glycosyl halides and "harder" nucleophiles such as azide anions (24, 25). In none of the literature surveyed, anomeric mixtures have been observed, suggesting that oxonium ion intermediates were not formed. Therefore, phase transfer catalyzed anomeric nucleophilic substitutions can be best generalized as occurring with inversion of configuration by a clean bimolecular S_N2 mechanism.

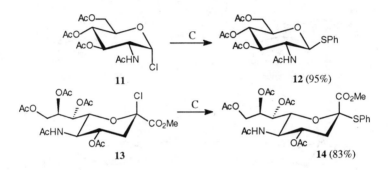

Scheme 5. Stereochemical outcome for the PTC transformation of 2-acetamido-2-deoxy (**11**) and 3-deoxy-2-ketoside (**13**) into phenyl 1-thio-glycopyranosides.

To further assess the stereochemical outcome of phase transfer catalyzed anomeric nucleophilic substitutions, anomeric pairs of 2,3,4-tri-*O*-acetyl-α/β-D-xylopyranosyl chlorides (**15, 16**) were independently prepared (Scheme 6). Each anomer was then treated with either thiophenol or sodium azide under optimized PTC conditions (EtOAc, 1M NaHCO₃, r.t., <30 min) using tetrabutylammonium hydrogen sulfate (TBAHS) as catalyst (33). In each case, the resulting peracetylated phenyl 1-thio-xylopyranosides (**17, 19**) or xylopyranosyl azides (**18, 20**) were obtained with complete anomeric inversion (Scheme 6). No trace of the other anomer could be detected from the crude reaction mixture, although trace amount of orthothioester was obtained from the 1,2-*trans* diequatorial β-D-xylopyranosyl chloride (**16**) and thiophenol. This is in sharp contrast to a previously reported reaction using β-D-glucopyranosyl chloride (18). This approach was then successfully applied to the syntheses of a large number of β-D-xylopyranosyl derivatives starting from the slightly more reactive α-bromide (33). It is also noteworthy to mention that, when treated under our mild PTC conditions in the absence of any external nucleophile, the β-chloro anomer (**16**) was isomerized to the more stable α-chloro anomer (**15**) in the presence of tetrabutylammonium chloride (TBAC) as phase transfer catalyst. The reaction showed an α:β anomeric mixture of 1:1.2 after one hour. The above experiment was used to demonstrate that the starting material can anomerize during PTC reactions and thus, care must be taken to avoid catalysts with halide anions. These results also imply that nucleophiles of poor lipophilicity have to compete against the catalyst counter anions for their efficient transfer into the organic phase.

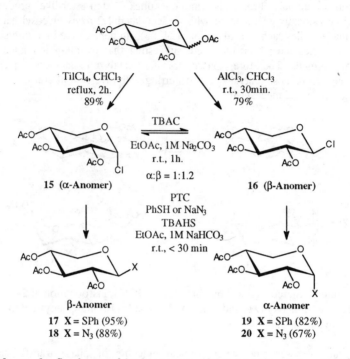

Scheme 6. Syntheses of both anomers of 2,3,4-tri-*O*-acetyl-α−(**15**)/β−(**16**) xylopyranosyl chlorides and their stereospecific transformation into anomerically inverted xylopyranosyl derivatives.

Anomeric Group Transformations

Having established the scopes and limitations of potential glycosyl halides in the above phase transfer catalyzed anomeric nucleophilic substitutions, a number a qualitative experiments were set to evaluate the other usual parameters. First, the efficiency of a number of ammonium salts were evaluated (22). Evidently, none of the reactions described above proceeded in the absence of phase transfer catalyst. Not surprisingly, the more lipophilic tetrabutylammonium cation (TBA) was slightly more effective (~2 times) than the corresponding benzyltriethylammonium (BTEA) chloride or bromide in methylene chloride. When phosphonium salts, such as ethyltriphenylphosphonium bromide, were used, the reactions were usually faster than when ammonium salts were used. However, irreversible formation of by-products, such as diphenylphosphine oxide or ylides, consumed the catalysts and complicated the purification processes. When benzene was used as solvent in place of dichloromethane with TBA bromide, the reaction rates were found to be much slower, even under refluxing conditions. More of the side-reactions described above were also produced. In the case of thiolates as nucleophiles and under stoichiometric conditions, it was found that some of the thiolates were consumed by reacting with dichloromethane (23). In those cases, EtOAc was found to be a good alternative solvent. In fact, in most of the reactions described in this chapter, EtOAc gave faster reactions and thus, is highly recommended for nucleophilic displacements of glycosyl halides.

There were no tremendous differences in rates and yields when a slight molar excess NaOH was used as the aqueous phase. In fact, it gave best results with aryloxides. However, in few circumstances, hydrolysis of anomeric halides and acetate protecting groups, together with 2-acetoxy-glycal formation (E2) did occur (26). For these reasons, 1M NaHCO$_3$ or 1M Na$_2$CO$_3$ were later preferred. The next section will describe the results obtained with a wide variety of nucleophiles under the optimum conditions just discussed.

O-Glycosyl Derivatives. Our early investigations on anomeric nucleophilic substitutions performed under PTC conditions have focused on N-acetylglucosamine derivative **11** and its transformation into a large number of *O*-aryl β-D-N-acetylglucosaminides (**21**) (21, 22). These numerous aryl glycosides with electron-donating and electron withdrawing substituents were prepared in order to probe potential electronic contributions originating from their binding to a plant lectin, wheat germ agglutinin (WGA), recognized to bind with better affinity to aryl glycosides than alkyl glycosides (34). In most cases, anchimeric group participation resulting in oxazoline formation (**22**) did not occur to any appreciable extent. Table 1 and Scheme 7 below illustrates few of the transformations achieved and the high yields obtained. The PTC conditions were also successfully applied for the syntheses of estrone pro-drug (**23**), to glycohydrolase enzyme substrates 7-hydroxy-4-methylcoumarin (**24**) and the new chromogenic substrate Fat Brown B® (**25**) (34). Anomeric esters (**26, 27**) could also be obtained in excellent yields (34). The PTC reaction was general and applicable to many other glycosyl halides. Besides aryl glycosides, other O-glycosyl derivatives were similarly prepared from acids (**26, 27**), including amino acids (**28**), N-hydroxy-succinimide (**29**) (32), dibenzylphosphate (**30**) (35), hydroxybenzotriazole (**31**) (33), and the enolates (**32, 33**) (34) taken as representative examples (Scheme 8).

Table 1. Results from the PTC Glycosidation of α-Chloride **11** with Aryloxides

Cpd 21			Cpd 21		
Aryl	X	Yields (%)	Aryl	X	Yields (%)
HO—⟨ ⟩—X	*p*-H	93	HO—⟨ ⟩—X	*p*-CN	80
	p-Me	84		*p*-CO$_2$Me	55
	p-F	77		*p*-OMe	85
	p-Cl	66		*p*-t-Bu	73
	p-Br	88		*p*-Ph	89
	p-I	83		*m*-Me	86
	p-NO$_2$	73		α-Naphthol	80
	p-CHO	67		β-Naphthol	70

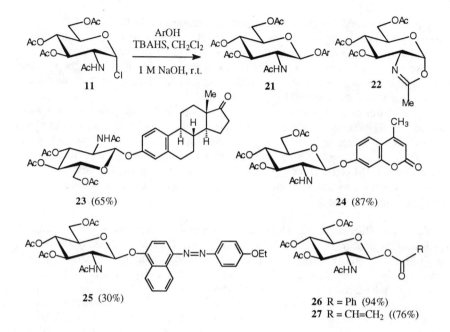

Scheme 7. Useful transformations of 2-acetamido-3,4,6-tri-*O*-acetyl-2-deoxy-α-D-glucopyranosyl chloride (**11**) into *O*-aryl glycosides and other derivatives under PTC.

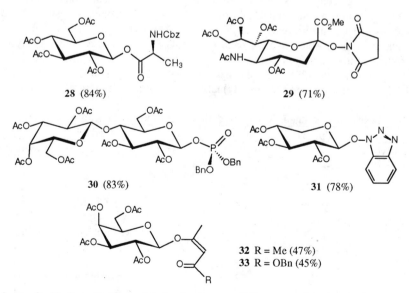

Scheme 8. Various *O*-glycosyl derivatives successfully prepared under PTC.

It is also worthy to mention that non-aromatic O-glycosides have been reported to occur under PTC conditions (10, 13). However, starting from a mixture of anomeric derivatives provided, as expected, anomeric mixtures of O-glycosides were produced.

S-and Se-Glycosyl Derivatives. Owing to the importance of hydrolytically stable glycosides towards the action of glycohydrolases and the need for potent glycosyl donors in stepwise and blockwise oligosaccharide syntheses, thio- and seleno-glycosides have triggered widespread interest. Although such glycosides were traditionally synthesized from peracetylated sugars using thiols and Lewis acids, the stereochemical outcome of these transformations were not always predictable. Based on previous successes obtained by Polish (18) and German chemists (14), we seeked to used our improved mild PTC conditions for the syntheses of both thio- and seleno-glycosides (23-25, 27-31, 34, 36).

The higher acidities and "softness" of thiols and selenols make them ideal candidates for liquid-liquid two phase PTC reactions using mildly basic conditions. To this end, we found that 1M $NaHCO_3$ or 1M Na_2CO_3, together with EtOAc as solvent and TBAHS, allowed fast and stereospecific entries into a large number of aryl thio- and seleno-glycosides, including the cumbersome sialic acid derivatives. The reaction was equally applicable to disaccharides (23, 36) and thioether-linked disaccharides (16, 17, 19). Although dichloromethane was still efficient in the above transformations, it was found that almost stoichiometric amounts of thiols and selenols could be used when EtOAc was used. We have unambiguously demonstrated that a large proportions of the thiols reacted with dichloromethane to provide solvent adducts such as bis(4-nitrophenyl)methane (23). Since thiols are prone to oxidation (disulfide formation), lower quantities of thiols can be used if the PTC reactions are performed under a nitrogen atmosphere. It was also found that thioacetic acid (28-30) and potassium xanthate (31) constituted other valuable entries into 1-thio-glycosyl derivatives (30). Using sialic acid as a representative example, a list of typical thio- and seleno-glycosides are illustrated in Scheme 9. Table 2 below also provides other examples.

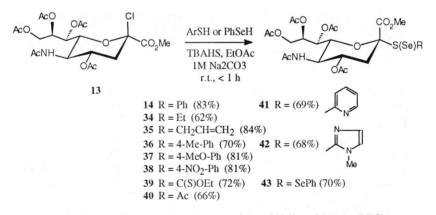

14 R = Ph (83%)   41 R = (69%)
34 R = Et (62%)
35 R = $CH_2CH=CH_2$ (84%)
36 R = 4-Me-Ph (70%)   42 R = (68%)
37 R = 4-MeO-Ph (81%)
38 R = 4-NO_2-Ph (81%)
39 R = C(S)OEt (72%)   43 R = SePh (70%)
40 R = Ac (66%)

Scheme 9. Synthesis of thio- and seleno-glycosides of sialic acid under PTC.

Glycosyl thiocyanates have also been recently reported as useful glycosyl donors by Kochetkov *et al.* (37). In their approach, they were formed by treatment of 1,2-cis glycosyl bromides with potassium thiocyanate in dry acetone and 18-crown-6 ether. Glycosyl thiocyanates were isolated as kinetic products in 55-70% yields. Using the above PTC conditions, acetobromoglucose (2) was reacted with a 10 molar excess of potassium thiocyanate. After one hour at room temperature, an 82% yield of a 10:1 mixture of β-D-glucosyl thiocyanate and isothiocyanate was obtained (34). The reaction did not proceed in CH_2Cl_2. Unlike Kochetkov's synthesis, the isothiocyanate produced by PTC had a 1,2-*trans* β-D-configuration. Thus, it must have been produced by nucleophilic substitution of bromide by isothiocyanate anion, which exists as a resonance form of thiocyanate (-S-C≡N ↔ -S=C=N-) and not by a rearrangement process as previously determined.

N-Glycosyl Derivatives. N-Glycosides constitute an important class of carbohydrate derivatives. For example, nucleosides and many glycopeptides are N-linked glycosides. A PTC synthesis of N-linked glycosides and precursors was therefore envisaged to further expand the usefulness of PTC glycosylations and potentially provide stereospecific access to this class of derivatives. The use of PTC for the syntheses of nucleosides with ribose, deoxy-ribose, and analogs has been reviewed previously and this topic will not be covered herein (1). Many useful applications have been noticed in this field. However, as a general trend, the glycosylation reactions were not stereospecific. As en entry to asparagine-linked glycopeptides, the pioneering work of Kunz *et al.* (15) has elegantly described the first use of PTC conditions for the synthesis of glycosyl azide. Under the same conditions reported above, we also found that PTC represents a versatile and high yielding entry into this class of derivatives (24, 33, 38). The chemistry involved provided stereospecific entry into glycosyl azides from glycosyl halides. All of the reactions occurred with complete anomeric inversion, as opposed to Lewis acid catalyzed treatment of peracetylated sugars with trimethylsilyl azide. Some of the carbohydrate analogs successfully transformed into glycosyl azides are listed in Table 2.

Table 2. Few Examples of PTC Transformations with Glycosyl Halides

Sugar	Nucleophiles (% Yield)					
	PhS⁻	p-NO₂PhS⁻	AcS⁻	EtO(S)CS⁻	N₃⁻	(BnO)₂P(O)O⁻
Glc	90	-	-	98	93	83
Gal	82	92	89	91	96	73
GlcNAc	93	73	45	91	98	-
Xyl	95	82	74	84	88	65
L-Fuc	91	-	-	-	94	-
Man	84	-	-	-	50	-
L-Rha	92	-	-	-	41	-
NeuAc	86	81	66	72	94	45
Lactose	92	91	-	-	98	83
Cellobiose	85	89	-	-	-	-
Maltose	89	71	-	-	97	-
Gentiobiose	77	70	-	-	-	-

Additionally, barbital (5,5-diethyl barbituric acid, pKa 7.4) and other barbiturates, represent attractive targets for PTC glycosylations. Indeed, N-glucosylation (39) and glucuronidation (40) have been claimed to be important metabolic pathways for the clearance of these drugs. Thus, in initial model studies, N-acetylglucopyranosyl chloride (11) and its analogous peracetylated galactopyranosyl bromide (44) were treated with barbital (Scheme 10) (34). However, bisglycosylation of barbital was observed in all reactions. Monoglycosylation products could be observed by thin layer chromatography when an excess of barbital (1.5 equiv) was used but seemed to disappear over the extended reaction time (~4 h) required for complete transformation of the glycosyl halides. Extensive de-O-acetylation occurred (1M Na₂CO₃) before the halides were depleted. Good yields (54-69%) of the bis glycosylated barbital 45 and 46 could be easily achieved when 2.5 molar excess of glycosyl halides to barbital was used with TBAHS in a CH_2Cl_2/ 2M Na_2CO_3 PTC system at room temperature.

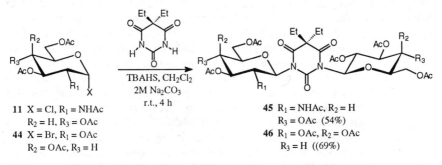

11 X = Cl, R₁ = NHAc
    R₂ = H, R₃ = OAc
44 X = Br, R₁ = OAc
    R₂ = OAc, R₃ = H

45 R₁ = NHAc, R₂ = H
    R₃ = OAc (54%)
46 R₁ = OAc, R₂ = OAc
    R₃ = H ((69%)

Scheme 10. Synthesis of N-linked glycosylated barbiturates.

C-Glycosyl Derivatives. C-Glycoside formation has triggered considerable attention in recent years. Although there is a wide variety of methods for their synthesis, no successful application of PTC has been reported to date. Our initial attempts with compounds containing active methylene groups (MeNO₂, CH₃CN, malonates, etc.) and peracetylated sugars were all met with failure. It is believed that concomitant de-O-acetylation was the source of the difficulties encountered with these smaller and "harder" nucleophiles. Attempts with 2,4-pentanedione and benzyl 3-keto-propanoate and acetobromogalactose (44) provided O-alkylation products (32, 33, ~45%) (Scheme 8) rather than the desired C-glycosides.

We did however obtained some success (39% yield) in the formation of 2-acetamido-3,4,6-tri-O-acetyl-2-deoxy-β-D-glucopyranosyl cyanide (47) when α-chloride 11 was treated with 5 molar equivalent of KCN and TBAHS (1 equiv) in a mixture of dichloromethane and 1M Na₂CO₃ at room temperature for 3.5 h (Scheme 11) (25). The low yield in this case was attributed to the exhaustive formation of oxazoline 22 and substantial de-O-acetylation. It is well known that this anion is an effective reagent for deesterification reactions. Since we have previously established that oxazoline 22 does not constitute an intermediate in PTC transformations (22), it is likely that it originated from the displacement of the cyanide anion in 47 by a favorable anchimeric group participation.

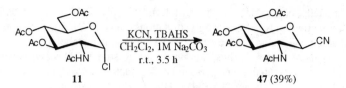

Scheme 11. Synthesis of C-glycoside under PTC conditions.

Biological Applications

Efficient access to a wide variety of glycosyl derivatives, made stereospecifically through PTC, offers practical entries into biologically useful precursors. Many of the above derivatives have been successfully transformed into more complex glycoconjugates (reviewed in 41). A large number of chromogenic substrates have been glycosidated with various sugars, including sialosides, to provide useful enzyme substrates. For instance, 4-nitrophenol and 7-hydroxy-4-methylcoumarin containing O-glycosides constitute highly efficient glycohydrolase substrates which liberate colored or fluorogenic aglycones. The same 4-nitrophenyl glycosides, whether O- or S-linked, have been transformed into antigenic protein or polymer conjugates (42). Similar 4-formylphenyl glycosides have also been directly transformed into neoglycoproteins by reductive amination with $NaBH_3CN$ (26). Syntheses of drug metabolites and pro-drug derivatives offer other interesting applications (2, 4). The syntheses of a number of plants flavonoids have also greatly benefited from high yielding PTC protocols (20). It is also obvious that the synthesis of a number a natural products containing some of the functionality discussed in this chapter will use PTC as key steps. For instance, glycosyl azides are now routinely used in solid phase glycopeptide syntheses (15).

Based on a number of available thioglycosides prepared under PTC, we recently described a new glycosylation strategy which utilized the concept of "active" and "latent" thioglycosyl donors (25, 27). The strategy is simply based on the modulation of various arylthio glycosides reactivity towards thiophilic promotors. Electron-rich (EDG) substituents (ex. H, OMe, NHAc) increase the sulfur nucleophilicity of the glycosyl donors, while electron withdrawing (EWG) substituents (ex. NO_2) render essentially inactive the corresponding glycosyl acceptors under the reaction conditions. The resulting di- or oligo-saccharides, containing EWG substituents, can be "reactivated" by transformation into EDG substituents (ex. $NO_2 \rightarrow NHAc$). The strategy is thus reiterative and allows blockwise syntheses of complex oligosaccharides.

Future prospects

Although PTC has been very useful in many circumstances, there is still numerous opportunities for improvements and fundamental studies. There is yet no systematic kinetic investigations on anomeric nucleophilic substitutions of glycosyl halides. Such studies would be very helpful to demonstrate if anchimeric group participation can occur to some extent under certain conditions. This is particularly valid for 1,2-*cis* peracetylated glycosyl halides. Many other families of carbohydrate derivatives have not been explored in details. For instance, it would be of interest to investigate the behavior of 2-deoxy aldohexoses and pentoses starting from anomerically pure glycosyl halides. The efficient syntheses of C-glycosides is still an open challenge. As more phase transfer catalysts are being developed, previously unsuccessful transformations may need to be reexamined. Kinetically versus thermodynamically controlled reactions should similarly be explored with new techniques such as solid-liquid PTC reactions. Other nucleophiles are also worth exploring. Access to glycosyl fluorides under PTC conditions would constitute a valuable achievement. Many of these topics are the subject of ongoing activities.

Conclusion

In summary, anomeric nucleophilic substitutions under PTC conditions do occur by an S_N2 pathway, irrespective of the nature of the neighboring protecting groups. Some specific cases may however rely on anchimeric group participation. Elimination and hydrolysis proceed to only a limited extent, although some nucleophiles such as 4-formylphenoxide can provide substantial E2 elimination to occur, particularly with glucosyl and galactopyranosyl halides (26). Acetate protecting groups are compatible to PTC reactions even when 1-2M NaOH is used as secondary liquid phase. Oxazolines formation may be accounted for the major by-products resulting from the treatment of 2-acetamido-2-deoxy sugars with poor nucleophiles or when the nucleophiles possess substantial nucleofugal characters. EtOAc and dichloromethane provide optimum results and allow the PTC reactions to proceed at room temperature. EtOAc is preferred over dichloromethane for thiols while the latter is the solvent of choice for hydroxysuccinimide and few other nucleophiles. So far, tetrabutylammonium hydrogen sulfate (TBAHS) has constituted the most efficient phase transfer catalyst. Its use avoids anomeric scrambling of the starting glycosyl halides and it offers a proper balance of hydrophobicity for efficient and rapid transfer of the accompanying anions into the organic phase. A brief schematic summary of all the transformation discussed in this chapter is presented in Scheme 12 below.

Typical Procedure

To a solution of the glycosyl halides (1 equiv.) and tetrabutyl-ammonium hydrogen sulfate (1 equiv.) in ethyl acetate (1.0 mL/100 mg of sugar) (or dichloromethane) were added the nucleophiles (1.2-3 equiv.) and 1M sodium carbonate (1.0 mL/100 mg of sugar). The reaction mixture was vigorously stirred at room temperature for 1 h or until the starting material was completely consumed (<3 h) as judged by TLC monitoring using a mixture of ethyl acetate and hexane as eluent. The solution was then diluted with ethyl acetate and the organic phase separated. The organic solution was washed with saturated sodium bicarbonate (2 × 20 mL), water (1 × 20 mL), and brine (20 mL), then dried over anhydrous sodium sulfate and concentrated. The crude residues were purified by silica gel column chromatography using a mixture of ethyl acetate and hexane as eluent. Pure glycosyl derivatives were obtained after silica gel column chromatography and usually recrystallised from ethanol (or Et_2O-hexane).

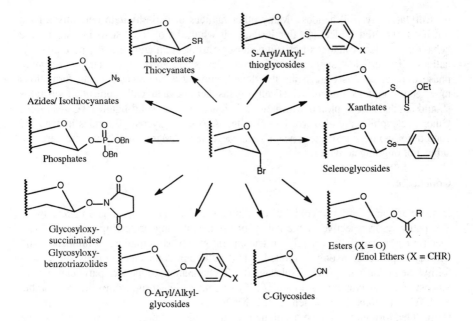

Scheme 12. General transformation describing the usefulness of PTC in anomeric nucleophilic substitutions.

References

1.   Roy, R. *Phase Transfer Catalysis in Carbohydrate Chemistry.* In: Handbook of Phase Transfer Catalysis; Sasson, Y, Neumann R., Eds. Chapman & Hall, Glasgow. In press.
2.   Kim, K. S.; Szarek, W. *Synthesis* **1978**, 48-50.
3.   Garegg, P. J.; Kvarnström, I.; Niklasson, G.; Svensson, S. C. T. *J. Carbohydr. Chem.* **1993**, *12*, 933-953.
4.   Szeja, W.; Kokt, I.; Grynkiewicz, G. *Recl. Trav. Chim. Pays-Bas* **1989**, *108*, 224-226.
5.   Zhdanov, Yu. A.; Alekseev, Yu. E.; Palui, I. N.; Serebrennika, E. V.; *Dokl. Akad. Nauk SSSR* **1983**, *268*, 883-885.
6.   Valpuesta, F. M.; Durante-Lanes, P.; López-Herrera, F. J. *Tetrahedron* **1990**, *46*, 7911-7922.
7.   Morris, P. E.; Kiely, D. E.; Vigee, G. S. *J. carbohydr. Chem.* **1990**, *9*, 661-673.
8.   Bessodes, M.; Antonakis, K. *Tetrahedron Lett.* **1985**, *26*, 1305-1306.
9.   Hansson, C.; Rosengren, E. *Acta Chem. Scand., Ser. B* **1976**, *30*, 871-875.
10. Di Cesare, P.; Gross, B. *Carbohydr. Res.* **1977**, *58*, C1-C3.
11. Brewster, K.; Harrison, J. M.; Inch, T. D. *Tetrahedron Lett.* **1979**, 5051-5054.

12. a) Dess, D.; Kleine, H. P.; Weinberg, D. V.; Kaufman, R. J.; Sidhu, R. S. *Synthesis* **1981**, 883-885; b) Kleine, H. P.; Weinberg, D. V.; Kaufman, R. J.; Sidhu, R. S. *Carbohydr. Res.* **1985**, *142*, 333-337; c) Kleine, H. P.; Sidhu, R. S. *Carbohydr. Res.* **1988**, *182*, 307-312.

13. Szeja, W. *Synthesis* **1988**, 223-224.

14. a) Rothermel, J.; Faillard, H. *Biol. Chem. Hoppe-Seyler* **1989**, *370*, 1077-1084; b) Rothermel, J.; Weber, B.; Faillard, H. *Liebigs Ann. Chem.* **1992**, 799-802; c) Rothermel, J.; Faillard, H. *Carbohydr. Res.* **1990**, *208*, 251-254; d) Rothermel, J.; Faillard, H. *Carbohydr. Res.* **1990**, *196*, 29-40; e) Reinhard, B.; Faillard, H. *Liebigs Ann. Chem.* **1994**, 193-203.

15. Kunz, H.; Waldmann, H. *Angew. Chem. Int. Ed. Engl.* **1985**, *24*, 883-885.

16. Hamacher, K. *Carbohydr. Res.* **1984**, *128*, 291-295.

17. Bogusiak, J.; Szeja, W. *Carbohydr. Res.* **1985**, *141*, 165-167.

18. Bogusiak, J.; Szeja, W. *Polish J. Chem.* **1985**, *59*, 293-298.

19. Chrétien, F.; Di Cesare, P.; Gross, B. *J. Chem. Soc., Perkin Trans. 1,* **1988**, 3297-3300.

20. Demetzos, C.; Skaltsounis, A.-L.; Tillequin, F.; Koch, M. *Carbohydr. Res.* **1990**, *207*, 131-137.

21. Roy, R.; Tropper, F. *Synthetic Commun.* **1990**, *20*, 2097-2102.

22. Roy, R.; Tropper, F. D. *Can. J. Chem.* **1991**, *69*, 817-821.

23. Tropper, F. D.; Andersson, F. O.; Grand-Maître, C.; Roy, R. *Carbohydr. Res.* **1992**, *229*, 149-154.

24. Cao, S.; Roy, R. *Carbohydr. Lett.* **1996**, *2*, 27-34.

25. Cao, S., *Ph. D. Dissertation*, **1996**, Department of Chemistry, University of Ottawa.

26. Roy, R.; Tropper, F. D.; Romanowska, A.; Letellier, M.; Cousineau, L.; Meunier, S. J.; Boratynski, J. *Glycoconjugate J.* **1991**, *8*, 75-81.

27. Cao, S.; Meunier, S. J.; Andersson, F. O.; Letellier, M.; Roy, R. *Tetrahedron: Asymmetry* **1994**, *5*, 2303-2312.

28. Roy, R.; Zanini, D.; Meunier, S. J.; Romanowska, A. *ACS Symposium Series* **1994**, *560*, 104-119.

29. Roy, R.; Zanini, D.; Meunier, S. J.; Romanowska, A. *J. Chem. Soc., Chem. Commun.* **1993**, 1869-1872.

30. Park, W. K. C.; Meunier, S. J.; Zanini, D.; Roy, R. *Carbohydr. Lett.* **1995**, *1*, 179-184.

31. Tropper, F. D.; Andersson, F. O.; Cao, S.; Roy, R. *J. Carbohydr. Chem.* **1992**, *11*, 741-750.

32. Cao, S.; Tropper, F. D.; Roy, R. *Tetrahedron* **1995**, *51*, 6679-6686.

33. Kim, J. M.; Roy, R. *J. Carbohydr. Chem.* **1996**, *15*, in press.

34. Tropper, F. D., *Ph. D. Dissertation*, **1992**, Department of Chemistry, University of Ottawa.

35. Roy, R.; Tropper, F. D.; Grand-Maître, C. *Can. J. Chem.* **1991**, *69*, 1462-1467.

36. Tropper, F. D.; Andersson, F. O.; Grand-Maître, C.; Roy, R. *Synthesis* **1991**, 734-736.

37.   Kochetkov, N. K.; Klimov, E. M.; Malysheva, N. N.; Demchenko, A. V. *Carbohydr. Res.* **1991**, *212*, 77-91.
38.   Tropper, F. D.; Andersson, F. O.; Braun, S.; Roy, R. *Synthesis* **1992**, 618-620.
39.   Yu, C.-F.; Soine, W. H.; Thomas, D. *Med. Chem. Res.* **1992**, *12*, 410-418.
40.   Neighbors, S. M.; Soine, W. H.; Paibir, S. G. *Carbohydr. Res.* **1995**, *269*, 259-272.
41.   Roy, R. Design and Syntheses of Glycoconjugates. In *Modern Methods in Carbohydrate Synthesis*; Khan, S. H., O'Neil, R., Eds.; Harwood Academic, Amsterdam, Netherlands, **1995**; pp. 378-402.
42.   Roy, R.; Tropper, F. D.; Romanowska, A. *Bioconjugate Chem.* **1992**, *3*, 256-261.

Chapter 14

Michael Addition of 2-Phenylcyclohexanone to Chalcone Under Phase-Transfer Catalysis Conditions

Enrique Diez-Barra, Antonio de la Hoz, Sonia Merino, and
Prado Sánchez-Verdú

Department of Organic Chemistry, Faculty of Chemistry,
University of Castilla–La Mancha, E–13071 Ciudad Real, Spain

The regioselectivity in the Michael addition of 2-phenylcyclohexanone to chalcone can be controlled by the appropriate choice of PTC conditions. In the absence of solvent, and using ephedrinium salts as catalyst, 2-substituted derivatives are predominant, while solid-liquid conditions and TBAB afford mainly 6-substituted derivatives. A stabilizing π–π interaction between the ephedrinium salt and the thermodynamic enolate is proposed to explain the results. A transition state with the participation of the metal ion is suggested to explain the obtained diastereoselectivity.

Michael addition is one of the most useful reactions in organic synthesis (1,2). The required basic catalysis for this reaction can efficiently be carried out by PTC methods. In the last twenty years, addition of acetamidomalonate (3,4), 2,4-pentanodione (5), methyl acetoacetate (5), fluorene (5), substituted indanones (6), nitroalkanes (7), thiolate (7), cyanide (8), and methyl phenylacetate (8) anions to a large number of Michael acceptors, mainly α,β-unsaturated ketones, under PTC have been reported.

Several approaches to the stereochemical control in the Michael addition and the alkylation of enolates have been envisaged. One of these is the use of chiral phase transfer catalysts. In this field, chiral crown ethers (8), quininium (7), ephedrinium (9-11), cinchoninium and cinchonidinium (13-20) salts have been used.

In most cases a π–π interaction between the catalyst and one of the reactants has been proposed to justify the enantiomeric and diastereomeric excesses achieved. Two types of interactions have been suggested:

    i) enolate-catalyst (figure 1, a and b)
    ii) substrate-catalyst (figure 1, c).

In the first case the catalyst blocks preferentially one face of the enolate, while the second case represents a higher accessibility of one face of the Michael acceptor.

The enolate-catalyst interaction has been studied in the alkylation of benzophenone Schiff bases (14-17) and 2-substituted indanones (6, 12, 19, 20). In both cases enantiomeric excesses up to 66% were obtained. A theoretical study establishing where the enolates bind to the catalyst (cinchoninium and cinchonidinium salts) was reported (21). Both theoretical and chemical studies showed how and why the use of chiral catalysts bearing π-systems benefit stereoselective reactions.

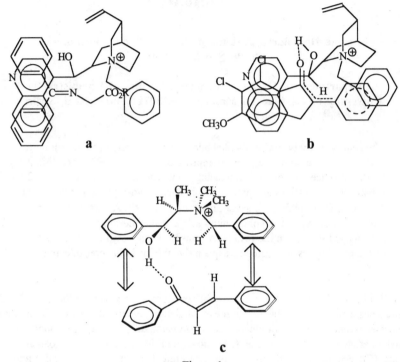

Figure 1

The substrate-catalyst interaction has been studied in the Michael addition of ethyl acetamidomalonate to chalcone (9, 10). A wide set of substituted ephedrinium and cinchoninium salts were used to show the influence of the π–π interactions on the stereoselectivity. Ephedrinium salts were found to be more efficient than cinchoninium salts. In the absence of solvent (solid-liquid PTC in solvent-free conditions) a reinforcement of the π–π interaction was suggested as indicated by the increase of the enantiomeric excess.

Although both interactions, enolate-catalyst and substrate-catalyst, have been described, no examples of competition between them have been reported.

Here the Michael addition of 2-phenylcyclohexanone 1 to chalcone 3 has been considered. In this case both reactants, the enolate and the Michael acceptor, have a π-system, so it may be possible to study which interaction, enolate-catalyst and Michael acceptor-catalyst, predominates.

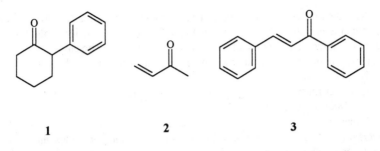

1   **2**   **3**

Regioselectivity

Methylvinylketone **2** was used first as a Michael acceptor. Reactions afforded in all PTC conditions, liquid-liquid (l-l), solid-liquid (s-l) and solid-liquid without solvent (ws), and tetrabutylammonium (TBAB) and N-benzylephedrinium bromides as catalysts, the expected 2-(3'-oxobutyl)-2-phenylcyclohexanone **4** as a racemic mixture. However, using chalcone **3**, mixtures of 2-phenyl-6-(1',3'-diphenyl-3'-oxopropyl)cyclohexanone (**A** products) and 2-phenyl-2-(1',3'-diphenyl-3'-oxopropyl)cyclohexanone (**B** products) were obtained. The formation of products **A**, not observed in the reaction with methylvinylketone, requires the attack from the carbon-6, i.e. from the kinetic enolate. Table I shows that the product distribution also depends on both the PTC conditions and the catalyst used.

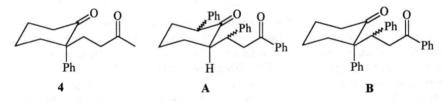

4   **A**   **B**

Table I. Reaction of **1** and **3** under PTC conditions. Base: KOH, temperature: 20°C, reaction time: 24 h

entry	PTC	Q⁺	yield	A	B	A/B ratio
1	l-l	TBAB	20	12	8	60/40
2		Ephedrinium	27	13	14	48/52
3	s-l	TBAB	46	42	4	91/9
4		Ephedrinium	44	31	13	70/30
5	ws	TBAB	42	17	25	40/60
6		Ephedrinium	48	13	35·	27/73

A isomers are favored in solid-liquid conditions and using TBAB. The combination of both effects gives a high regioselectivity (91/9). B isomers are favored in solvent-free conditions and using ephedrinium salts, combination of both gives a good regioselectivity (27/73). The lower regioselectivity was observed in liquid-liquid conditions.

As indicated before, formation of regioisomer **A** requires the attack from the kinetic enolate but in order to analyze the effect of the reaction conditions, it is necessary to know the reaction pathway. Two possibilities may be envisaged:

i) An initial attack from the kinetic enolate together with attack from the thermodynamic enolate

ii) Initial formation of regioisomer **B** and subsequent equilibration of enolates after a retroMichael reaction from **B**.

Blank experiments show that transformation from isomers **B** into isomers **A**, and the reversal transformation from **A** to **B**, under the reaction conditions shown in Table I, do not occur. Thus, it can be confirmed that the two enolates are present (figure 2) and the approach to the kinetic enolate, in position 6 is less hindered that of the thermodynamic enolate, in position 2. The larger size of chalcone *vs* methylvinylketone may explain the change in regioselectivity.

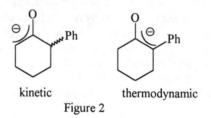

kinetic   thermodynamic

Figure 2

However, in the reactions with chalcone, using N-benzylephedrinium bromide the ratio of **B** regioisomers increases in all PTC conditions. So, it must be concluded that the presence of ephedrinium salts favors the thermodynamic enolate and the change in regioselectivity is produced by a stabilizing interaction between the catalyst bearing π-moieties and the thermodynamic enolate. In the thermodynamic enolate, the phenyl group and the enolate are planar and the interaction with the ephedrinium salt is more effective (figure 3). In solvent-free conditions the effect must be enhanced (*9*).

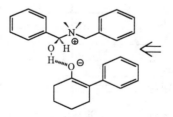

Figure 3

As a preliminary conclusion, ephedrinium salts produce an effect not previously described, a change in regioselectivity. These changes in the regioselectivity cannot be explained considering a chalcone-enolate interaction, shown in figure 1c, because this interaction does not change the accessibility of both enolates.

By both the enolate-catalyst interaction, shown in figure 3, and the chalcone-enolate interaction, enantiomeric excesses were expected. However, no ee were obtained, even when using large amounts of catalyst (up to 30 %). These results suggest that the enolate-catalyst interaction does not require the hydrogen bond between the oxygen of the enolate and the hydroxy group of the catalyst, as proposed by O'Donnell (*16*). In fact, O-benzylated ephedrinium salts produced the same regio and stereoselectivity. In this way both faces of the enolates are similarly favored.

It can be concluded that the enolate-catalyst interaction predominates over the chalcone-catalyst interaction and that this enolate-ephedrinium salt interaction stabilizes the thermodynamic enolate.

The effect of temperature, cation and the strength of the base are shown in tables II and III.

Table II. Effect of the temperature and base on the regioselectivity

entry	catalyst	base	T (°C)	yield	A	B	A/B ratio
1	TBAB	KOH	20	42	17	25	40/60
2	Ephedr.	KOH	20	48	13	35	27/73
3	TBAB	KOH	60	49	41	8	84/16
4	Ephedr.	KOH	60	42	26	16	60/40
5	TBAB	KOBu^t	20	36	14	22	39/61
6	Ephedr.	KOBu^t	20	43	13	30	30/70

Table III. Effect of the cation effect on the regio and diastereoselectivity. Q$^+$, ephedr.; T=60°C; t=24h

entry	base	yield	A1/overall A (yield)	B1/overall B (yield)	A/B ratio
1	LiOBu^t	49	87/100 (45)	50/100 (4)	92/8
2	NaOH	47	74/100 (42)	0/100 (5)	89/11
3	KOH	42	61/100 (26)	31/100 (16)	60/40
4	Cs$_2$CO$_3$	32	40/100 (10)	50/100 (22)	31/69

High temperatures favor the kinetic enolate and diminish the catalyst effect (entries 1 and 2 *vs* 3 and 4), while the strength of the bases does not change the selectivity (entries 1 and 2 *vs* 5 and 6).

The nature of the cation has also a strong influence on the regioselectivity (table III). **B** isomers are obtained from the thermodynamic enolate that produces a more crowded transition state while **A** isomers are obtained from the kinetic enolate that produces a less crowded transition state. This effect is enhanced with small cations such as Li$^+$. Thus, the ratio **A/B** increases as the size of the cation is reduced, i.e. from Cs$^+$ to Li$^+$ (table III).

Diastereoselectivity

A careful separation by column chromatography, and an accurate structural assignment by ^{1}H and ^{13}C-NMR spectroscopy and molecular mechanics calculation, permits the identification of four **A** diastereoisomers (2S,6R,1'S, **A1**; 2S,6R,1'R, **A2**; 2R,6R,1'R, **A3**; 2R,6R,1'S, **A4**) and two **B** diastereoisomers (2R,1'R, **B1**; 2R,1'S, **B2**). Yields of each diastereoisomer have been determined by HPLC.

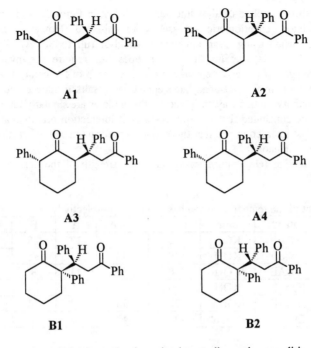

Diastereoisomer **A1** is predominant in almost all reaction conditions (table IV).

Table IV.

reaction conditions	A1/overall A (yield)	B1/overall B (yield)
ws, KOH, RT, no Q$^+$	69/100 (23)	33/100 (15)
ws, KOH, RT, TBAB	59/100 (17)	36/100 (25)
ws, KOH, RT, Ephed.	46/100 (13)	50/100 (34)
ws, KOH, 60°, TBAB	50/100 (41)	25/100 (8)
ws, KOH, 60°, Ephed.	61/100 (26)	31/100 (16)

The approaches of enolate-chalcone with and without coordination of the cation, can be considered (22). Figure 4 shows both possibilities and represents the less hindered approaches for isomers **A**, with the phenyl group equatorial, an equatorial approach of the chalcone and the chalcone phenyl group *out* of the cyclohexane ring. The approach *out* leads to 2S,6R,1'R regioisomer (**A2**), while the *out-M$^+$* leads to regioisomer 2S,6R,1'S (**A1**), i.e. coordination of the cation by the enolate and chalcone has an important role on diastereoselectivity. As **A1** is the predominant diastereoisomer the *out-M$^+$* approach produces the favored transition state. The effect of the cation in the diastereoselectivity can be deduced from the data in table III. Again the size of the cation affects the diastereoselectivity favoring the less crowded transition state, that produces isomer **A1**, thus the ratio **A1**/overall **A** increases from Cs$^+$ to Li$^+$. The approach *in*, with the phenyl group over the cyclohexane ring (figure 5a), which produces isomer **A2**, is more crowded and is observed in a lesser extent.

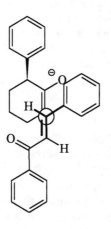
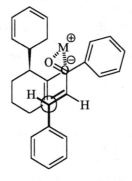

out   out-M⁺

Figure 4

a)

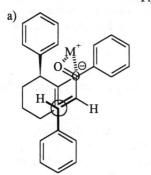
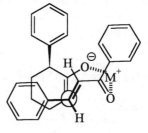

out   *in*

b)

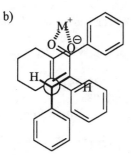
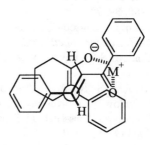

Figure 5

Regarding **B** isomers (figure 5b) there is not a clearly favored approach: in the *in* approach, which produces isomer **B2**, the chalcone phenyl group is over the ring, while in the *out* approach, which produces isomer **B1**, a *pseudo* 1,3-diaxial phenyl-phenyl interaction exists. From the data in table III, it can be deduced that the second interaction is more important and that, in this particular case the *in* approach is less hindered.

Conclusions

The regioselectivity in the Michael addition of 2-phenylcyclohexanone to chalcone can be selected using the adequate PTC technique and in solvent-free conditions by a careful choice of the reaction conditions. Products coming from the thermodynamic enolate are favored by the use of and large cations, such as Cs^+, and ephedrinium salts as catalysts. In the first case, the use of a large cation minimize the steric interactions in the transition state. In the second, a $\pi-\pi$ interaction between catalyst and enolate must stabilize the thermodynamic enolate. The absence of solvent reinforces this interaction. Under these conditions, the catalyst-enolate interaction is preferred to the catalyst-chalcone interaction. A transition state with the participation of the metal ion is suggested in order to explain the obtained diastereoselectivity. Isomer **A1**, produced from a stable conformation of the enolate and the less hindered approach, is always the major product.

Acknowledgments

Financial support from Spanish DGICYT (PB-94-0742) is gratefully acknowledged.

References

1.   Pelmutter, P. *Conjugate Addition Reactions in Organic Synthesis*; Tetrahedron Organic Series No 9; Pergamon Press: Oxford, UK, 1992.
2.   Lee, V. J. In *Comprehensive Organic Synthesis*; Trost, B. M.; Fleming, I.; Pergamon Press: Oxford, UK, 1991, Vol. 4; Chapter 1.2.
3.   Bram, G.; Sansoulet, J.; Galons, H.; Miocque, M. *Synth. Commun.* **1988**, *18*, 367.
4.   Galons, H.; Labidalle, S.; Miocque, M.; Ligniere, B. *Phosphorus and Sulfur* **1988**, *39*, 73.
5.   Bram, G.; Sansoulet, J.; Galons, H.; Bensaid, Y.; Combet-Farroux, C. Miocque, M. *Tetrahedron Lett.* **1985**, *26*, 4601.
6.   Conn, R.S.E.; Lovell, A. V.; Karady, S.; Weinstock, L. M. *J. Org. Chem.* **1986**, *51*, 4710.
7.   Colonna, S.; Re, A. *J. Chem. Soc., Perkin Trans. 1* **1981**, 547.
8.   Dehmlow, E. V.; Knufinke *Liebigs Ann. Chem.* **1992**, 283.
9.   Loupy, A.; Sansoulet, J.; Zaparucha, A.; Merienne, C. *Tetrahedron Lett.* **1989**, *30*, 333.
10.   Loupy, A.; Zaparucha, A. *Tetrahedron Lett.* **1993**, *34*, 473.
11.   Fiaud, J. C.; *Tetrahedron Lett.* **1975**, 3495.
12.   Dolling, Ulf-H.; Davis, P.; Grabowski, J. J. *J. Am. Chem. Soc.,* **1984**, *106*, 446.
13.   Colonna, S.; Hiemstra, H.; Wynber, H. *J. Chem. Soc., Chem. Comm.* **1978**, 238.

14.   O'Donnell, M. J.; Bennett, W. D.; Wu, S. *J. Am. Chem. Soc.,* **1989,** *111,* 2353.
15.   Gasparski, C. M.; Miller, M. J.*Tetrahedron* **1991,** *47,* 5367.
16.   O'Donnell, M. J.; Wu, S.; Huffman, J. C. *Tetrahedron* **1994,** *50,* 4507.
17.   O'Donnell, M. J.; Wu, S. *Tetrahedron Asymmetry* **1992,** *3,* 591.
18.   Nerinchx, W.; Vandewalle, M. *Tetrahedron Asymmetry* **1990,** *1,* 265.
19.   Hughes, D. L.; Dolling, Ulf H.; Ryan, K. M.; Schoenewaldt, E. F.; Grabowski, E. J. J. *J. Org. Chem.* **1987,** *52,* 4752.
20.   Bhattacharya, A.; Dolling, Ulf H. *Angew. Chem., Int. Ed. Eng.* **1986,** *25,* 476.
21.   Lipkowitz, K. B.; Cavanaugh, M. W.; Baker, B.; O'Donnell, M. J. *J. Org. Chem.,* **1991,** *56,* 5181.
22.   Ouvard, N.; Rodriguez, J; Santelli, M. *Angew. Chem., Int. Ed. Eng.* **1992,** *31,* 1651.

Chapter 15

Synthesis of Sulfides, Thiol Esters, and Cyclic Polythiaethers from Thioiminium Salts with a Phase-Transfer Catalyst

Toshio Takido[1], Takayoshi Fujihira[2], Manabu Seno[1], and Kunio Itabashi[1]

[1]Department of Industrial Chemistry, College of Science and Technology, Nihon University, 1–8 Kanda Surugadai, Chiyoda-ku, Tokyo 101, Japan
[2]Technical Department and Research Laboratory, Japan Sugar Refiners' Association, 5–7 Sanbancho, Chiyoda-ku, Tokyo 102, Japan

In general, the synthesis of unsymmetrical sulfides, thiolesters, and cyclic polythiaethers have been achieved by the reactions of thiol or dithiol with alkyl or acyl halides. However, thiol compounds are hard to handle, due to their unpleasant smell, and not always simple to prepare. We have explored the synthetic method of these sulfur-containing compounds from 1-(alkylthio)ethaniminium salts, 1-(acylthio)ethaniminium salts or 1,1'-(dithioalkane)diethaniminium salts rather than the common thiols or dithiols as starting materials, under phase-transfer conditions.

Many methods regarding the preparation of unsymmetrical sulfides (*1-3*) and thiolesters (*4-8*) have been reported. Organic sulfides are useful intermediates in organic synthesis and the thiolesters are useful acylating agents in biochemical reactions (*9-10*). Furthermore, in recent years several new synthetic methods for preparing cyclic polythiaethers have been reported (*11-12*), especially because cyclic polythiaethers prefer to complex and extract soft metal ions, as opposed to cyclic polyethers which selectively complex and extract hard metal ions (*13*). We have explored the synthesis of these sulfur-containing compounds starting with 1-(alkylthio)ethaniminium salts, 1-(acylthio)ethaniminium salts or 1,1'-(dithioalkane)diethaniminium salts as starting materials, instead of thiols or dithiols, under phase-transfer conditions.

1-(Alkylthio)ethaniminium salts are prepared by the reaction of thioacetamide with alkyl halides. 1-(Acylthio)ethaniminium salts are similarly prepared from thioacetamide with acyl halides, but since these compounds are unstable to moisture they are not isolated. 1,1'-(Dithioalkane)diethaniminium salts are synthesized by the reaction of dihaloalkane with two-fold excess of thioacetamide. These thioiminium salts are utilized to generate a variety of otherwise unavailable thiolate and dithiolate ions under mild conditions.

We present here an efficient method for the synthesis of various unsymmetrical sulfides, thiolesters and cyclic polythiaethers starting with 1-(alkylthio), 1-(acylthio) ethaniminium salts or 1,1'-(dithioalkane)diethaniminium salts in a two phase system and in the presence of a phase-transfer catalyst (PTC), instead of unpleasant thiols or dithiols as starting materials.

Synthesis of Unsymmetrical Sulfides and Nitriles

Freshly prepared 1-(alkylthio)ethaniminium salt **1** reacts with alkyl halide **2** to afford unsymmetrical sulfide **3** in a liquid-liquid two-phase system consisting of benzene, an aqueous solution of 30% sodium hydroxide and a catalytic amount (0.03 molar ratio) of tetrabutylammonium bromide (TBAB) as PTC for 15 minutes at room temperature under nitrogen. At the end of the reaction, the organic layer is dried and evaporated. The residual product is distilled under reduced pressure to afford pure unsymmetrical sulfide **3** (*14*) .

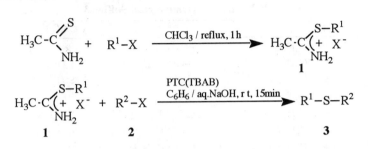

We have examined three different kinds of PTC, TBAB, methyltrioctylammonium chloride (MTOAC) and benzyltriethylammonium chloride (BTEAC), among which TBAB is shown to be the most effective. It is frequently pointed out that symmetrical ammonium ions are efficient as PTC (*15*). The reaction proceeds smoothly under these conditions and the resulting unsymmetrical sulfides are obtained in very good yields, except for **3g**, **3k** and **3l** which require 90 rather than 15 minutes to obtain similar results (Table I). The synthetic methods of unsymmetrical sulfide by using PTC are known for the reactions of S,S-dialkyl dithiocarbonates with organic halides under liquid-liquid conditions (*16*) and those of 1-(alkylthio)alkaniminium salts with organic halides under liquid-solid conditions (*17*). However, the former require extreme conditions such as boiling temperatures, and the latter give only moderate yield due to the formation of disulfide as by-product.

Furthermore, 1-(alkylthio)phenylmethaniminium salt, which is prepared from thiobenzamide with alkyl halide, reacts with alkyl halide to afford the by-products benzonitrile as well as the corresponding dialkylsulfide. In order to convert the benzonitrile into main product we have investigated the reaction of primary thioamide with benzyl chloride under phase-transfer conditions (*18*).

$$R^3-C\overset{\displaystyle S}{\underset{\displaystyle NH_2}{\big\diagup}} + 2\ PhCH_2Cl \xrightarrow[\text{C}_6\text{H}_6\,/\,\text{aq.NaOH, 30\,°C, 2h}]{\text{PTC}} R^3-CN + (PhCH_2)_2S$$
$$\qquad\qquad\qquad\qquad\qquad\qquad\qquad\qquad\qquad\qquad\qquad\mathbf{4}$$

The reaction of thioamide with two equivalents of benzyl chloride was carried out at 30°C for 2 h under two phase system consisting of benzene and an aqueous solution of 30% sodium hydroxide and TBAB(molar ratio 0.05) as PTC to yield the corresponding nitrile **4** and dibenzyl sulfide (Table II). Amides or aldoximes react with dichlorocarbene or carbon disulfide under phase transfer conditions to give corresponding nitriles, but the yields of nitriles from thioamides are not so good (*19,20*).

Table I. Yields of Unsymmetrical Sulfide 3

3	R^1	R^2	yield(%)[a]
a	C_4H_9	i-C_4H_9	88
b	C_4H_9	p-$CH_3C_6H_4CH_2$	82
c	C_8H_{17}	C_2H_5	90
d	C_8H_{17}	C_4H_9	91
e	C_8H_{17}	$C_6H_5CH_2$	91
f	C_8H_{17}	p-$CH_3C_6H_4CH_2$	95
g	$H_2C=CHCH_2$	$C_2H_5OCOCH_2$	77 [b]
h	$C_6H_5CH_2$	C_2H_5	80
i	$C_6H_5CH_2$	C_4H_9	97
j	$C_6H_5CH_2$	C_8H_{17}	92
k	$C_6H_5CH_2$	$H_2C=CHCH_2$	80 [b]
l	$C_6H_5CH_2$	$C_2H_5OCOCH_2$	86 [b]
m	$C_6H_5CH_2$	p-$CH_3C_6H_4CH_2$	96
n	$C_6H_5CH_2$	p-$ClC_6H_4CH_2$	95
o	$C_6H_5CH_2$	p-$NO_2C_6H_4CH_2$	100
p	p-$CH_3C_6H_4CH_2$	p-$ClC_6H_4CH_2$	100
q	p-$CH_3C_6H_4CH_2$	p-$NO_2C_6H_4CH_2$	100
r	p-$CH_3C_6H_4CH_2$	p-$CH_3OC_6H_4CH_2$	92
s	acetyl glucopyranosyl	CH_3	90 [c]
t	acetyl glucopyranosyl	acetyl galactopyranosyl	86 [c]

[a] Yield of purely isolated product, based on thioacetamide.
[b] Reaction time: 1.5h.
[c] $(C_4H_9)_4PBr$ as PTC, 50% NaOH.
SOURCE: Adapted from refs. 14 and 28.

Table II. Yields of Nitrile 4 [a]

	R³	yield (%)[b]	
		4	dibenzyl sulfide
a	p-CH$_3$C$_6$H$_4$	92(95)	93
b	o-CH$_3$C$_6$H$_4$	80(88)	81
c	p-ClC$_6$H$_4$	90(92)	91
d	C$_6$H$_5$	91(97)	90
e	C$_6$H$_5$CH$_2$	86(90)	90
f	CH$_3$	- (80)[c]	78
g	CH$_3$CH$_2$	- (94)[c]	92
h	CH$_3$(CH$_2$)$_2$	75(83)	80
i	CH$_3$(CH$_2$)$_6$	80(92)	88

[a] Products were identified by direct comparison with authentic specimens.
[b] Yield of purely isolated product, based on thioacetamide. Parenthesis: by GLC data.
[c] 4 could not isolated by distillation.
SOURCE: Adapted from ref. 18.

Synthesis of Thiolesters

In a similar way to the synthesis of unsymmetrical sulfides, we have attempted to react 1-(alkylthio)ethaniminium salt **1** with acyl halide instead of alkyl halide. However, the corresponding thiolester could not be obtained due to the prefered formation of N-acylthioimidate **5**, which is produced from the reaction of 1-(alkylthio)ethaniminium salt with acyl halide under basic conditions without PTC (*21*).

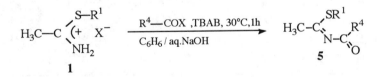

Method A

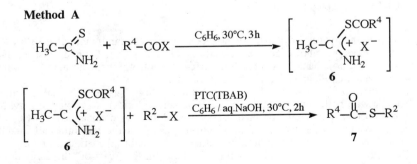

Thiolesters are prepared by hydrolysis of 1-(alkylthio)alkaniminium salts in an acidic medium (22). This method is not applicable for a general use (23). Therefore, we have developed a more general method for the synthesis of thiolester 7 using 1-(acylthio)alkaniminium salt 6 as a source of thiolate ion (**Method A**) (24). 1-(Acylthio)ethaniminium salts 6 are produced by the reaction of thioacetamide with acyl halides in benzene at 30°C for 3h, but are not isolated because they are air sensitive. Therefore, the subsequent reaction with alkyl halide was carried out on the crude reaction mixture using a two-phase system consisting of benzene and an aqueous solution of sodium hydroxide with PTC.

When this reaction was carried out using an aqueous solution of 30% sodium hydroxide, a small amount of thiolester was obtained together with the corresponding carboxylic acid as a by-product resulting from the decomposition of the thiolester. On the other hand, upon using an aqueous solution of 10% sodium hydroxide the reaction proceeded smoothly to afford the desired thiolesters in good yields. The reaction was almost complete within 2h at 30°C as shown by monitoring on GLC. The reaction of 1-(acylthio)ethaniminium salt 6 with alkyl halide did not afford thiolester in the absence of PTC, and corresponding thio acid was obtained as a main product. Various PTC were examined. For example, the yields of S-octylthiobenzoate 7f obtained from the reaction of 1-(benzoylthio)ethaniminium bromide with octyl bromide are 87%(TBAB), 71%(MTOAC), and 3%(BTEAC), respectively. In the case of thiolesters, the effect of PTC is identical to that of the unsymmetrical sulfides.

In the case of the synthesis of p-substituted octylthiobenzoate($7i$,$7j$), 1-(p-methoxybenzoylthio)ethaniminium bromide reacts easily with octyl bromide to form $7i$ in 95% yield, and 1-(p-nitrobenzoylthio)ethaniminium bromide reacts with octyl bromide to afford $7j$ in 63% yield (Table III).

Table III. Yields of Thiolester 7 (Method A)

7	R^4	R^2	Yield(%)[a]	7	R^4	R^2	Yield(%)[a]
a	C_5H_{11}	C_8H_{17}	89	g	C_6H_5	$CH_3CH=CHCH_2$	88
b	C_5H_{11}	$C_6H_5CH_2$	86	h	C_6H_5	$C_6H_5CH_2$	82
c	C_7H_{15}	$C_6H_5CH_2$	86	i	p-$CH_3OC_6H_4$	C_8H_{17}	95
d	t-C_4H_9	C_8H_{17}	83	j	p-$NO_2C_6H_4$	C_8H_{17}	63
e	C_6H_5	C_4H_9	91	k	$C_6H_5CH=CH$	$C_6H_5CH_2$	93
f	C_6H_5	C_8H_{17}	87				

[a] Yield of purely isolated product, based on thioacetamide.
SOURCE: Adapted from ref. 24.

Moreover, the hydrolysis of 1-(p-methoxybenzoylthio)ethaniminium salt at 30°C for 30 minutes gives p-methoxythiobenzoic acid in 89% yield, while the hydrolysis of 1-(p-nitrobenzoylthio)ethaniminium bromide affords p-nitrothiobenzoic acid and $p$nitrobenzoic acid in 59% and 38% yields, respectively.

Furthermore, we examined the competition reaction of 1-(octylthio)ethaniminium bromide with both butyl bromide and p-toluic acid chloride. N-(p-toluoyl) octylthioimidate, octyl p-methylthiobenzoate and butyl octyl sulfide were obtained in 78%, 17% and 5% yields, respectively. This result suggests that the iminium carbon-sulfur linkage of 1-(alkylthio)ethaniminium salt is more stable than that of 1-(acylthio) ethaniminium salt in an alkaline solution. Thus, in this reaction, the formation of thiolester 7 is less favored than that of N-acylthioimidate 5 which is formed by electro-

Method B

$$H_3C-C\begin{matrix} \nearrow S-R^1 \\ (+ \quad X^- \\ \searrow NH_2 \end{matrix} \quad \xrightarrow[\text{2) } R^4-COX, 30°C, 1h]{\text{1) } C_6H_6 / aq.NaOH, TBAB, 30°C, 0.5h} \quad R^1-S-\overset{\overset{\displaystyle O}{\|}}{C}-R^4$$

$$\textbf{1} \hspace{8cm} \textbf{7}$$

philic attack of the carbonyl carbon of acyl halide on the nitrogen atom of 1-(alkylthio)ethaniminium salt.   It is, therefore, assumed that the generation of a sufficient amount of alkanethiolate anions by decomposition of 1-(alkylthio) ethaniminium salt in a basic solution is a key point to the preparation of thiolester from 1-(alkylthio) ethaniminium salt and acyl halide.

Then, we continued to study an effective procedure for the conversion of 1-(alkylthio)ethaniminium salt with acyl halide to thiolester **7** (**Method B**) (*25*).   The decomposition of 1-(alkylthio)ethaniminium bromide was carried out under vigorous stirring at 30 °C for 30 or 60 minutes in the presence of TBAB (0.03 molar ratio) in a two-phase system consisting of benzene and an aqueous solution of 20% sodium hydroxide.   The alkanethiolate anions formed in the aqueous phase are transferred to the organic phase by the PTC.   Acyl halide was added to the organic phase over 30 minutes.   The mixture was stirred for an additional 30 minutes.   After the reaction was over, the unreacted acyl halide could not be detected by gas chromatography and two new components were detected in the organic phase.   After the separation, these two compounds were identified as thiolester and dialkyl disulfide, respectively, by instrumental analysis (*25*).   N-Acylthioimidate **5** could not be detected.   The solvent was evaporated under reduced pressure and the residue was separated on a silica gel column.   Upon using hexane as an eluent only a very small amount of dialkyl disulfide was recovered.   The silica gel was extracted with acetone to give pure thiolester **7** in a very good yield (Table IV).

Table IV. Yields of Thiolester 7 (Method B)

7	R^1	R^4	Yield (%)[a]	7	R^1	R^4	Yield (%)[a]
d	C_8H_{17}	t-C_4H_9	79 (86)[b]	o	C_8H_{17}	C_2H_5	85 (96)[b]
e	C_4H_9	C_6H_5	81 (85)	p	C_8H_{17}	$C_4H_9CH(C_2H_5)$	84 (90)
l	C_4H_9	C_2H_5	94 (98)[b]	q	t-C_4H_9	$C_4H_9CH(C_2H_5)$	83 (94)
m	C_4H_9	C_9H_{19}	87 (98)	r	$C_6H_5CH_2$	C_2H_5	62 (72)[b]
n	C_4H_9	p-$CH_3C_6H_4$	82 (96)	s	$CH_3CH=CHCH_2$	p-$CH_3C_6H_5$	84 (92)

[a] Yield of purely isolated product, based on thioacetamide. Parenthesis: by GLC data.
[b] Decomposition conditions of **1**: 60min, 30 °C.
SOURCE: Adapted from ref. 25.

The yields of thiolesters depends on the decomposition time of 1-(alkylthio)-ethaniminium salt.   Thus, the yields of **7d**, **7l**, **7 o**, and **7r** increase with an increase of the decomposition time from 30 to 60 minutes.   When these reactions are carried out in the absence of PTC, the yield of thiolester decreases substantially.

Consequently, the reaction of 1-(acylthio)ethaniminium salt with alkyl halide is a useful synthetic route for the preparation of thiolesters **7** having normal, branched and unsaturated alkyl groups, and the synthetic method of thiolesters **7** from 1-(alkylthio)ethaniminium salts and acyl halides has the following advantages: 1-(alkylthio)ethaniminium salts are air stable and easily handled; the method is applicable to a wide range of 1-(alkylthio)ethaniminium salts and acyl halides under mild conditions; the reaction proceeds almost selectively with very good yields of the pure thiolesters after the simple purification by column chromatography on silica gel.

Synthesis of Cyclic Polythiaether

Cyclic polythiaethers are known to favor their complexing ability with soft Lewis acids (*13*), because the sulfur atoms in the thioether groups are acting as soft Lewis bases. We developed a simple method for the synthesis of cyclic polythiaethers using the reaction of 1,1'-(dithioalkane)diethaniminium salt **9** with alkyl or aryl dihalides **10** under phase-transfer conditions. The reaction of dihaloalkane **8** with two-fold excess of thioacetamide affords 1,1'-(dithioalkane)diethaniminium salt **9** in boiling temperature of chloroform for 2-3h or at 70-80°C for 2h without any solvent. The yields of 1,1'-(dithioalkane)diethaniminium salts **9** are ranging between 70-90%.

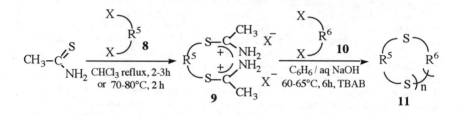

Alkyl or aryl dihalide **10** is added dropwise to a stirred reaction mixture containing 1,1'-(dithioalkane)diethaniminium salt **9** in a two-phase system consisting of benzene and an aqueous solution of alkali hydroxide at 60-65°C for 6h in the presence of TBAB. After the reaction is over, the organic phase is analyzed by a gel permeation chromatography(GPC) using chloroform as an eluent. The main products of this reaction are cyclic polythiaethers and linear polymers having sulfur atoms. Two types of cyclic polythiaethers **11** were obtained; 1:1 ratio (n=1) and the 2:2 ratio (n=2) which are separated by column chromatography on silica gel using hexane / EtOAc (9:1) as an eluent. The yields and physical properties of the cyclic polythiaethers are summarized in Table V.

The yields for the cyclic polythiaethers depend on the kind of alkali cations in the aqueous phase. While, the reaction time and the alkali concentration of the aqueous solution exhibit only a slight dependence on the yield. When the reaction of **9**(R^5=(CH$_2$)$_4$) with **10**(R^6=(CH$_2$)$_4$) in a benzene / 30% aqueous sodium hydroxide system produces 1:1 stoichiometrical product **11g**, dithia-10-crown-2, and 2:2 stoichiometrical product **11h**, tetrathia-20-crown-4, in 8 and 14% yields, respectively. The use of cesium hydroxide in place of sodium hydroxide increases the yield of 2:2 stoichiometrical product **11h**(17%), together with a decrease of the 1:1 stoichiometrical product **11g**(2%), and a small amount of 3:3 stoichiometrical product **11i**(5%), hexathia-30-crown-6, as a new product. Similarly, the reaction of **9**(R^5=(CH$_2$)$_3$) with **10**(R^6=(CH$_2$)$_3$) in a benzene / sodium or potassium hydroxide system produces 1:1 stoichiometrical compound **11c** as a main product but the similar reaction in a benzene / rubidium or cesium hydroxide system affords the 2:2 stoichiometrical compound **11d** as the main product.

These results shows that the unit number (n) of cyclic polythiaether increases with an increase in the ionic radius of alkali cation. The alkali cation plays an important roles as a template in the formation of cyclic polythiaethers **11**.

The extractability of alkali, transition and heavy metal ions by the cyclic polythiaethers is evaluated. An aqueous solution(3 mL) containing 1×10^{-2}M metal nitrate, the 3×10^{-3}M picric acid was placed in a 10 mL glass cylindrical tube equipped with a glass stopper. After the addition of 3 mL of the 3×10^{-4}M polythiaether(Host) solution of 1,2-dichloroethane, the mixture was vigorously shaken for 30 min at 25°C. The concentration of the metal ion in the aqueous phase was determined by means of ultraviolet spectrophotometry. The typical results are given in Table VI.

All of the 2:2 stoichiometrical products show a high selectivity toward Ag^+(*26*), but the 1:1 stoichiometrical product shows only about half the extractability of 2:2 product. Cyclic polythiaethers **11b**, **11d**, **11f** and **11h** have 20 to 50% extractability for ions other than Ag^+ and **11k** to **11u** show only a weak extractability except for Ag^+ ion.

Conclusion

This reactions of thioiminium salts in the presence of PTC are developed for the preparation of sulfur-containing compounds such as unsymmetrical sulfides, thiolesters, and cyclic polythiaethers. In the absence of PTC, the reaction either do not proceed at all or only slowly.

The present method offers many advantages; First, unpleasant smell is reduced by the use of thioiminium salts as a thiolate ion source instead of thiols or dithiols; secondly, the reaction proceeds almost to completion under mild conditions; thirdly, the selectivity of the reaction is high, except for the synthesis of the cyclic polythiaethers, when the products can be easily separated; and lastly, this method can be applied to the synthesis of a wide range of unsymmetrical sulfides and thiolesters.

Experimental Section

1-(Benzylthio)ethaniminium Bromide **1h**; A mixture of thioacetamide (3.75g, 50mmol) and benzyl bromide (8.55g, 50mmol) in chloroform (50mL) was refluxed for 1h. After cooling, the product was isolated by suction and washed with ether; yield: 11.04g(90%); mp 174 - 176°C (mp 174 - 176°C (*27*)). If, using other alkyl halides instead of benzyl bromide, the product did not crystallize from the mixture, some amount of ether was added.

Unsymmetrical Sulfides **3**; A mixture of the freshly prepared 1-alkylthioethaniminium salts **1** (10mmol), alkyl halides **2** (10mmol), tetrabutylammonium bromide (TBAB; 97mg, 0.3mmol) as a phase-transfer catalyst, benzene (50 mL), and 30 wt% aqueous sodium hydroxide (50 g) was vigorously stirred at room temperature for 15 min (1.5h in the cases of **3g**, **3k**, and **3l**) under nitrogen atmosphere. The organic layer was separated, and the aqueous layer was extracted with benzene. The combined organic layer was washed with water (3x50mL), dried, and evaporated. The residual product was distilled under reduced pressure to afford the pure unsymmetrical sulfide **3**.

Nitriles **4**; A mixture of thioamides (10mmol), benzyl chloride (2.53g, 20mmol), tetrabutylammonium bromide (TBAB; 161mg, 0.5mmol), benzene (70mL) and 30 wt% aqueous sodium hydroxide (50 g) was vigorously stirred at 30°C. The reaction was complete in 2h as shown by GLC monitoring. The organic layer was separated, and the aqueous layer was extracted with benzene. The combined organic layer was washed with water (3x50 mL), dried, and evaporated. The residual product was distilled under reduced pressure to afford nitrile **4** and dibenzyl sulfide.

Table V. Yields and Physical Properties of Cyclic Polythiaether 11

11	R^5	R^6	n	yield[a] (%)	m.p. (°C)	Mass spectra(70eV)[c] m/z(M^+)	Element	^{13}C-NMR(CDCl$_3$/TMS)[d] δ(ppm) : CH$_2$
a	$(CH_2)_3$	$(CH_2)_2$	1	8	45-47	134.0216	$C_5H_{10}S_2$	31.8, 32.7, 38.6
b			2	10	118-119	268.0458	$C_{10}H_{20}S_4$	30.1, 30.4, 31.7
c	$(CH_2)_3$	$(CH_2)_3$	1	37	oil	148.0377	$C_6H_{12}S_2$	30.2, 30.8
d			2	10	51-53	296.0780	$C_{12}H_{24}S_4$	29.6, 30.7
e	$(CH_2)_4$	$(CH_2)_3$	1	22	56-58	162.0531	$C_7H_{14}S_2$	21.7, 25.9, 28.8, 32.2
f			2	24	59-60	324.1075	$C_{14}H_{28}S_4$	28.4, 29.7, 30.6, 31.5
g	$(CH_2)_4$	$(CH_2)_4$	1	8(2)[b]	90-92	176.0679	$C_8H_{16}S_2$	25.5, 31.5
h			2	14(17)[b]	31-32	352.1383	$C_{16}H_{32}S_4$	28.3, 31.3
i			3	-(5)[b]	71-73	528.2066	$C_{24}H_{48}S_6$	28.7, 31.6
j	$(CH_2)_6$	$(CH_2)_6$	1	12	78-79	232.1320	$C_{12}H_{24}S_2$	25.6, 27.2, 30.1
k			2	22	71-73	464.2636	$C_{24}H_{48}S_4$	28.5, 29.6, 32.1
l	$(CH_2)_8$	$(CH_2)_4$	1	10	65-67	232.1300	$C_{12}H_{24}S_2$	25.5, 26.7, 27.0, 29.8, 30.3
m			2	9	47-48	464.2634	$C_{24}H_{48}S_4$	28.5, 28.6, 28.9, 29.6, 31.6, 32.0

	Structure		Yield	mp	HRMS	Formula	^{13}C NMR
n	$(CH_2)_8$	1	25	51-52	288.1925	$C_{16}H_{32}S_2$	27.7, 27.8, 29.0, 31.0
o	$(CH_2)_8$	2	9	53-56	576.3847	$C_{32}H_{64}S_4$	
p	$(CH_2)_{10}$	1	21	69-71	274.1768	$C_{15}H_{30}S_2$	27.5, 27.1, 27.6, 29.1, 29.4, 31.9, 32.0
q	$(CH_2)_5$	2	4	57-59	548.3569	$C_{30}H_{60}S_4$	28.0, 28.7, 29.0, 29.2, 29.3, 29.7, 31.9, 32.1
r	$(CH_2)_{10}$	1	19	48-50	344.2557	$C_{20}H_{40}S_2$	28.1, 28.5, 28.8, 29.3, 31.4
s	$(CH_2)_{10}$	2	18	70-72	688.5156	$C_{40}H_{80}S_4$	29.0, 29.2, 29.5, 29.7, 32.2
t		1	9	120-121	272.0702	$C_{16}H_{16}S_2$	37.8, (-CH=)127.0, 128.5, 131.8, (-C=)137.0
u		2	18	92-94	544.1356	$C_{32}H_{32}S_4$	

[a] The yield is based on dithioiminium halides **9**. [b] Parentheses refer to the case using of aq.CsOH, otherwise aq.NaOH was used. [c] Recorded on a Hitachi M-80B spectrometer. [d] Obtained on a JEOL GX-400 spectrometer.

Table VI. Extraction of Alkaline,Transition Metal and Heavy Metal Ions

Host	Extraction (%)								
	Li^+	Na^+	K^+	Co^{2+}	Ni^{2+}	Cu^{2+}	Ag^+	Cd^{2+}	Pb^{2+}
11g	4.2	2.6	5.2	2.6	0.9	2.8	44	5.0	4.8
11l	2.0	5.6	2.9	2.2	2.9	3.8	66	3.6	2.4
11b	1.7	3.8	20	7.6	20	8.4	99	28	22
11d	23	2.9	4.5	6.2	6.8	5.8	99	6.0	5.0
11f	6.8	4.1	8.3	6.6	50	5.4	98	7.0	9.0
11h	1.3	1.8	8.4	17	5.0	36	99	2.0	3.0
11k	2.2	1.3	2.5	3.0	3.0	8.4	99	11	6.0
11m	3.9	8.9	3.9	5.8	7.4	6.4	92	9.2	6.0
11o	1.1	2.0	3.4	3.0	3.6	4.4	98	4.8	6.1
11q	2.2	6.0	4.9	4.8	9.0	11	91	12	13
11s	4.1	7.2	3.8	6.2	5.4	4.4	97	5.0	5.2
11u	1.5	1.4	1.9	2.8	2.6	6.6	96	3.0	5.2
none	0.6	0.7	1.5	1.5	1.1	1.4	1.8	0.6	1.8

Org. phase (CH_2ClCH_2Cl) ; [Host] $=3.0 \times 10^{-4}$ M
Aq. phase ; [Pic^-H^+] $=3.0 \times 10^{-5}$ M , [Metal Nitrate] $=0.01$M

N-Acylthioimidate **5**; A mixture of the freshly prepared 1-alkylthioethaniminium salts **1** (10mmol), acyl halide (10mmol), tetrabutylammonium bromide (TBAB; 97mg, 0.3mmol) as a phase-transfer catalyst, benzene (50mL), and 30 wt% aqueous sodium hydroxide (50g) was vigorously stirred at 30°C for 15 min under nitrogen atmosphere. The organic layer was separated, and the aqueous layer was extracted with benzene. The combined organic layer was washed with water (3x50mL), dried, and evaporated. The residual product was distilled under reduced pressure to afford the pure N-acylthioimidate **5**.

Thiolesters **7**; The reaction of 1-acylthioethaniminium salts with alkyl halides (**Method A**) : A mixture of thioacetamide (1.83g, 30mmol) and acyl halides (30mmol) in benzene (50mL) was stirred at 30°C for 3h under nitrogen atmosphere. Then, alkyl halide (30 mmol), tetrabutylammonium bromide (TBAB; 0.29g, 0.9mmol) as a phase-transfer catalyst, and 10 wt% aqueous sodium hydroxide (50g) were added and the mixture was vigorously stirred at 30°C for 2h. The organic layer was separated, and the aqueous layer was extracted with benzene. The combined organic layer was washed with water (3x50 mL), dried, and evaporated. The residual product was column chromatographed on silica gel with hexane/EtOAc (4:1) as eluent to afford the pure thiolesters **7**.

The reaction of 1-alkylthioethaniminium salts with acyl halides (**Method B**) :
A mixture of the freshly prepared 1-alkylthioethaniminium salts **1** (10mmol), tetrabutyl-ammonium bromide (TBAB; 97mg, 0.3mmol) as a phase-transfer catalyst, benzene (50mL), and 20 wt% aqueous sodium hydroxide (50g) was vigorously stirred at 30°C for 30 min under nitrogen atmosphere. Acyl halide in benzene (10mL) was then added dropwise to the solution over 30 min. The resulting solution was stirred for a further 30 min. After the reaction, the organic layer was separated, and the aqueous layer was extracted with benzene. The combined organic layer was washed with water (3x50 mL), dried, and evaporated. The residual was chromatographed on a silica gel column with hexane and then acetone. A very small amount of disulfide was eluted first with hexane. Then the silica gel in the column was extracted with acetone to afford the pure thiolesters **7**.

1,1'-(Dithioalkane)diethaniminium salt **9**; A mixture of thioacetamide (15g, 0.2mol) and alkyl dibromides **8** (0.1mol) in chloroform (100mL) was refluxed for 2-3h or at 70-80°C for 2h without any solvent. After cooling, the product **9** was isolated by suction and washed with ether.

Cyclic polythiaethers **11**; A mixture of the freshly prepared 1,1'-(dithioalkane) diethaniminium salt **1** (20mmol), tetrabutylammonium bromide (TBAB; 0.64g, 2mmol) as a phase-transfer catalyst, benzene (50mL), and 30 wt% aqueous sodium hydroxide (100 g) was vigorously stirred at 60-65°C for 1h under nitrogen atmosphere. The alkyl dibromide **10** (20mmol) in benzene (50 mL) was then added dropwise to the solution over 4h. The resulting solution was stirred for a further 1h. After the reaction, the organic layer was separated, and the aqueous layer was extracted with benzene. The combined organic layer was washed with water (3x100mL), dried, and evaporated. The cyclic polythiaethers including 1:1 and 2:2 stoichiometric products **11** were isolated from residual product by column chromatography on silica gel using hexane / EtOAc (9:1) as an eluent.

Acknowledgments

This work was supported in part by a grant from Tokyo Ohka Foundation for the Promotion of Science and Technology, to which we greatly acknowledge. We thank Mr. Aimi for measurements of mass spectra and Ms. Nasu and Ms. Fujita for measurements of NMR spectra.

Literature Cited
(*1*) Sih, J.C.; Graber, D.R. *J Org. Chem.* **1983**, *48*, 3842.
(*2*) Guindon, Y.; Frenette, R.; Fortin, R.; Rokach, J. *J. Org. Chem.* **1983**, *48*, 1357.
(*3*) Luzzio, F.A.; Little, A.D. *Synth. Commun.* **1984**, *14*, 209.
(*4*) Nowicki, T.; Markowska, A. *Synthesis* **1986**, 305.
(*5*) Cardellicchio, C.; Fiandanese, V.; Marchese, G.; Ronzin, L. *Tetrahedron Lett.* **1985**, *26*, 3595.
(*6*) Payne, N.G.; Peach, M.E. *Sulfur Lett.* **1986**, *4*, 217.
(*7*) Gauthhier, J.Y.; Bourden, F.; Young, R.N. *Tetrahedron Lett.* **1986**, *27*, 15.
(*8*) Dellaria, J.F.; Nordeen, C.; Swett, L.R. *Synth. Commun.* **1986**, *16*, 1047.
(*9*) Bruce,T.; Benkovic, S. *Bioorganic Mechanisms*; Benjamin,W.A.; New York, **1966**; Vol. 1, p.259.
(*10*) Janssen, M. *The Chemistry of Carboxylic Acids and Esters*; Patai, S., Ed.; John Wiley & Sons: New York, **1969**, p.705.
(*11*) Singh, P.; Jain, A. *Indian J. Chem.* **1987**, *26B*, 707.
(*12*) Singh, P.; Kumar, M.; Singh, H. *Indian J. Chem.* **1987**, *26B*, 861.
(*13*) Ahrland, S.; Chatt, J.; Davies, N.R. *Quart. Rev.*, London **1958**, *12*, 265.
(*14*) Takido, T.; Itabashi, K. *Synthesis* **1987**, 817.
(*15*) Herriott, A.W.; Picker, D. *J. Am. Chem. Soc.* **1975**, *97*, 2345.
(*16*) Degani, I.; Fochi, R.; Regondi, V. *Synthesis* **1983**, 630.
(*17*) Singh, H; Batra, S.M.; Singh, P. *Indian J. Chem. Sect.B* **1985**, *24*, 131.
(*18*) Funakoshi,Y.; Takido, T.; Itabashi, K. *Synth. Commun.* **1985**, *15*, 1299.
(*19*) Saraie,T.; Ishiguro,T.; Kawashima, K.; Morita, K. *Tetrahedron Lett.*. **1973**, 2121.
(*20*) Shinozaki, H.; Imaizumi, M.; Tajima, M. *Chem. Lett.*.**1983**, 929.
(*21*) Walter, W.; Krohn, J. *Ann. Chem.* **1973**, 443.
(*22*) Chaturvedi, R.K.; McMahon, A.E.; Schmir, G.L. *J. Am. Chem. Soc.* **1978**, *100*, 6125.
(*23*) Walter, W.; Krohn, J. *Chem, Ber.* **1969**, *102*, 3786.
(*24*) Takido, T.; Toriyama, M.; Itabashi, K. *Synthesis* **1988**, 404.
(*25*) Takido, T.; Sato, K.; Nakazawa, T.; Seno, M. *Sulfur Letter* **1995**, *19*, 67.
(*26*) Saito, K.; Masuda, Y.; Sekido, E. *Analytica Chimica Acta.* **1983**, *151*, 447.
(*27*) Walter, W.; Krohn, J. *Libigs Ann. Chem.* **1973**, 443.
(*28*) Fujihira,T.; Sakurai,M.; Saito,S.; Miki,T.; Takido,T.;Nakazawa,T.; Seno,M. *Proc. Research Soc. Japan Sugar Ref. Tech.* **1994**, *42*, 53.   C.A. 123: 9779b.

Chapter 16

Microwave-Irradiated Alkylations of Active Methylenes Under Solid–Liquid Phase-Transfer Catalytic Conditions

Yaozhong Jiang, Yuliang Wang, Runhua Deng, and Aiqiao Mi

Chengdu Institute of Organic Chemistry, Chinese Academy of Sciences, Chengdu 610041, People's Republic of China

Convenient procedures are described for a series of microwave promoted and phase transfer catalyzed alkylations on the active methylene compounds, including ethyl phenylsulfonylacetate, methyl N-benzylidene glycinate, ethyl acetoacetate, ethyl phenylmercaptoacetate, diethyl malonate, and acetylacetone. These useful reactions can be carried out within several minutes in good yields, and most of the procedures are solvent-free.

Alkylation of the enolate anions of acidic methylene groups with alkyl halides or other alkylating agents has been profusely studied for many years, due to the importance of carbon-carbon bond formation in organic synthesis (1). Organic strong base were used in typical method to derive the carbanion, later inorganic base were used in phase transfer catalytic method. We have reported the solid-liquid phase transfer catalyzed (PTC) synthesis of amino acids via the alkylation of N-benzylidene glycinate (2-6). In the synthesis of phenylalanine, the alkylation with benzyl bromide could be carried out smoothly in good yield, but the alkylation with benzyl chloride could not take place under the same condition. For a large scale preparation of phenylalanine, benzyl chloride is much cheaper than benzyl bromide as alkylating agent. So it is reasonable to use some methods to increase the reactivity of reactants.

In recent years microwave irradiation has been used in the organic synthesis to increase the activity and shorten the reaction time (7-11). So we have the idea to use this new method to develop a series of procedures for the alkylation of active methylenes, improving the synthesis of amino acids and other useful compounds. Here we would like to describe the summary of 28 alkylations on 6 methylene compounds.

Results and Discussion

The alkylation of sulfonylacetate by typical method had been used in the synthesis of 11,12-secoprostaglandins (12); and the phase transfer catalyzed alkylation has been carried out by stirring the reaction mixture for 1-4 hs at 40°C *(13)*. Here we give an improved procedure to perform the rapid and efficient alkylation of ethyl phenylsulfonylacetate (compound 1) in microwave oven, using TEBA (triethylbenzyl ammonium chloride) as phase transfer catalyst.

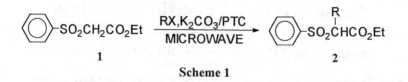

1   **2**

Scheme 1

Most of the alkylation of **1** was simply achieved by microwave irradiation to the mixture of ethyl phenylsulfonylacetate, alkyl halide, potassium carbonate and TEBA for 2-3 minutes. But we failed to mix the substrate with bromide well in the synthesis of **2b** and **2c**, so we added 1ml of toluene to solve this problem. Followed by isolation and purification gave monoalkylated products (**2a-e**) in 76-86% yield (Table I). All the products were characterised by IR, ^{1}H-NMR and MS.

Table I. alkylation of ethyl phenylsulfonylacetate

product	alkyl halide	mol. ratio (RX:Subst.)	irradiation time (min)	isol. yield (%)
2a	$C_6H_5CH_2Cl$	1.4:1	3	76
2b	$p\text{-}ClC_6H_4CH_2Br$	1:1	2	76
2c	$2\text{-}C_{10}H_7CH_2Br$	1;1	2	86
2d	n-OctylBr	1:1	3	79
2e	n-BuBr	2:1	3	83

Phase transfer catalyzed alkylation of alkyl N-benzylidene glycinate (compound 3) has been successfully used in the synthesis of amino acids *(2-6)*, and we improve the synthesis by using microwave irradiation. The synthetic route is as follows:

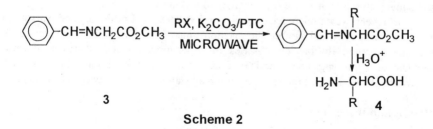

3

Scheme 2

New method dramatically enhance the activity of the alkylation with benzyl chloride and other alkyl halide. All the alkylations are rapidly carried out in 1-2 minutes and give the monoalkylated products, which are stirred with 1N HCl at room temperature for 14 hs, then with 6N HCl at 60^0C for 8-10 hs to offer amino acids (**4a**-**e**). The results are summarized (Table II). All the products were characterized by IR, ^{1}H-NMR and elemental Analyses.

Table II. The synthesis of amino acids

product	alkyl halide	irradiation time (min)	overall yield (%)
4a: Phenylalanine	$PhCH_2Cl$	1	62.5
4b: Aspartic acid	$BrCH_2CO_2Et$	1	61.4
4c: Serine	$PhCH_2OCH_2Cl$	1	43.6
4d: Leucine	$CH_3(CH_2)_3Br$	2	54.5

The alkylation of acetoacetate (compound **5**) were used in the typical synthesis of ketone and carboxylic acid, and the PTC alkylation could be carried out *(14-18)*. The improved procedure by microwave irradiation can perform a rapid and   efficient alkylation of ethyl acetoacetate, using TBAC (tetra-butyl ammonium chloride) as phase transfer catalyst without solvent.

$$CH_3COCH_2CO_2Et \xrightarrow[\text{MICROWAVE}]{RX,KOH,K_2CO_3/PTC} \overset{R}{\underset{}{CH_3COCHCO_2Et}}$$

$$\quad\quad\quad 5 \quad\quad\quad\quad\quad\quad\quad\quad\quad\quad\quad\quad\quad\quad\quad\quad 6$$

Scheme 3

The alkylation of **5** was simply achieved by microwave irradiation to the mixture of ethyl acetoacetate, alkyl halide, potassium hydroxide-potassium carbonate (1:4) and PTC for 3-4.5 minutes. Followed by isolation and purification gave monoalkylated products (**6a-e**) in 59-82% yield. The microwave power scale and the irradiation time were changed according to the activity of alkyl halide (Table III). All the products were characterised by IR, ^{1}H-NMR and MS.

Table III. Alkylation of ethyl acetoacetate

product	alkyl halide	mol. ratio (RX:Subst.)	power level	irradiation time (min)	isol. yield (%)
6a	$CH=CHCH_2Br$	1:1	2	3	81
6b	$PhCH_2Cl$	1:1	3	4	69
6c	m-$CH_3OC_6H_4CH_2Cl$	1:1	5	3.5	82
6d	p-$ClC_6H_4CH_2Br$	1:1	4	4	59
6e	$CH_3(CH_2)_3Br$	1.2:1	6	4.5	61

B. M. Trost and co-workers *(19)* have led the development and utilization of the sulphenylation-oxidation-dehydrosulphenylation sequence to provide 2-unsaturated esters. Here we develop a alternative procedure for a rapid and convenient synthesis of 2-phenylmercaptoesters via the alkylation of phenylmercaptoacetate (compound 7) in the absence of solvent in microwave oven, using TBAC (tetra-butyl ammonium chloride) as phase transfer catalyst.

Scheme 4

The rapid PTC alkylations of 7 with a series of halides were easily performed in 650W domestic microwave oven to yield the α-alkylated products (8a-e) in 58-83% isolated yield. The microwave power scale and the irradiation time were changed according to the activity of alkyl halide (Table IV). All the products were characterised by IR, ^{1}H-NMR and MS.

When p-chlorobenzyl bromide was used, we fail to mix it with substrate well, so we add 1 mL of toluene to solve this problem. We used 2 equivlent of alkyl halide, and obtained the dialkylated product.

Table IV. Alkylation of ethyl phenylmercaptoacetate

product	alkyl halide	mol. ratio (RX:Subst.)	power level	irradiation time (min)	isol. yield (%)
8a	PhCH$_2$Cl	1:1	3	4.5	83
8b	CH$_2$=CHCH$_2$Br	1:1	2	3.5	67
8c	p-ClC$_6$H$_4$CH$_2$Br	2:1	5	4	61
8d	CH$_3$(CH$_2$)$_3$Br	1:1	8	4.5	59
8e	m-MeOC$_6$H$_4$CH$_2$Cl	1:1	6	4.5	58

The alkylation of malonate (compound 9) were used for the synthesis of carboxylic acids in typical method, and the advanced PTC alkylation could be carried out (18,20,21). Our improved procedure can perform the alkylation more conveniently by microwave irradiation, using TBAB (tetra-butyl ammonium bromide) as phase transfer catalyst.

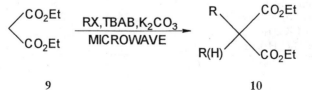

Scheme 5

The rapid alkylations of **9** with a series of halides were easily performed in 650W domestic microwave oven by irradiation to the mixture of diethyl malonate, alkyl halide, potassium and PTC for 2 minutes. Five alkylated products (**10a-e**) were synthesized in 64-86% isolated yield (Table V). All the products were characterised by IR, ^{1}H-NMR and MS. In the synthesis of **10e**, we fail to mix alkyl halide with substrate well, so we add 1 mL of toluene to solve this problem; In this case we used 2 equivlent of alkyl halide, and obtained the dialkylated product.

Table V. Alkylation of diethyl malonate

product	alkyl halide	mol.ratio (RX:Subst.)	power level	irradiation time (min)	isol. yield (%)
10a	PhCH$_2$Cl	1.2:1	1	2	72
10b	m-MeOC$_6$H$_4$CH$_2$Cl	1.05:1	2	2	71
10c	CH$_2$=CHCH$_2$Br	2:1	1	2	75
10d	CH$_3$(CH$_2$)$_3$Br	2.5:1	6	2	86
10e	p-ClC$_6$H$_4$CH$_2$Br	2.5:1	5	2	64

Also we have improved the dialkylation of acetylacetone *(14,22)* with some active alkyl halides by microwave irradiated synthetic method, using TBAB as phase transfer catalyst. Most of the procedures are solvent-free.

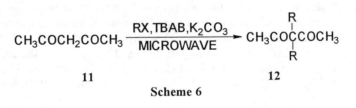

$$CH_3COCH_2COCH_3 \xrightarrow[\text{MICROWAVE}]{RX,TBAB,K_2CO_3} CH_3CO\overset{\displaystyle R}{\underset{\displaystyle R}{C}}COCH_3$$

              11   **12**

Scheme 6

The rapid dialkylations of acetylacetone (compound **11**) with a series of halides were carried out easily by microwave irradiation to the mixture of acetylacetone, alkyl halides, potassium and PTC within 2 min. When p-chlorobenzyl bromide was used, we fail to mix it with substrate well, so we add 1 mL of toluene to solve this problem. Four compounds (**12a-d**) were prepared in 68-84% isolated yield (Table VI). All the products were characterised by IR, ^{1}H-NMR and MS.

Table VI. Dialkylation of acetylacetone

product	alkyl halide	mol.ratio (RX:Subst.)	power level	irradiation time (min)	isol. yield (%)
12a	PhCH$_2$Cl	2.5:1	1	2	75
12b	m-MeOC$_6$H$_4$CH$_2$Cl	2.5:1	2	2	72
12c	p-ClC$_6$H$_4$CH$_2$Br	2:1	5	2	68
12d	CH$_2$=CHCH$_2$Br	3:1	1	2	84

Conclusion. Microwave promoted alkylations on active methylenes under PTC conditions can be carried out in good yield within several minutes, offering a series of more convenient route to synthesize some useful compounds. When the substrate can be mixed with halide well, the solvent-free method is facile and efficient; When the substrate cannot be mixed with halide completely, toluene is added and a higher power level is used because the energy which the reactants obtain from microwave irradiation can transfer partially to the solvent.

Experimental Section

General Considerations: 2 types of microwave oven were used: 1). FEIYUE 960W domestic microwave oven, cook scale and defrost scale; 2). S. M. C E70E domestic microwave oven, the maxium power is 650W, 1-10 scales. IR spectra were recorded on a Nicolet infrared spectrometer; ^{1}H-NMR spectra were recorded on a Varian FT-80A spectrometer, using TMS as internal standred; MS data were obtained using a HP-5988A instrument; Elemental data were obtained using a Carlo Erba-1106 elemental analyser; K_2CO_3 was grounded and dried at 500^0C for 3 hs.

Ethyl 2-(phenylsulfonyl)-3-phenylpropionate (2a): ethyl phenylsulfonylacetate (1.14g, 5.0mmol), benzyl chloride (0.89g, 7.0mmol) and TEBA (0.23g, 1.0mmol) were added into a 50 mL open container and mixed well, then potassium carbonate (3.0g, 22mmol) was added and the mixture was introduced into a FEIYUE 960W domestic microwave oven. Microwave irradiation was carried out for 3 minutes at defrost power. Then 50 mL of ether was poured into cooled mixture, filtration, concentration and followed by purification through column chromatography (silica gel 100-200 mesh; eluent: petroleum ether:EtOAc=4:1) to give **2a**. Yield:1.25g (76%). IR (cm^{-1}): 1740, 1306, 1140. ^{1}H-NMR (CDCl$_3$): 7.96-7.26 (5H, m); 7.07 (5H, m); 4.10 (1H, m); 3.82 (2H, q, J=7.0Hz); 3.30 (2H, d, J=6.0Hz); 0.87 (3H, t, J=7.0Hz). MS [m/z (%)]: 273 (M$^+$-OEt, 1.06).

Ethyl 2-(phenylsulfonyl)-3-(p-chlorophenyl)propionate (2b): synthesized as **2a** according to the reaction condition listed in table I. We failed to mix the alkyl bromide with the substrate, so added 1 mL toluene as solvent. Yield: 76%. IR (cm^{-1}): 1738, 1322, 1146. ^{1}H-NMR (CDCl$_3$): 8.15-6.60 (9H, m); 4.10 (1H, m); 3.82 (2H, q, J=7.0Hz); 3.78 (2H, d, J=6.2Hz); 0.88 (3H, t, J=7.0Hz). MS [m/z (%)]: 353 (M$^+$+1, 0.34); 355 (M$^+$+3, 0.18); 307 (M$^+$-OEt, 7.03).

Ethyl 2-(phenylsulfonyl)-3-(2'-naphthyl)propionate (2c): synthesized as **2a** according to the reaction condition listed in table I. We failed to mix the alkyl bromide with the substrate, so added 1 mL toluene as solvent. Yield: 86%. IR (cm^{-1}): 1738, 1322, 1148. ^{1}H-NMR (CDCl$_3$): 8.18-6.95 (12H, m); 4.50-3.06 (3H, m); 3.78 (2H, q, J=7.0Hz); 0.82 (3H, t, J=7.0Hz). MS [m/z (%)]: 369(M$^+$+1, 1.86); 323 (M$^+$-OEt, 1.87).

Ethyl 2-(phenylsulfonyl)-decanoatate (2d): synthesized as **2a** according to the reaction condition listed in table I. Yield: 79%. IR (cm^{-1}): 1740, 1328, 1150. ^{1}H-NMR

(CDCl$_3$): 7.86-7.26 (5H, m); 4.00 (2H, q, J=7.0Hz); 3.80 (1H, m); 1.95-1.65 (2H, m); 1.65-0.62 (18H, m). MS [m/z (%)]: 295 (M$^+$-OEt, 0.70).

Ethyl 2-(phenylsulfonyl)-hexanoateate (2e): synthesized as **2a** according to the reaction condition listed in table I. Yield: 83%. IR (cm^{-1}): 1738, 1311, 1150. ^{1}H-NMR (CDCl$_3$): 7.82-7.20 (5H, m); 3.95 (2H, q, J=7.0Hz); 3.78 (1H, m); 2.05-1.65 (2H, m); 1.05 (3H, t, J=7.0Hz); 1.35-0.60 (7H, m). MS [m/z (%)]: 285(M$^+$+1, 0.56); 239 (M$^+$-OEt, 6.14).

Phenylalanine (4a): The mixture of methyl N-benzylidene glycinate (5.0g, 26 mmol), benzyl chloride (4.0g, 31mmol), TEBA and anhydrous potassium carbonate (15.0g, 108mmol) was put into a domestic microwave oven, irradiated for 1 minute at 225W power. Final temperature is 141^{0}C. The reaction mixture was cooled to room temperature and 100 mL of ether was added. The solid was removed by filtration, and the solvent was evaporated to give alkylated product, which was dissolved in a solution of 50 mL of ether and 70 mL of 1N hydrochloric acid, and stirred at room temperature for 14h. After removing the ether, the reaction mixture with 58mL of 12N hydrochloric acid was stirred at 60^{0}C for 8h. After evaporation , 20mL of ethanol and 15mL of propene oxide were added to the residue, the solution was heated to reflux for 15 minutes and the precipitate was filtered, recrystallized with water to give 2.7g of **4a** (62.5%). m.p. 249.5-251^{0}C. Analysis for C$_9$H$_{11}$NO$_2$: C, 65.48; H, 6.59; N, 8.44. Cacld: C, 65.48; H, 6.64; N, 8.47. IR (cm^{-1}, KBr): 1620, 1530, 1500, 1410, 745, 700. ^{1}H-NMR (D$_2$O): 7.3-7.0 (5H, m); 4.25-4.0 (1H, m); 3.1-2.9 (2H, m).

Aspartic acid (4b): synthesized as **4a** according to the reaction condition listed in Table II. Yield: 61.4%. m.p. 250-251.5^{0}C. Anal. For C$_4$H$_7$NO$_4$: C, 35.82; H, 5.44; N, 10.44. Cacld: C, 36.06; H, 5.26; N, 10.52. IR (cm^{-1}, KBr): 1600, 1500, 1070. ^{1}H-NMR (D$_2$O): 3.87 (1H, t, J=6.0Hz); 2.81 (2H, D, J=6.0Hz).

Serine (4c): synthesized as **4a** according to the reaction condition listed in Table II. Yield: 43.6%. m.p. 216-219 ^{0}C (dec.). Anal. For C$_3$H$_7$NO$_3$: C, 33.93; H, 7.20; N, 13.12. Cacld: C, 34.28; H, 6.71; N, 13.33. IR (cm^{-1}, KBr): 3400, 3100-2400, 1650, 1500. ^{1}H-NMR (D$_2$O): 4.2-3.6 (3H, m).

Leucine (4d): synthesized as **4a** according to the reaction condition listed in Table II. Yield: 54.5%. m.p. 291-293 ^{0}C (dec). Anal. For C$_6$H$_{13}$NO$_2$: C, 54.84; H, 10.19; N, 10.60. Cacld: C, 54.88; H, 10.73; N, 10.67. IR(cm^{-1}, KBr):3100-2400, 1620, 1520. ^{1}H-NMR (D$_2$O): 3.7 (1H, t, J=6.0Hz); 2.0-1.5 (2H, m); 1.50-0.97 (7H, m).

Ethyl 2-acetyl-4-pentenoate (6a): The mixture of ethyl acetoacetate (0.65g, 5.0mmol), allyl bromide (0.6g, 5.0mmol), potassium hydroxide-potassium carbonate (1:4, 4.0g, 6.5mmol) and TBAC (0.15g, 0.5mmol) was introduced into a S.M.C. E 70E 650W domestic microwave oven in an open container, microwave irradiation was carried out for 3 minutes at power level 2, final temperature was 75^{0}C. Then 50 mL of ether was poured into cooled mixture, filtration, concentration and followed by purification through column chromatography (silica gel 100-200 mesh; eluent:

petroleum ether:EtOAc=6:1) to give **6a**. Yield: 0.69g (81%). IR (cm^{-1}): 1745, 1720. ^{1}H-NMR (CDCl$_3$): 5.70-4.85 (3H, m); 4.10 (2H, q, J=7.0Hz); 3.40 (1H, m); 2.50 (2H, m); 2.15 (3H,s); 1.20 (3H, t, J=7.0Hz). MS [m/z (%)]: 273 (M$^+$, 10.90); 43(100).

Ethyl 2-acetyl-3-phenyl-propionate(6b): synthesized as **6a** according to the reaction condition listed in Table III. Yield: 69%. IR (cm^{-1}): 1740, 1718. ^{1}H-NMR (CDCl$_3$): 7.20 (5H, s);.); 4.10 (2H, q, J=7.0Hz); 3.65 (1H, t, J=8.0Hz); 3.20 (2H, d, J=8.0Hz); 2.18 (3H, s); 1.20 (3H, t, J=7.0Hz). MS [m/z (%)]: 220 (M$^+$, 19.05); 177(M$^+$-CH$_3$CO, 100).

Ethyl 2-acetyl-3-(m-methoxyphenyl)-propionate(6c): synthesized as **6a** according to the reaction condition listed in Table III. Yield: 82%. IR (cm^{-1}): 1736, 1713. ^{1}H-NMR (CDCl$_3$): 7.10-6.65 (4H, m); 4.00-3.65 (3H, m); 3.65 (3H, s); 3.10 (2H, d, J=8.0Hz); 2.10 (3H, s); 1.10 (3H, t, J=7.0Hz). MS [m/z (%)]: 250 (M$^+$, 74.24); 207(M$^+$-CH$_3$CO, 100).

Ethyl 2-acetyl-3-(p-chlorophenyl)-propionate(6d): synthesized as **6a** according to the reaction condition listed in Table III. Yield: 59%. IR (cm^{-1}): 1740, 1715. ^{1}H-NMR (CDCl$_3$): 7.25-6.90 (4H, m); 4.20 (2H, q, J=7.0Hz); 3.60 (1H, m); 3.10 (2H, d, J=8.0 Hz); 2.18 (3H, s); 1.20 (3H, t, J=7.0Hz). MS [m/z (%)]: 254 (M$^+$, 4.12); 211(M$^+$-CH$_3$CO, 100).

Ethyl 2-acetyl-hexanoate (6e): synthesized as **6a** according to the reaction condition listed in Table III. Yield: 61%. IR (cm^{-1}): 1740, 1718. ^{1}H-NMR (CDCl$_3$): 4.00 (2H, q, J=7.0Hz); 3.25 (1H, t, J=8.0Hz); 2.10 (3H, s); 1.70 (2H, m); 1.55-1.05 (7H, m); 0.80 (3H, t, J=7.0Hz). MS [m/z (%)]: 186 (M$^+$, 2.28); 143 (M$^+$-CH$_3$CO, 18.89), 73 (100).

Ethyl 2-phenylthio-3-phenyl-propionate (8a): The mixture of ethyl phenyl-mercaptoacetate (0.98g, 5.0mmol), benzyl chloride (0.65g, 5.0mmol), potassium hydroxide-potassium carbonate (1:2, 4.0g, 13mmol) and TBAC (0.15g, 0.5mmol) was introduced into a S.M.C. E 70E 650W domestic microwave oven in an open container, microwave irradiation was carried out for 4.5 minutes at power level 3. Then 50 mL of ether was poured into cooled mixture, filtration, concentration and followed by purification through column chromatography (silica gel 100-200 mesh; eluent: petroleum ether:EtOAc=6:1) to give **8a**. Yield: 81%. IR (cm^{-1}): 1730, 1580, 1380, 1260, 740. ^{1}H-NMR (CDCl$_3$): 7.35-7.05 (10H, m); 4.45 (2H, q, J=7.0Hz); 3.45 (3H, m); 1.20 (3H, t, J=7.0Hz). MS [m/z (%)]: 286 (M$^+$, 100).

Ethyl 2-phenylthio-4-pentenoate(8b): synthesized as **8a** according to the reaction condition listed in Table IV. Yield: 67%. IR (cm^{-1}): 1725, 1640, 1380, 920, 700. ^{1}H-NMR (CDCl$_3$): 7.65-7.15 (5H, m); 6.10-4.90 (3H, m); 4.10 (2H, q, J=7.0Hz); 3.61 (1H, t, J=7.0Hz); 2.55 (2H, d, J=7.0Hz); 1.20 (3H, t, J=7.0Hz). MS [m/z (%)]: 236 (M$^+$, 83.45); 161 (100).

Ethyl 2-phenylthio-2-(p-chlorobenzyl)-3-(p-chlorophenyl)-propionate (8c): synthesized as **8a** according to the reaction condition listed in Table IV, using 1 mL of toluene as solvent. Yield: 61%. IR (cm^{-1}): 1730, 1600, 1490, 1100, 700. ^{1}H-NMR (CDCl$_3$): 7.10-6.75 (13H, m); 3.90 (2H, q, J=7.0Hz); 3.00 (4H, s); 1.10 (3H, t, J=7.0Hz). MS [m/z (%)]: 444 (M$^+$, 29.14); 110 (100).

Ethyl 2-phenylthio-hexanoate(8d): synthesized as **8a** according to the reaction condition listed in Table IV. Yield: 59%. IR (cm^{-1}): 1730, 1580, 1380, 690. ^{1}H-NMR (CDCl$_3$): 7.50-7.00 (5H, m); 4.05 (2H, q, J=7.0Hz); 3.61 (1H, m); 1.75 (2H, m); 1.65-0.75 (4H, m); 1.20 (3H, t, J=7.0Hz); 0.85 (3H, t, J=6.0Hz). MS [m/z (%)]: 252 (M$^+$, 42.56); 224 (100).

Ethyl 2-phenylthio-3-(m-methoxyphenyl)-propionate(8e): synthesized as **8a** according to the reaction condition listed in Table IV. Yield: 58%. IR (cm^{-1}): 1730, 1600, 1380, 1270, 700. ^{1}H-NMR (CDCl$_3$): 7.40-6.75 (9H, m); 4.50 (2H, d, J=6.0Hz); 3.85 (3H, s); 3.55 (3H, m); 1.20 (3H, t, J=7.0Hz). MS [m/z (%)]: 316 (M$^+$, 12.09); 122 (100).

Diethyl 2-benzyl-malonate(10a): The mixture of diethyl malonate (1.62g, 10.0mmol), benzyl chloride (1.52g, 12.0mmol), potassium carbonate (5.48g, 40.0mmol) and TBAB (0.32g, 1.0mmol) was introduced into a S.M.C. E 70E 650W domestic microwave oven in an open container, microwave irradiation was carried out for 2 minutes at power level 1. Then 50 mL of ether was poured into cooled mixture, filtration, concentration and followed by purification through column chromatography (silica gel 100-200 mesh; eluent: petroleum ether:EtOAc=4:1) to give **10a**. Yield: 2.10g (75%). IR (cm^{-1}): 1735, 755, 700. ^{1}H-NMR (CDCl$_3$): 7.16 (5H,s); 4.03 (4H, q, J=7.0Hz); 4.03-3.80 (1H, m); 3.09 (2H, d, J=7.0Hz); 1.10 (6H, t, J=7.0Hz). MS [m/z (%)]: 251(M$^+$-1); 91 (100).

Diethyl 2-(m-methoxybenzyl)-malonate(10b): synthesized as **10a** according to the reaction condition listed in Table V. Yield: 71%. IR (cm^{-1}): 1735, 872, 795,708. ^{1}H-NMR (CDCl$_3$): 7.20-6.50 (4H, m); 4.05(4H, q, J=7.0Hz); 4.02 (1H, m); 3.72 (3H, s); 3.05 (2H, d, J=7.0Hz); 1.12 (6H, t, J=7.0Hz). MS [m/z (%)]: 282(M$^+$-1); 122 (100); 121 (16.34).

Diethyl 2-allyl-malonate(10c): synthesized as **10a** according to the reaction condition listed in Table V. Yield: 75%. IR (cm^{-1}): 3030, 1735, 1642. ^{1}H-NMR (CDCl$_3$): 6.00-4.80 (3H, m); 4.07 (4H, q, J=7.0Hz); 4.25-3.88 (1H, m); 3.02 (2H, m); 1.18 (6H, t, J=7.0Hz). MS [m/z (%)]: 202(M$^+$).

Diethyl 2-(n-butyl)-malonate (10d): synthesized as **10a** according to the reaction condition listed in Table V. Yield: 86%. IR (cm^{-1}): 2960, 2936, 2876, 1735. ^{1}H-NMR (CDCl$_3$): 4.08(4H, q, J=7.0Hz); 4.22-3.90 (1H, m); 1.95-1.60 (2H, m); 1.20 (6H, t, J=7.0Hz). 1.40-0.65 (7H, m). MS [m/z (%)]: 218(M$^+$); 160 (100).

Diethyl 2,2-di(p-chlorobenzyl)-malonate(10e): synthesized as 10a according to the reaction condition listed in Table V, using 1 mL of toluene as solvent. Yield: 64%. IR (cm^{-1}): 1730, 830. ^{1}H-NMR (CDCl$_3$): 7.15 (4H, d, J=9.0Hz); 6.95 (4H, d, J=9.0Hz); 3.98 (4H, q, J=7.0Hz); 3.05 (4H, s); 1.07 (6H, t, J=7.0Hz). MS [m/z (%)]: 408(M$^+$); 125(100).

3,3-dibenzyl-2,4-pentanedione(12a): The mixture of acetoacetone (1.00g, 10.0mmol), benzyl chloride (3.16g, 25.0mmol), potassium carbonate (6.00g, 43.4mmol) and TBAB (0.32g, 1.0mmol) was introduced into a S.M.C. E 70E 650W domestic microwave oven in an open container, microwave irradiation was carried out for 2 minutes at power level 1. Then 50 mL of ether was poured into cooled mixture, filtration, concentration and followed by purification through column chromatography (silica gel 100-200 mesh; eluent: petroleum ether:EtOAc=4:1) to give 12a. Yield: 1.81g (72%). IR (cm^{-1}): 1691,1600, 1580, 1495, 1455. ^{1}H-NMR (CDCl$_3$): 7.30-6.70 (10H, m); 3.18 (4H, s); 2.06 (6H, s). MS [m/z (%)]: 280(M$^+$, 15.57); 43 (CH$_3$CO$^+$, 100).

3,3-di(m-methoxybenzyl)-2,4-pentanedione (12b): synthesized as 12a according to the reaction condition listed in Table VI. Yield: 72%. IR (cm^{-1}): 1697, 1601, 1584, 1491, 1454. ^{1}H-NMR (CDCl$_3$): 7.20-6.85 (8H, m); 3.68 (6H, s); 3.18 (4H, s); 2.05 (6H, s). MS [m/z (%)]: 307(M$^+$-1, 76); 43 (CH$_3$CO$^+$, 100).

3,3-di(p-chlorobenzyl)-2,4-pentanedione (12c): synthesized as 12a according to the reaction condition listed in Table VI, using 1 mL of toluene as solvent. Yield: 68%. IR (cm^{-1}): 1690, 1491, 1452. ^{1}H-NMR (CDCl$_3$): 7.06 (4H, d, J=8.0Hz); 6,69 (4H, d, J=8.0Hz); 3.01 (4H, s); 1.96 (6H, s). MS [m/z (%)]: 348 (M$^+$, 23.83); 305 (M$^+$-CH$_3$CO, 100).

3,3-diallyl-2,4-pentanedione (12d): synthesized as 12a according to the reaction condition listed in Table VI. Yield: 68%. IR (cm^{-1}): 3085, 1700. ^{1}H-NMR (CDCl$_3$): 5.70-4.65 (6H, m); 2.57 (4H, d, J=7.0Hz); 2.05 (6H, s). MS [m/z (%)]: 178(M$^+$-2, 42.36); 121 (M$^+$-1-2C$_2$H$_4$, 100).

Acknowledgement

We are grateful to the financial support of the National Natural Science Foundation of P. R. China.

Literature Cited:

1. House, H. O., "*Modern Synthetic Methods*", W. A. Benjamin, Menlo Park, 1972.
2. Jiang, Yaozhong; Chen, Daimo; Li guangnian; *Acta. Chimica Sinica*, 1985, *43*, 275.
3. Jiang, Yaozhong; Ma, Youan; Li, Guangnian; *Chinese J. Amino Acids*, 1987, *33*, 12.
4. Jiang, Yaozhong; Ma, Youan; Chen, Zhenwan; Li, Guangnian; *Chinese J. Appl. Chem.*, 1987, *4(1)*, 44

5. Wu, Shengde; Liu, Guilan; Chen, Daimo; Jiang, Yaozhong; *Youji Huaxue,* **1987,** *5,* 357.
6. Jiang, Yaozhong; Zhou, Changyou; Wu, Shengde et al; *Tetrahedron,* **1988,** *44,* 5343.
7. Gedye, R., Smith, F.; Westaway, K., Ali, H., Baldisera, L.; Rousell, J.; *Tetrahedron Lett.,*1986, *27(3),* 279.
8. Giguere, R. J.; Bray, T. L.; *Tetrahedron Lett.,* **1986,** *27(41),* 4945.
9. Giguere, R. J.; Namen, A. M.; Lopez, B. O.; Arepally, A.; Ramos, D. E.; *Tetrahedron Lett.,* **1987,** *28(52),* 6553.
10. Gedye, R. N., Smith, F. E.; Westaway, K. C.; *Can. J. Chem.,*1988, *66,* 17.
11. Bram, G.; Loupy, A.; Majdoub, M.; *Synth. Commun.,* **1990,** *20(1),* 125.
12. Smith R. L. et al, *J. Med. Chem.,* **1977,** *20(4),* 540.
13. Zhang, Zhen; Liu, Guangjian; Wang, Yuliang; Wang, Yong; *Synth. Commun.,* **1989,** *19(7&8),* 1167.
14. Babayan, A. T., Indzhikyan, M. G., *Tetrahedron,* **1964,** *20,* 1371.
15. Bandstrom, A., Junggren, U., *Acta. Chem. Scand.,* **1969,** *23,* 2204.
16. Kurst, A. K., Dem'yanor, P. I., Beletskaya, I. P. et al, *Zhur. Org. Khim.,* **1973,** *9,* 1313.
17. Durst, H. D., Liebeskind, L.; *J. Org. Chem.,*1974, *39,* 3271.
18. Singh, R. K., Danishetsky, S.; *J. Org. Chem.,*1975, *40,* 2969.
19. Trost, B. M., Salzmann, T. N., Hiroi, K., *J. Amer. Chem. Soc.,* **1976,** *98,* 4887, and references therein.
20. Brandstrom, A., Junggren, U.; *Tetrahedron Lett.,* **1972,** *13,* 473.
21. Jon'czyk, A., Ludwikow, M., Makosza, M., *Rocz. Chem.,* **1973,** *47,* 89.
22. Bandstrom, A., Junggren, U.; *Acta. Chem. Scand.,* **1969,** *23,* 3585.

Chapter 17

Chemical Modification of Polymers via a Phase-Transfer Catalyst or Organic Strong Base

Tadatomi Nishikubo

Department of Applied Chemistry, Faculty of Engineering, Kanagawa University, Rokkakubashi, Kanagawa-ku, Yokohama 221, Japan

It is well-known that phase-transfer catalysis (PTC) is a very convenient and useful method for organic synthesis. This method can also be used for the chemical modification of polymers to synthesize various functional polymers. That is, functional polymers can be synthesized ordinarily by two different methods. One method is synthesis of monomers with functional groups and selective polymerization of the monomers to give the corresponding polymers with pendant functional groups. Although polymers containing mostly 100 mol% of the functional group can usually be obtained, synthesis and isolation of the monomers are not easy, and polymerization of these monomers is strongly affected or inhibited some times by the functional groups in the monomers. The other method is synthesis of functional polymers by the chemical modification of polymers containing reactive groups such as pendant chloromethyl groups. Although synthesis of polymers containing fully 100% of functional groups is very difficult by this method, many synthetic routes can be used for the synthesis of a functional polymer. Furthermore, many commercial monomers and polymers can be used as starting materials for the synthesis of targeted functional polymers. This means that the PTC method is a very convenient and powerful method for chemical modification of polymers to synthesize functional polymers.

Over the years, progress from the chemical modification of polymers using the classical method to the PTC method has occurred. Substitution reactions of polymers containing haloalkyl groups with some nucleophilic reagents were found to proceed smoothly in aprotic polar solvents such as DMF, DMSO, and HMPA without use of phase transfer catalysts. Okawara et al. reported[1,2] substitution reactions of poly(vinyl chloride) with potassium dithiocarbamate and sodium azide in DMF in 1966 and 1969, respectively. Minoura and his co-workers reported[3] various substitution reactions of poly(bischloromethyl oxetane) (PBCMO) in DMSO to give the corresponding polymers with high conversions in 1967. Nishikubo et al. reported[4] substitution reactions of poly(vinyl chloroacetate) (PVCA) and poly(2-chloroethyl acrylate) (PCEA) with potassium cinnamate to synthesize new photosensitive polymers in 1972.   These reactions proceeded very smoothly to give the targeted polymers with high conversion in HMPA.   Gibson and Bailey also reported[5] substitution reaction of poly-(chloromethylstyrene) with potassium carbazole in DMF.

When the reaction of poly(epichlorohydrin) (PECH) with potassium cinnamate to obtain a new and highly photo-responsive polymer was investigated [6,7] in the early 1970s, it was found that the degree of esterification of the polymer was only 82% even at 100 °C in HMPA.   However, the reaction of PECH proceeded smoothly to give poly(glycidyl cinnamate) with 97% conversion by the addition of small amounts of methyltriethylammonium iodide (MTEAI) as a catalyst .   In this reaction, the solubility of potassium cinnamate in HMPA was increased, and the rate of the reaction enhanced by the addition of MTEAI.   MTEAI was recognized as a very good **"catalyst"** to accelerate this reaction.   However, the authors did not see Starks's papers[8,9], because our research work on the reaction of PECH with potassium cinnamate was finished by May, 1970 as shown in the patent[6].   This research work was also reported[7] in Nippon Kagaku Kaishi in 1973.   Okawara and Takeishi also reported[10,11] in 1973 that the substitution reaction of PVC with sodium azide in organic solvents and in solid-liquid two-phase reaction systems was strongly enhanced by the addition of tetrabutylammonium bromide (TBAB), which was labeled as **"surfactant"** by them.   Roovers investigated[12] the substitution reaction of chloromethylated polystyrene with potassium acetate in a mixture of benzene and acetonitrile using dicyclohexyl-18-crown-6 (DCHC) as a catalyst in 1976.   Roeske and Gesellchen also reported[13] that the substitution reaction of the crosslinked poly(chloromethylstyrene) with boc-amino acid derivatives proceeds very smoothly using 18-crown-6 (CR6) in various organic solvents in 1976.   That is, although the excellent catalytic effect of quaternary onium salts and crown ethers was found by a number of us, none of us ever referred to the term "PTC" in our work in this field.   Boileau et al. examined[14] the substitution reaction of PECH with some nucleophilic reagents such as sodium carbazole in 1978.   They suggested that the reaction of PECH with sodium carbazole did not proceed in chloroform even using tetrabutylammonium hydrogensulfate (TBAH), although the reaction proceeded in DMF with the addition of TBAH, of DCHC, or of cryptand[2,2,2].   Referring to this reaction system, they used the name "phase transfer catalyst" (PTC).   They also reported[15] in 1979 that the substitution reaction of partly-chloromethylated soluble poly(styrene) with some nucleophilic reagents proceeded very smoothly in DMF using "PTC".   Frechet et al. reported[16,17] successful reactions of the crosslinked chloromethylated poly(styrene) beads with some nucleophilic reagents using a certain PTC in a liquid-liquid two-phase reaction system in 1978 and 1979.   Gozdz also reported[18,19] very good results on the substitution reaction of the crosslinked polystyrene bead with potassium iodide and potassium thiocyanate using CR6 as a PTC.

However, there had not been, up to that time, any systematic research work or publication on the chemical modification of polymer containing pendant chloromethyl groups with various reagents using PTCs, which had already become such an important method for the synthesis of functional polymers.   Therefore, in 1978, our research group started[20,21] working again with chemical modification of polymers containing pendant chloromethyl groups using phase transfer catalyst and commercial organic solvents such as toluene, because such a reaction system promised to be a better method from the viewpoint of economics than the classical reaction in aprotic polar solvents such as DMF, DMSO and HMPA.

Chemical Modification of Polymers with Pendant Chloromethyl Groups or Arylchloride Groups using PTC

The substitution reaction of poly(chloromethylstyrene) (PCMS) with various reagents such as potassium acetate, potassium benzoate, potassium thioacetate, potassium thiocyanate, sodium azide, and potassium phthalimide was examined[22,23] using certain PTCs (Scheme 1). The reaction of PCMS with potassium acetate did not proceed in non-polar solvents such as toluene and diglyme without PTC at room temperature for 24 h. However, the reaction did proceed with 77% conversion in DMF under the same conditions. When CR6 was added as a PTC, the solid-liquid two phase reaction proceeded with 24 and 33% conversions in toluene and diglyme, respectively. Furthermore, the reaction proceeded with nearly 100% conversion in DMF by the addition of CR6. This means that the substitution reaction of PCMS proceeds even in a nonpolar solvent under mild reaction conditions in the presence of PTC and the reaction was enhanced by the addition of PTC in an aprotic polar solvent. As summarized in Table 1, the reaction of PCMS with various O-anions such as potassium acetate, sodium acetate, potassium benzoate was performed in toluene using various PTCs. Tetrabutylammonium bromide (TBAB) and tetrabutylphosphonium bromide (TBPB) showed higher catalytic activity than crown ethers in solid-liquid and liquid-liquid two-phase systems.

When several symmetric tetraalkylammonium bromides were used in the reaction of PCMS with potassium acetate, tetrapentylammonium bromide, having a good balance of hydrophobicity and steric hindrance, showed the highest catalytic activity (Figure 1). Meanwhile, when the reaction of PCMS with potassium acetate was carried out in liquid-solid and liquid-liquid two-phase systems using various tetrabutylammonium salts as the catalysts, TBAB showed higher catalytic activity than other onium salts (Figure 2). This means that the counter anions also affected the catalytic activity of the PTC.

As summarized in Table 2, when the reaction of PCMS with certain S-anions such as potassium thiophenoxide, potassium thioacetate, and potassium thiocyanate was carried out in toluene using various phase transfer catalysts, DCHC showed higher catalytic activity than TBAB and TBPB in solid-liquid two-phase reaction systems. The same results on the reaction of PCMS with S-anions were also obtained in the liquid-liquid two-phase reaction systems. In the case of the solid-liquid two-phase reaction of PCMS with potassium thiocyanate, tetrabutylammonium chloride (TBAC), TBAB, tetrabutylammonium iodide (TBAI) and tetrabutylammonium perchlorate (TBAP) showed nearly the same catalytic activity, although TBAB had the highest activity on the reaction with potassium acetate (Figure 3). This means that the effect of the counter anions of the quaternary ammonium salts varied with the kind of nucleophilic reagents used.

When the reaction of PCMS with certain N-anions was examined, the catalytic activity of PTC was found to depend on the combination of PTC and anions (Table 3). That is, in the reaction of PCMS with sodium azide , TBAB and TBPB showed higher activity than crown ethers. However, in the reaction of PCMS with potassium azide, crown ethers showed higher catalytic activity than quaternary onium salts. Furthermore, in the reaction with potassium phthalimide , DCHC showed extremely lower activity than CR6, TBAB and TBPB.

From these results, the following interesting conclusions can be drawn. 1) Appropriate quaternary onium salts such as TBAB and TBPB have higher catalytic activity than crown ethers such as CR6 or DCHC on the reaction of PCMS with O-anions. 2) Lipophilic crown ethers such as DCHC have higher catalytic activity than TBAB and TBPB on the reaction of PCMS with S-anions. 3) The catalytic activity of phase transfer catalysts on the reaction of PCMS with N-anions lies between that of the reaction of PCMS with S-anions and with O-anions. 4) Phase transfer catalysts

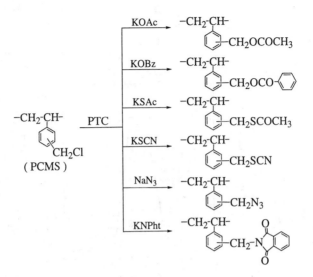

Scheme 1.

Table 1. Substitution Reaction of PCMS with Salts of O-Anion Using PTC [a)

Run	Nucleophilic reagent	Solvent	PTC	Degree of substitution (mol%)
1	KOAc	Toluene	None	0
2	KOAc	Toluene	CR5	Trace
3	KOAc	Toluene	CR6	24
4	KOAc	Toluene	DCHC	4
5	KOAc	Toluene	TBAB	74
6	KOAc	Toluene	TBPB	79
7	KOAc	Toluene / Water[b)	CR5	Trace
8	KOAc	Toluene / Water[b)	CR6	Trace
9	KOAc	Toluene / Water[b)	DCHC	6
10	KOAc	Toluene / Water[b)	TBAB	66
11	KOAc	Toluene / Water[b)	TBPB	83
12	NaOAc	Toluene	CR5	14
13	NaOAc	Toluene	CR6	Trace
14	NaOAc	Toluene	DCHC	0
15	NaOAc	Toluene	TBAB	38
16	NaOAc	Toluene	TBPB	46
17	KOBz	Toluene	CR5	Trace
18	KOBz	Toluene	CR6	11
19	KOBz	Toluene	DCHC	7
20	KOBz	Toluene	TBAB	64
21	KOBz	Toluene	TBPB	55

a) The reaction was carried out with PCMS (4 mmol) and the reagent(4mmol) using PTC (0.4 mmol) in toluene (10 ml) at 30°C for 24 h at 300 rpm. b) the reaction was carried out with 5 mg saturated aqueous solution of the reagent.

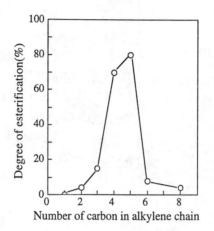

Figure 1. Effect of alkylene chain length of symmetric tetraalkylammonium bromide on a solid-liquid two-phase reaction of PCMS with KOAc.

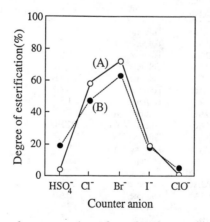

Figure 2. Effect of counter anion of tetrabutylammonium salt on a solid-liquid (A) and a liquid-liquid (B) two-phase reaction of PCMS with KOAc.

Table 2. Substitution Reaction of PCMS with Salts of S-Anion Using PTC[a]

Run	Nucleophilic reagent	Solvent	PTC	Degree of substitution (mol%)
1	KSPh	Toluene	None	0
2	KSPh	Toluene	CR6	99
3	KSPh	Toluene	TBAB	99
4	KSPh	Toluene	TBPB	99
5	KSAc	Toluene	None	Trace
6	KSAc	Toluene	CR6	99 (56)[b]
7	KSAc	Toluene	DCHC	99 (88)[b]
8	KSAc	Toluene	TBAB	99 (36)[b]
9	KSAc	Toluene	TBPB	99 (64)[b]
10	KSCN	Toluene	None	Trace
11	KSCN	Toluene	CR5	13
12	KSCN	Toluene	CR6	Trace
13	KSCN	Toluene	DCHC	77
14	KSCN	Toluene	TBAB	63
15	KSCN	Toluene	TBPA	64

a) The reaction was carried out with PCMS (4 mmol) and the reagent (4mmol) using PTC (0.4 mmol) at 30 °C for 24 h. b) The reaction was carried out for 60 min.

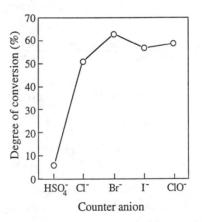

Figure 3.   Effect of counter anion of tetraalkylammonium salt on a solid-liquid two-phase reaction of PCMS with KSCN.

Table 3. Substitution Reaction of PCMS
with Salts of N-Anion Using PTC [a]

Run	Nucleophilic reagent	Solvent	PTC	Degree of substitution (mol%)
1	NaN_3	Toluene	None	0
2	NaN_3	Toluene	CR5	9
3	NaN_3	Toluene	CR6	28
4	NaN_3	Toluene	DCHC	16
5	NaN_3	Toluene	TBAB	95
6	NaN_3	Toluene	TBPB	98
7	KN_3	Toluene	CR5	70
8	KN_3	Toluene	CR6	85
9	KN_3	Toluene	DCHC	89
10	KN_3	Toluene	TBAB	54
11	KN_3	Toluene	TBPB	51
12	KNPht	Toluene	None	0
13	KNPht	Toluene	CR6	56
14	KNPht	Toluene	DCHC	12
15	KNPht	Toluene	TBAB	54
16	KNPHt	Toluene	TBPB	66

a) The reaction was carried out with PCMS (4 mmol) and the reagent
(4 mmol) using PTC (0.4 mmol) in toluene (10 ml) at 30 °C for 24 h.

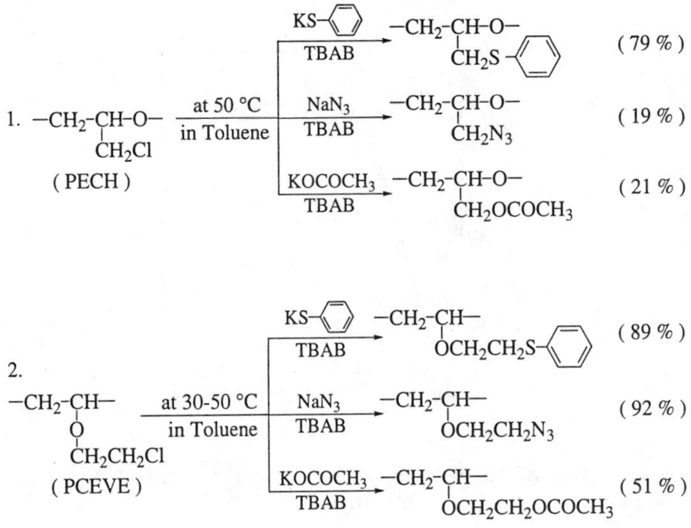

Scheme 2.

should be therefore classified into two categories: Soft phase transfer catalysts such as crown ethers without electric charge and hard phase transfer catalysts such as quaternary onium salts with electric charge. Soft phase transfer catalysts are effective ordinarily for the reaction of soft nucleophilic reagents such as S-anions, and hard phase transfer catalysts are effective for the reaction of hard nucleophilic reagents such as O-anions. 5) Although the substitution reaction proceeds ordinarily with high conversion by the liquid-solid two-phase reaction, polymer with high conversion can be also obtained by the liquid-liquid two-phase reaction, when the reaction is performed with a soft reagent (S-anion) using a soft phase transfer catalyst such as DCHC.

Chemical modifications of PECH and poly(2-chloroethyl vinyl ether) (PCEVE) were also tried[24-26] using various PTCs (Scheme 2). The substitution reaction of PECH with potassium thiophenoxide, sodium azide, and potassium acetate produced the corresponding polymers with 78, 19, and 21% conversions, respectively, using TBAB at 50 °C for 24 h. The reaction of PCEVE with the same reagents gave the corresponding polymers with 89, 92, and 51% conversions, respectively, under the same conditions.

It was found that a correlation existed between the degree of conversion in the reaction of the polymers with potassium acetate and the number of carbon atoms in the alkyl chain of the symmetric quaternary ammonium bromide (Figure 4). This means that the catalytic activity of the quaternary ammonium salt was related to the kind of polymer chain used. That is, although TBAB showed the highest catalytic activity in the reaction of PECH with a shorter spacer chain from the polymer main chain, tetrahexylammonium bromide had the highest activity in the reaction of PCEVE with a longer spacer chain from the polymer main chain. PCEVE with a longer spacer chain from the polymer main chain had higher reactivity than PECH with shorter spacer chain. This means that the polymer main chain acts as a bulky groups exerting steric hindrance on the substitution reaction.

Elimination reactions of PECH and PCEVE with certain bases were performed using PTCs at room temperature (Scheme 3). The elimination reaction of polymers proceeded smoothly even at 30 °C (Table 4). From this result, it seems that crown ethers have high catalytic activity in the elimination reaction of these polymers. Correlation between degree of elimination and number of carbon atoms in the alkyl chain of symmetric quaternary ammonium bromide was observed in the elimination reaction of PECH and of PCEVE. It was also shown that TBAB had the highest catalytic activity on the reaction of PECH, and tetrahexylammonium bromide had the highest activity on the reaction of PCEVE (Figure 5). It was found as well that PECH produced the corresponding vinyl ether polymer with higher conversion on the elimination reaction than PCEVE under the same conditions, although PCEVE showed higher reactivity than PECH in the substitution reaction. The effect of counter anions of quaternary ammonium salts on the substitution and elimination reactions was examined. TBAB with bromine as counter anion showed the highest activity in reactions such as substitution reaction with potassium acetate (Figure 6).

Substitution reactions of pendant arylchloride of 4-chloro-3-nitrobenzoate group in the polymer side chain were investigated[27] using PTC (Scheme 4). Although the reaction of the polymer with potassium thiophenolate and potassium thioacetate proceeded to give the corresponding polymers even without PTC in anisole, the reaction was enhanced by the addition of PTCs such as TBPB and DCHC. The reaction with sodium azide did not occur without PTC. However, this reaction proceeded with 82 and 43% conversions by the addition of TBPB and DCHC, respectively (Table 5). The effect of the number of carbon atoms in the alkyl chain of symmetric ammonium bromide was examined on the reaction of this polymer with sodium azide, and it was found that tetrapropylammonium bromide (TPAB) showed the highest activity in the reaction with sodium azide (Figure 7).

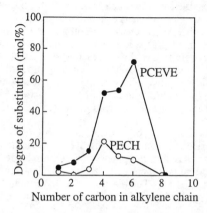

Figure 4. Effect of alkylene chain length of symmetric tetraalkylammonium bromide on two-phase reactions of PCEVE and PECH with KOAc.

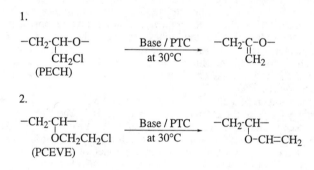

Scheme 3.

Table 4. Elimination Reaction of PECH and PCEVE with Base[a]

Run	Polymer	Solvent	Base	PTC	Degree of elimination (mol%)
1	PECH	Toluene	KOH	TBAB	19
2	PECH	Toluene	KOH	CR6	45
3	PECH	Toluene	KOH	DCHC	35
4	PECH	Diglyme	KOH	TBAB	70
5	PCEVE	Toluene	KOH	None	Trace
6	PCEVE	Toluene	KOH	TBAB	9
7	PCEVE	Toluene	KOH	CR6	49
8	PCEVE	Toluene	KOH	DCHC	31
9	PCEVE	Toluene	KOBut	None	Trace
10	PCEVE	Toluene	KOBut	CR6	66

a) Elimination reaction of 4 mmol of the polymer with 4 mmol of base was carried out in the presence of 0.4 mmol of PTC in 10 mL of solvent for 24 h.

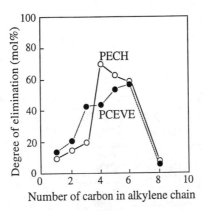

Figure 5. Effect of alkylene chain length of symmetric tetraalkylammonium bromide on elimination reactions of PCEVE and PECH with KOH.

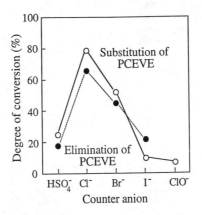

Figure 6. Effect of counter anion of tetrabutylammonium salt on a substitution reaction of PCEVE with KOAc and an elimination reaction of PCEVE with KOH.

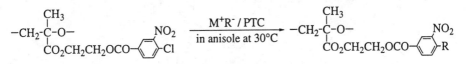

Scheme 4.

Table 5. Substitution Reaction of Polymer with Certain
Reagents Using PTC[a]

Run	Reagent	PTC	Degree of conversion (%)
1	KSC_6H_5	None	89
2	KSC_6H_5	TBPB	90
3	KSC_6H_5	DCHC	97
4	$KSCOCH_3$	None	28
5	$KSCOCH_3$	TBPB	46
6	$KSCOCH_3$	DCHC	83
7	NaN_3	None	0
8	NaN_3	TBPB	82
9	NaN_3	DCHC	43

a) The reaction was carried out with polymer (4 mmol) and
the reagent (4 mmol) in anisole (10 mL) using 10 mol% of
the PTC at 30°C for 24 h.

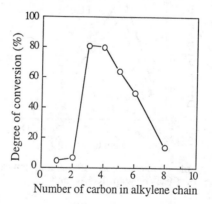

Figure 7. Effect of alkylene chain length of symmetric tetraalkylammonium
bromide on a substitution reaction of polymer with NaN_3.

These results also show that the precise combination of reacting polymer, reagents, and PTC is a very important factor in obtaining the targeted polymers with high degree of conversion, although the reaction solvent also strongly affects the reaction.   That is, the selection of a suitable PTC is the most important factor leading to successful chemical modification of polymers using the PTC method.

Synthesis of Functional Polymers using Phase Transfer Catalysis
Although systematized research on phase transfer catalysis is very important for the advancement of chemical modification of polymers, the synthesis of functional polymers using PTC under mild reaction conditions is producing further fruitful results in the field of polymer synthesis.   New multifunctional photosensitive polymers with both pendant photosensitizing groups and photosensitive moieties were synthesized[28,29] by reactions of PCMS and PCEVE with some reagents using TBAB, which is the most popular PTC with high activity (Scheme 5).   The reaction of PCMS with potassium salts of photosensitizing compounds proceeded very smoothly in DMF using TBAB at 30 °C, and the targeted polymer with pendant chloromethyl groups and photosensitizing moieties were obtained with good yield.   The reaction of the polymers with potassium salts of photosensitive compounds also proceeded very smoothly under the same conditions, and new multifunctional polymers with high photosensitivity were obtained.   Multifunctional photosensitive polymers with both pendant photosensitizing groups and photosensitive moieties were also prepared by the reaction of PCEVE with potassium salts of photosensitizing compounds followed by the reaction with potassium salts of photosensitive compound using TBAB in DMF. However, to achieve the high conversions, the reaction was carried out at 100 °C, because the reactivity of PCEVE was lower than PCMS.

Multifunctional polymeric photosensitizers with both pendant photosensitizing groups and substrate attracting groups were prepared[30,31] by the reaction of soluble PCMS and the crosslinked PCMS beads with the corresponding reagents.   Typical procedures for the synthesis of the polymeric photosensitizers were showed in Scheme 6.   This photosensitizer can be synthesized very smoothly by three different ways. 1) The substitution reaction of PCMS with potassium salt of the photosensitizer using PTC followed by the addition reaction of the resulting polymer with tributylphosphine (TBP).   2) The addition reaction of PCMS with TBP followed by the substitution reaction of the polymer with the potassium salt of the photosensitizer, in which the addition of PTC is not necessary because the pendant quaternary phosphonium salt acts as a polymer-supported PTC for the reaction.   3) Synthesis of the photosensitizing monomer using PTC and synthesis of substrate-attracting monomer followed by the radical copolymerization of these monomers.   The photosensitizing efficiency was strongly affected by the contents of the photosensitizing group and substrate attracting groups, and these multifunctional polymeric photosensitizers with appropriate amounts of the functional groups showed about ten times higher efficiency than the corresponding low molecular weight photosensitizers.   A typical application of multifunctional photosensitizers prepared by the PTC method is photochemical valence isomerization of norbornadiene (NBD) derivatives to quadricyclane derivatives for solar energy storage and exchange systems[32].

New photoresponsive polymers with pendant NBD moieties, which show promise as solar-energy storage polymers, were also synthesized[33-35].   Typical soluble polymers with pendant NBD moieties were prepared in high conversions by the substitution reaction of PCMS with potassium carboxylate of NBD derivatives using TBAB as a PTC. The resulting NBD polymers were isomerized to QC polymers by photo-irradiation.   The resulting QC polymer could also be converted to the NBD polymer releasing about 90 kJ/mol of thermal energy by contact with a catalyst Co-TPP,

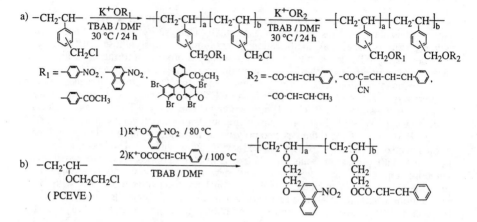

Scheme 5.

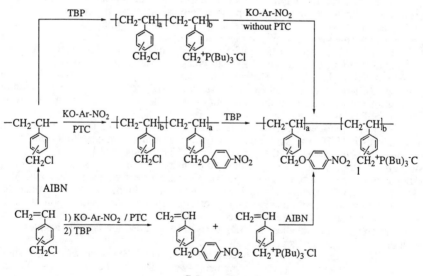

Scheme 6.

by heating, or by photo-irradiation with UV light (Scheme 7). The crosslinked polystyrene beads having pendant NBD moieties were also synthesized[36] by the reaction of the crosslinked chloromethylated polystyrene beads with some NBD derivatives in DMF using phase transfer catalyst TBAB.

Chemical Modification of Pendant Carboxylate Groups and Chloromethyl Groups using Organic Base

The substitution reaction of quaternary ammonium salt of poly(methacrylic acid) with some alkyl halides proceeded[37,38] in DMSO and DMF to give the corresponding poly(methacrylic esters). However, the reaction of potassium poly(methacrylate) and sodium poly(acrylate) with alkylbromides did not proceed[39,40] even in the presence of PTC under normal reaction conditions. This means that phase transfer catalysis is not a suitable method for the chemical modifications of polymers with pendant carboxylate anions. In 1987, we reported[39] finding that the reaction of poly(methacrylic acid) with propargyl bromide proceeded quantitatively using diazabicyclo[5.4.0]-7-undesen (DBU) as an organic base in DMSO even at room temperature for 10 min to give poly(propargyl methacrylate) (Scheme 8). The reaction of poly(methacrylic acid) with a spiro ortho ester compound having a bromomethyl group using DBU produced[41] a new thermo-setting resin with both pendant carboxylate and spiro ortho ester groups. Poly(nitrobenzyl methacrylates), which can be used as positive-type photo-resists, were synthesized[42] in 94-97% conversions by the reaction of poly(methacrylic acid) with o-, m-, or p-nitrobenzylbromides using DBU in DMSO at 30 °C for 3 h. A new self-sensitized multifunctional photo-resist with high photosensitivity was also prepared[43] by the reaction of poly(methacrylic acid) with propargyl bromide and bromomethylated anthraquinone using DBU as a base under similar conditions. These results show the DBU method to be an appropriate method for the chemical modification of polymeric anions and the synthesis of functional polymers from the reaction of poly(methacrylic acid) and poly(acrylic acid) under mild reaction conditions.

Chemical modification of pendant chloromethyl groups in PCMS were also examined[44] using the DBU for comparison with the PTC. The rate of the reaction of PCMS with acetic acid increased with reaction temperature, and the polymer with nearly 100% pendant ester groups was obtained at 80°C for 60 min (Figure 8). When the reactions of PCMS with potassium benzoate and benzoic acid were performed using TBAB and DBU in DMF at 60 °C, respectively, the rate of the reaction with benzoic acid using the DBU method was higher than that with potassium benzoate using TBAB as a PTC in DMF under the same conditions (Figure 9). These results indicate that DBU is a very useful and convenient reagent for the chemical modification of the polymers containing pendant carboxylic acid groups and pendant chloromethyl groups, although PTCs can not be used ordinarily for the chemical modification of the polymers with pendant carboxylic acid groups under normal conditions.

Conclusion

1. Phase transfer catalysts such as quaternary onium salts and crown ethers are very convenient and powerful reagents for the chemical modification of polymers, especially chemical modification of pendant chloromethyl groups in polymers and synthesis of various functional polymers. However, the selection of suitable PTC, and the combination of the catalyst, reagent and polymer are most important factors for successful reactions. 2. Although there are limitations on use of phase transfer catalysts in reactions involving pendant carboxylic acid groups in polymers, DBU is a very suitable reagent for the esterification reaction of pendant carboxylic acids with alkyl halides.

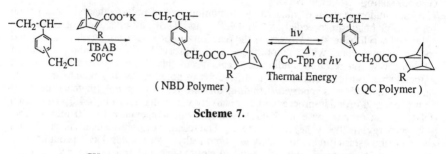

Scheme 7.

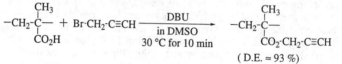

Scheme 8.

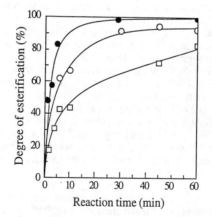

Figure 8.   Reaction of PCMS with acetic acid using DBU in DMF: (●) at 30 °C; (○) at 60 °C, (□) at 80 °C.

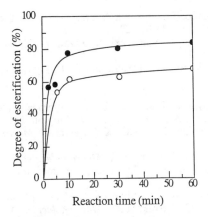

Figure 9. Reaction of PCMS with benzoic acid derivatives in DMF at 60 °C: (●) with benzoic acid using DBU; (○) with potassium benzoate using TBAB.

References
1. Okawara, M.; Morishita, K.; Imoto, E. *Kogyo Kagaku Zasshi* **1966**, 69, 761.
2. Takeishi, M.; Okawara, M. *J. Polym. Sci. Polym. Lett.* **1969**, 7, 201.
3. Minoura, Y.; Shiina, K.; Yoshikawa, K. *J. Polym. Sci. Part A-1* **1967**, 5, 2843.
4. Nishikubo, T.; Tomiyama, Y.; Maki, K.; Takaoka, T. *Kobunshi Kagaku* **1972**, 29, 295.
5. Gibson, H. W.; Bailey, F. C. *Macromolecules* **1976**, 9, 688.
6. NOK Co. Ltd. *Japan Patent* **48-3220**, January 30, 1973 (accepted; May 11, 1970).
7. Nishikubo, T.; Ichijyo, T.; Takaoka, T. *Nippon Kagaku Kaishi* **1973**, 35.
8. Starks, C, M. *J. Am. Chem. Soc.* **1971**, 93, 195.
9. Starks, C. M. *J. Am. Chem. Soc.* **1973**, 95, 3613.
10. Takeishi, M. Kawashima, R.; Okawara, M. *Makromol. Chem.* **1973**, 167, 261.
11. Takeishi, M. Kawashima, R.; Okawara, M. *Angew. Makromol. Chem.* **1973**, 28, 111.
12. Roovers, J. E. L. *Polymer*, **1976**, 17, 1106.
13. Roeske, R. W.; Gesellchen, P. D. *Tetrahedron Lett.* **1976**, (38), 3369.
14. N'Guyen, T. D.; Deffieux, A.; Boileau, S. *Polymer* **1978**, 19, 423.
15. N'Guyen, T. D.; Boileau, S. *Tetrahedron Lett.* **1979**, (28), 2651.
16. Farrall, M. J.; Frechet, J. M. J. *J. Am. Chem. Soc.* **1978**, 100, 7998.
17. Frechet, J. M. J.; Smet, M. D.; Farrall, M. J. *J. Org. Chem.* **1979**, 44, 1774.
18. Gozdz, A. S.; Rapak, A. *Makromol. Chem. Rapid Commun.* **1981**, 2, 359.
19. Gozdz, A. S. *Makromol. Chem. Rapid Commun.* **1981**, 2, 595.
20. Nishikubo, T.; Iizawa, T.; Kobayashi, K.; Okawara, M. *Makromol. Chem. Rapid Commun.* **1980**, 1, 765.
21. Nishikubo, T.; Iizawa, T.; Kobayashi, K.; Okawara, M. *Makromol. Chem. Rapid Commun.* **1981**, 2, 387.
22. Nishikubo, T.; Iizawa, T.; Kobayashi, K.; Masuda, Y.; Okawara, M. *Macromolecules*, **1983**, 16, 722.
23. Iizawa, T.; Nishikubo, T.; Masuda, Y.; Okawara, M. *Macromolecules*, **1984**, 17, 992.
24. Nishikubo, T.; Iizawa, T.; Mizutani, Y.; Okawara, M. *Makromol. Chem. Rapid Commun.* **1982**, 3, 617.
25. Nishikubo, T.; Iizawa, T.; Ichikawa, M.; Okawara, M. *Makromol. Chem. Rapid Commun.* **1983**, 4, 93.

26. Iizawa, T.; Nishikubo, T.; Ichikawa, M.; Sugawara, Y.; Okawara, M. *J. Polym. Sci. Polym. Chem. Ed.* **1985**, 23, 1893.
27. Nishikubo, T.; Iizawa, T.; Numazaki, N.; Okawara, M. *Makromol. Chem. Rapid Commun.* **1983**, 4, 187.
28. Iizawa, T.; Nishikubo, T.; Takahashi, E.; Hasegawa, M. *Makromol. Chem.* **1983**, 184, 2297.
29. Nishikubo, T. Iizawa, T.; Takahashi, E. ACS Symposium Series **266**, "Polymers for Microlithography", Am. Chem. Soc., Washington, D.C. (1984).
30. Nishikubo, T.; Uchida, J.; Matui, K.; Iizawa, T. *Macromolecules*, **1988**, 21, 1583.
31. Nishikubo, T.; Kondo, T; Inomata, K. *Macromolecules*, **1989**, 22, 3827.
32. Nishikubo, T.; Kawashima, T.; Inomata, K.; Kameyama, A. *Macromolecules*, **1992**, 25, 2312.
33. Nishikubo, T.; Shimokawa, T.; Shahara, A. *Macromolecules*, **1989**, 22, 8.
34. Iizawa, T.; Hijikata, C.; Nishikubo, T. *Macromolecules*, **1992**, 25, 21.
35. Nishikubo, T.; Kameyama, A.; Kishi, K.; Mochizuki, Y. *J. Polym. Sci. Part A. Polym. Chem.* **1994**, 32, 2765.
36. Nishikubo, T.; Kameyama, A.; Kishi, K.; Hijikata, C. Reactive Polymers, **1994**, 24, 65.
37. Bailey, D.; Tirrell, D.; Vogl, O. *J. Macromol. Sci.-Chem.* **1978**, A12, 661.
38. Maa, Y. F.; Chen, S. H. *Macromolecules*, **1989**, 22, 2036.
39. Shimokawa, T.; Nishikubo, T. *Kobunshi Ronbunshu*, **1987**, 44, 641.
40. Chen, S. H.; Maa, Y. F. *Macromolecules*, **1988**, 21, 904.
41. Isobe, N.; Numasawa, H.; Nishikubo, T.; Tagoshi, H.; Endo, T. *J. Polym. Sci. Part A. Polym. Chem.* **1989**, 27, 681.
42. Nishikubo, T.; Iizawa, T.; Takahashi, A.; Shimokawa, T. *J. Polym. Sci. Part A. Polym. Chem.* **1990**, 28, 105.
43. Shimokawa, T.; Suzuki, T.; Nishikubo, T. *Polym. J.* **1994**, 26, 967.
44. Kameyama, A.; Suzuki, M.; Ozaki, K.; Nishikubo, T. *Polym. J.* **1996**, 28, 155.

Chapter 18

Synthesis of Aromatic Polyesters Bearing Pendant Reactive Groups by Phase-Transfer Catalysis

Shigeo Nakamura and Chonghui Wang

Department of Applied Chemistry, Faculty of Engineering, Kanagawa University, Kanagawa-ku, Yokohama 221, Japan

Aromatic polyesters having reactive groups in the side chain have been prepared chemoselectively by a water phase/organic phase interfacial polycondensation using phase transfer catalyst. Polyesters bearing pendant carboxyl groups are prepared utilizing a higher nucleophilicity of phenolate compared to carboxylate, whereas high molecular-weight polyesters having pendant chlorohydrin groups are synthesized by controlling the molar ratio of bisphenolate to isophthaloyl chloride and the phase ratio of the water phase to the organic phase.

Since Schnell has first reported the preparation of polycarbonate by interfacial polycondensation (1), a variety of aromatic condensation polymers have been prepared by a water phase/organic phase interfacial polycondensation using phase transfer catalyst.

Many reactive polymers have been synthesized, and their applications have been investigated extensively. However, these investigations are almost restricted to the addition polymers, and those of condensation polymers are very limited.

The synthesis of polyesters bearing pendant reactive groups is very difficult due to the high reactivity of ester groups in the main chain. The synthesis of these polyesters involves the preparation of monomers having an additional functional group, the chemoselective reaction among the reactive groups, and the degradation reaction of the ester groups of resulting polyesters with the pendant groups.

Recently, attention has been focused on chemoselective synthesis of aromatic polyesters and polyamides with reactive groups using condensation agents (2). Aromatic polyesters bearing reactive groups can be used for the preparation of various functional polymers, which are expected for applications as polymer catalysts, amphoteric polyelectrolytes, chelating agents for metal ions, drug delivery systems and biodegradable polymers.

Aromatic polyesters bearing pendant reactive groups can be also used for composite materials with inorganic materials, metals and other organic polymers, and those having pendant bifunctional or multifunctional groups can be blended with other polymers to produce intermolecular complexes and interpenetrating polymer networks (IPNs) or semi-IPNs.

Most aromatic polyesters have been merely blended with condensation polymers such as other polyesters, polyimides and polycarbonates (3,4). Therefore, the polyesters bearing pendant reactive groups are expected to find wide applications in polymer blends due to specific interactions such as hydrogen bonding, ion-ion interaction and electron transfer (5).

A cooperative effect has been proposed in the reaction of partially ionized poly (methacrylic acid) with bromoacetic acid (6,7). The reaction is affected not only by the long-range Coulomb forces but also hydrogen bonding or hydrophobic bond (8). Therefore, polyesters bearing pendant reactive groups behave differently from addition polymers due to the polarizability and reactivity of the ester groups in the main chain.

This article describes the chemoselective synthesis of aromatic polyesters bearing reactive groups such as carboxyl groups and chlorohydrin moieties in the side chains by the water phase/organic phase interfacial polycondensation using phase transfer catalysts.

Polyesters Having Pendant Carboxyl Groups

Chemoselective Reaction of Phenolate. When a sodium carboxylate is reacted with an acid chloride in the presence of tertiary amine, an anhydride can be obtained almost quantitatively at room temperature (9). When sodium phenolate is reacted with an acid chloride at room temperature, an ester of phenol is obtained in a high yield even without a tertiary amine. Generally, the nucleophilicity of phenolate to acid chloride is much higher than that of carboxylate.

The higher nucleophilicity of phenolate than that of carboxylate can be utilized in the interfacial polycondensation and polyesters having pendant carboxyl groups can be obtained chemoselectively from aromatic diols having a carboxyl group, leaving the carboxylic group unreacted (*10*).

Phase Transfer Catalyst.   A low-molecular quaternary ammonium salt, *tert-n*-butyl-phosphonium bromide (TBPB) and a polyester having quaternary ammonium salt groups in the side chain and amino acid moieties in the main chain (poly-cat) were used as phase transfer catalysts (PTC).   The polymer catalyst, poly-cat was prepared as reported previously (*11*) by an organic phase/organic phase interfacial polycondensation of isophthaloyl chloride (IPC) with triethyl-2,3-propanediolammonium chloride and *N,N*-di- (2-hydroxyethyl)-3-aminopropionic acid.   The molar ratio of the quaternary ammonium salt groups to the amino acid groups in the poly-cat was 56/44.

Preparation of Polyesters.   Polyester bearing pendant carboxyl groups (PEA) was prepared from IPC and 4,4-bis(p-hydroxyphenyl)valeric acid (diphenolic acid) (DPA) according to Equation 1 of Scheme 1 (*10*).

A solution polycondensation of IPC with DPA was carried out in DMAc using TEA as an acid acceptor to compare with the phase transfer-catalyzed polycondensation. Although the yield is high (76%), the molecular weight of the product is very low (η_{sp}/ C: 0.09dL/g).   Thus, the polycondensation in an organic solvent without phase transfer catalyst is insuitable for the synthesis of high molecular-weight polyester PEA because the resulting oligomer precipitates at the initial stage of reaction.

0.02 Mole of IPC in 80 mL of organic solvent was added to 0.02 mole of DPA in 180 mL of aqueous 0.06 N NaOH solution in the presence of 0.2 g of PTC.   The mixture was stirred for 3 h at 30°C and then the pH of the solution was adjusted to about 3 by 2 N hydrochloric acid to precipitate polyester (PEA).

Poly(ester-amide-thioester) (PEATA) and copolyester (PEBA) were obtained by substituting 0.01 mole DPA with 0.01 mole of 2-aminothiophenol (ATP) or 0.01 mole of bisphenol A (BPA), respectively (Equations 2 and 3 of Scheme 1).   Constituent ratios of the resulting copolyesters were 59/41 of DPA/ATP for PEATA and 48/52 of DPA/BPA for PEBA.

In the IR spectra of polyester PEA, copolyesters PEATA and PEBA, the band due to anhydride groups is not observed at 1800-1850 cm^{-1}.   The carboxylic acid content of the resulting polymers was determined by titration and agrees well with the calculated values. Therefore, linear polyesters having pendant carboxyl groups are obtained by interfacial polycondensation.   Molecular weight and yield of the resulting polymer are remarkably affected by the phase transfer catalyst, base and reaction medium as seen in Table 1.

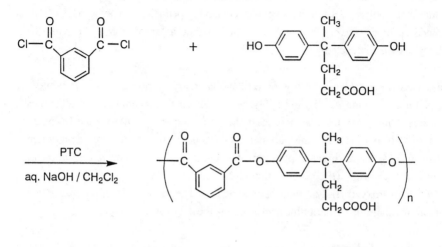

PEA

Table 1. Effects of PTC and Solvent on the Reduced Viscosity of Polyesters Bearing Pendant Carboxyl Groups

Run no	PTC	Reaction media	Yield %	$\eta_{sp/c}$ [1] dL/g
PEA-1	TBAB	H_2O/toluene	43	---[2]
2	TBAB	H_2O/CCl$_4$	36	---[2]
3	TBAB	H_2O/CHCl$_3$	74	0.30
4	TBAB	H_2O/CH$_2$Cl$_2$	84	0.41
5	TBAB+TEA	H_2O/CH$_2$Cl$_2$	17	0.19
6	Poly-cat	H_2O/CCl$_4$	62	---[2]
7	Poly-cat	H_2O/CHCl$_3$	76	0.62
8	Poly-cat	H_2O/CH$_2$Cl$_2$	87	0.83
PEATA	Poly-cat	H_2O/CH$_2$Cl$_2$	87	1.65
PEBA	poly-cat	H_2O/CH$_2$Cl$_2$	87	1.15

[1] Measured at 0.5g/dL in tetrachloroethane-phenol (2:3 by weight) containing 4.76% (by volume) of H_2SO_4 at 30°C.

[2] The value of $\eta_{sp/c}$ is very low.

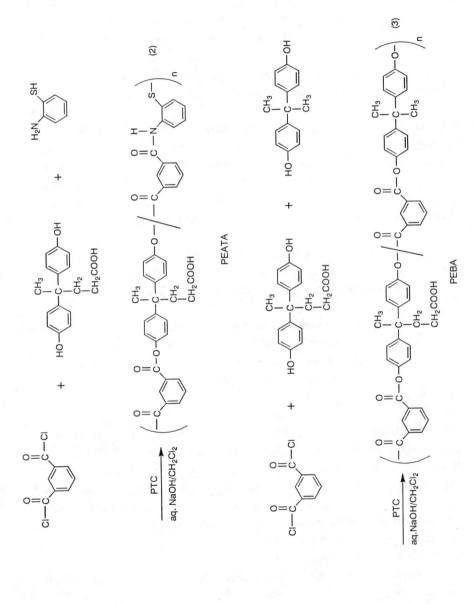

Effect of Phase Transfer Catalyst and Reaction Medium. When TBAB is used as a phase transfer catalyst in a H_2O/CCl_4 system, the yield of PEA is only 36% and the molecular weight is low because the complex of TBAB with phenolate is difficult to diffuse from the water phase to the CCl_4 phase. Both the yield and molecular weight of PEA are increased to 62% when poly-cat is used as a phase transfer catalyst.

In $H_2O/CHCl_3$ and H_2O/CH_2Cl_2 systems, the yield becomes higher when either TBAB or poly-cat is used.   However, the molecular weight is remarkably increased when poly-cat is used.   The phase transfer process of poly-cat may be different from that of TBAB because poly-cat contains amino acid moieties as nucleophilic catalyst.

Effect of Organic Base. When TBAB is used together with TEA as a complex catalyst, the yield is very low.   Aromatic acid chloride is considerably less reactive than aliphatic acid chlorides.   Therefore, when Schotten-Baumann reaction is used for preparation of polyester, aqueous sodium hydroxide or organic base must be added to the reaction system.   However, the use of organic base triethylamine (TEA) is not effective in water phase/organic phase interfacial polycondensation as seen in Table 1.

Catalytic Mechanism.   As shown in the catalytic mechanism of TEA (Equations 4, 5 and 6 of Scheme 2), the partial positive charge δ^+ on the carbonyl carbon of the intermediate I is increased and the nucleophilic attack of carboxylate and water occurs on the intermediate I.

In the IR spectra of reaction products, the characteristic absorption bands of anhydride groups are observed at 1770 and 1830 cm^{-1}.   The poly-cat is a complex catalyst and contains both quaternary ammonium groups and amino acid groups.   The quaternary ammonium groups acts as a phase transfer catalyst and the amino acid moiety as a nucleophilic catalyst.

Polymer Catalyst.   Previously, we have reported the synthesis of copolyesters having tertiary amine groups in the main chain from terephthaloyl chloride (TPC), N-ethyl diethanolamine (EDA) and bisphenol A (BPA) by an interfacial polycondensation (12). The resulting copolymer contains higher EDA residues when Na_2CO_3 is used as an acid acceptor.   If the self-nucleophilic catalysis of the tertiary amine groups in the main chain of the polymer occurs in the interfacial polycondensation, the content of EDA residues in the resulting copolymers is higher because aliphatic diols do not react with TPC when Na_2CO_3 is used as an acid acceptor.

As the same reason, the intermediate of nucleophilic catalysis from poly-cat and IPC is formed during the reaction.   Polymer effect is exerted by the intermediate of nucleophilic catalysis resulting from the amino acid moiety of the poly-cat and IPC; that is, the nucleophilic attack of carboxylate and water on the carbonyl carbon of the

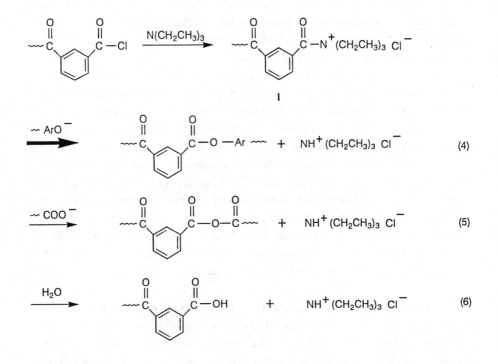

propagating chain end is disturbed by the hydrophobic nature of polymer chain segments in the vicinity of the intermediate.

It has been reported that high molecular-weight aromatic polyesters can be obtained in the presence of surfactants or quaternary ammonium salts (13). When a quaternary ammonium salt is added, it acts as a phase transfer catalyst and accelerates the transportation of reactants from the aqueous phase to the organic phase.

Phase Transfer of Reactants.   However, when the poly-cat is used as a complex catalyst, two types of transportation occur. The complex produced by bisphenolate with the quaternary ammonium salt group transfers from the aqueous phase to the organic phase and the intermediate from the reaction of IPC with the amino acid moiety of poly-cat transfers from the organic phase to the aqueous phase. The phase transfer mechanism is represented in Equations 7 and 8 of Scheme 3.

The resulting polyester is soluble in basic aqueous solution, and the intermediate transfers from the organic phase to the aqueous phase. The mechanism of the interfacial polycondensation differs from that of polyarylate formation because the rate of phase transfer of polymer chain segments from the aqueous phase to the organic phase is slower for polymers with phenolate end groups than that of the low-molecular bisphenolates.

The relation between the reduced viscosity and the reaction time is given in Figure 1 for the polycondensation of IPC with DPA at 20°C using poly-cat and TBAB as chain transfer catalysts keeping the molar ratio of IPC to DPA to unity. As expected, the rate of propagation of polyester PEA is faster for poly-cat than that for TBAB as a catalyst. The poly-cat contributes to the increase in the molecular weight due to the nucleophilic catalytic action of the amino acid moiety in the main chain and the phase transfer of the resulting intermediate from the organic phase to the aqueous phase.

Effect of the molar Ratio of Reactants.   The molecular weight of polyarylate is significantly affected by the molar ratio of acid chloride to BPA (14), because the molar ratio of acid chloride to BPA affects the mechanism of the interfacial polycondensation. Usually, monomer or dimer reacts with the propagating polymer in the interfacial polycondensation.

At the initial stage of reaction, the solubility of oligomer in the reaction media is governed by the end groups of the oligomer itself.   In the synthesis of polyarylate, therefore, very high molecular weight can be attained when the molar ratio of BPA to acid chloride is larger than unity (15), because the propagating oligomer or polymer having phenolate end groups is insoluble in both the organic phase and the aqueous phase. However, the molecular weight of PEA is remarkably affected by the molar ratio of reactants (Table 2).

When the molar ratio of DPA to IPC is 1.15 and 1.20, the reaction tends to occur in

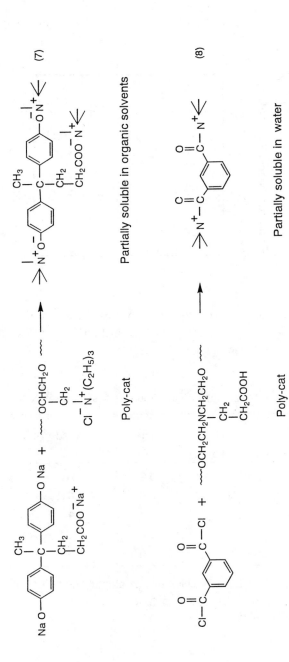

Partially soluble in organic solvents

Poly-cat

(7)

Partially soluble in water

Poly-cat

(8)

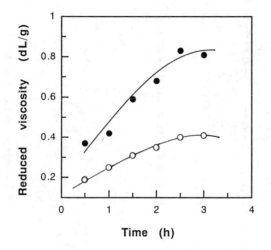

Figure 1 The relation between the reduced viscosity and the reaction time in the
interfacial polycondensation of 0.02 mole each of IPC with DPA at 20°C.
Catalyst: (○) TBPB, (●) poly-cat.

Table 2. Effect of Molar Ratio on the Reduced Viscosity of PEA

Run no.	DPA/IPC Molar ratio	Time h	Yield %	η_{sp}/c [1] dL/g
PEA-1'	0.83	0.25	76	0.45
2'	0.91	3.0	81	1.30
3'	1.05	3.0	84	0.41
4'	1.15	3.0	--	0.59
5'	1.20	3.0	--	0.55

[1] Measured at 0.5g/dL in tetrachloroethane-phenol (2:3 by weight) containing
4.76% (by volume) of H_2SO_4 at 30°C.

the aqueous phase because the resulting oligomer has pendant groups and phenolate groups at both ends. It is difficult for acid chloride to transfer from the organic phase to the aqueous phase or for oligomer having phenolate end groups and pendant carboxylate groups to transfer from the aqueous phase to the organic phase. Consequently, the high molecular weight of PEA is not very high.

For the molar ratio of 1.05, the oligomer obtained at the initial stage of reaction is partly soluble in the organic phase, and PEA with fairly high molecular weight is obtained.

For the molar ratio of 0.91, PEA with high molecular weight can be obtained, because the oligomer having pendant carboxylate groups in the side chain and acid chloride groups at the both ends is produced at the initial stage of reaction, which is neither soluble in the organic phase nor in the aqueous phase, but is located in the interface.

For the molar ratio of 0.83, the polycondensation is stopped because stirring becomes difficult after 0.25h of reaction.

For the same reason, copolymers PEATA and PEBA are not easily soluble in the organic phase nor in the water phase at the early stage of reaction when the molar ratio of DPA to IPC is unity, and the molecular weight of the resulting copolymers is very high (Table 1).

Polymer Reaction. The polyester PEA was reacted with 3-bromo-1-propanol, allyl bromide and 2-dimethylaminoethyl chloride in DMAc-H_2O containing K_2CO_3 using TBAB (*10*). The pendant carboxyl groups were converted to hydrazyl groups, carbon-carbon double bonds and tertiary amine groups, respectively.

Polyesters Having Pendant Chlorohydrin Moieties

Preparation of Polyesters. Condensation polymers having pendant hydroxyl groups have been reported (*16-19*) and polyester having pendant chlorine-containing groups has been also described (*20*). Reactive aromatic polyesters having pendant chlorohydrin moieties are also novel reactive polymer (*21*). Various functional polyesters can be obtained by substituting the hydroxyl or chlorine groups in the chlorohydrin moiety with other functional groups.

0.02 Mole of propyl chlorohydrin diphenolate (PCHDP) and 0.2 g of *tetra-n*-butyl-phosphonium bromide (TBPB) were dissolved in 180mL of water containing 0.04 mole of NaOH, and then 0.02 mole of IPC in 80ml of solvent was added. The mixture was stirred for 3h at 25°C. Precipitated polyester (PECH) was filtered off. Copolyester (PEACH) was prepared by substituting 0.01mole of PCHDP with 0.01mole of DPA in the feed.

Polyester PECH and copolyester PEACH having pendant chlorohydrin moieties were prepared according to Scheme 4.   The hydroxyl groups of aliphatic diols do not react with IPC in a water phase/organic phase reaction system, and the carbon linked to chlorine atom does not react with phenolate at room temperature.   The monomers, phenolate and acid chloride react preferentially with the oligomers when the concentration of oligomers in an organic liquid or organic solid phase is higher than that of acid chloride, and then the molecular weight of polymer is increased. Therefore, the pendant chlorohydrin moiety of PCHDP does not react with either IPC nor the phenolate groups of PCHDP itself.   Linear polyester and copolyester having pendant chlorohydrin moieties are obtained exclusively by the water phase/organic phase polycondensation.

Effect of Reaction Media.   The results of interfacial polycondensation are given in Table 3.   The effects of reaction media on the yield and molecular weight of polymers are very complicated in an interfacial polycondensation.

Generally, the interfacial polycondensation is divided into two types: Whether the oligomer resulted at the initial stage of reaction is soluble or insoluble in the organic phase.   At the initial stage of reaction, the oligomers of most aromatic polyesters are soluble in polar organic phases, such as nitrobenzene, chloroform and dichloromethane.

This local concentration effect contributes to the increase in molecular weight of polymer.   Therefore, high molecular-weight species is increased with the transfer of phenolate from the water phase to organic phase or the organic solid phase and of acid chloride from the organic phase to the organic solid phase.   The transfer of phenolate and acid chloride depends on the solubility or swellability of polymer in the organic phase and the water phase, the phase transfer catalyst, the organic solvent and the phase ratio of the water phase to the organic phase.

The interfacial polycondensation changes from the liquid/liquid into the liquid/ solid reaction when polymers with higher molecular weight precipitate from the organic phase and is located in the interface.   The liquid/solid reaction is very desirable for obtaining polymers with high molecular weight.   Therefore, a highly polar organic solvent is preferable to the liquid/solid reaction due to the polar ester groups in polyester.

Swain et al. designate the polarity of various solvents using a summation (A+B) of solvent parameters A and B (20).   The values A+B for cyclohexane, carbon tetrachloride, benzene, nitrobenzene and chloroform are 0.09, 0.43, 0.73, 1.14 and 1.15, respectively, and that for water is 2.00.   Consequently, when cyclohexane, carbon tetrachloride or benzene with low polarity is used as the organic phase, the molecular weight of polymer is not high (Table 3).   On the other hand, high molecular weight is attained when a more polar organic solvent such as nitrobenzene, chloroform and dichloromethane is used.   The value A+B for dichloromethane should be close to that of chloroform due to their similar chemical structure.

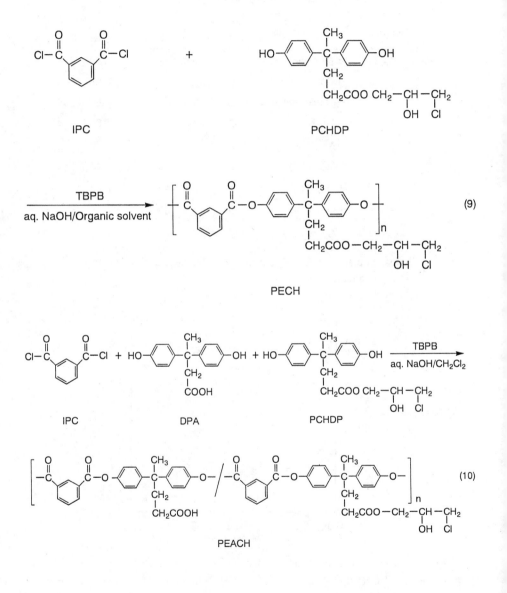

Table 3. Effect of Reaction Media on the Reduced Viscosity of Polyesters
Bearing Pendant Chlorohydrin Moieties

Run no.	Reaction media	Time h	Yield %	η sp/c [1] dL/g
PECH 1	H_2O/cyclohexane	2	56	0.36
2	H_2O/benzene	2	54	0.34
3	H_2O/nitrobenzene	2	66	0.60
4	H_2O/CCl_4	2	62	0.39
5	H_2O/$CHCl_3$	2	77	0.66
6	H_2O/CH_2Cl_2	1	80	0.49
7	H_2O/CH_2Cl_2	2	80	0.66
8	H_2O/CH_2Cl_2	3	80	0.66
PEACH	H_2O/CH_2Cl_2	2	79	0.63

[1] Measured at 0.5 g/dL in DMAc containing 30 vol.% of N,N-dimethyl-
1,3-propanediamine at 30°C.

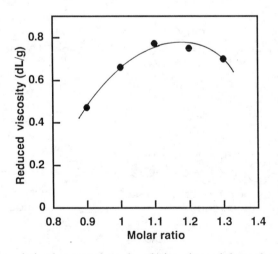

Figure 2 The relation between the reduced viscosity and the molar ratio of the
sodium salt of PCHDP to IPC in the H_2O/CH_2Cl_2 interfacial polycondensation at
20°C.

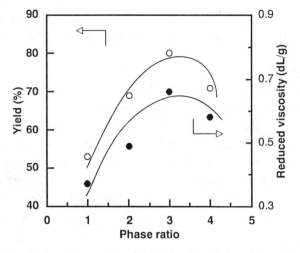

Figure 3 The relations between the yield and reduced viscosity and the phase ratio of H_2O to CH_2CCl_4 in the interfacial polycondensation of 0.02 mole of IPC with 0.023 mole of PCHDP at 20°C. (○) yield, (●) reduced viscosity.

Effect of the Molar Ratio of Reactants.   The relation between the reduced viscosity and the molar ratio of the sodium salt of to IPC is given in Figure 2 for the interfacial polycondensation of IPC with.   If the molar ratio of bisphenolate to IPC is higher than unity, the polymer or oligomer has a high concentration of phenolate end groups.   The ester groups in polyester are hydrophobic, whereas the phenolate end groups are hydrophilic.   Therefore, the propagating polymer chains contain both hydrophobic and hydrophilic moieties and are advantageously attacked by IPC from the organic phase and by bisphenolate from the water phase due to the surface activity of the polymer itself.   Accordingly, polymers with high molecular weight can be obtained when the molar ratio of the sodium salt of to IPC is higher than unity but lower than 1.15 as shown in Figure 2.

Swelling of Polymers.   The relations between the reduced viscosity and yield and the phase ratio of the water phase to the organic phase are given in Figure 3.   Swelling of the precipitated oligomer or polymer strongly depends on the phase ratio of the water phase to the organic phase, which in turn affects the transfer rates of both bisphenolate from the water phase and of IPC from the organic phase to the liquid/solid interface.

In fact, PEACH with lower molecular weight and phenolate end groups can swell obviously in both water and a polar organic solvent due to both hydrophilic and hydraphobic natures of pendant chlorohydrin moiety in the polyester.   The swelling will be varied when PEACH absorbs different quantities of water and organic solvent.   When PEACH absorbs excess water, the transfer of IPC from the organic phase to the liquid/solid interface is hindered. However, when it absorbs excess organic solvent, transfer of bisphenolate from the water phase to the interface is disturbed.   When the phase ratio is 2.0 to 3.5, PEACH with high molecular weight can be obtained in a higher yield.

Conclusions

High molecular-weight aromatic polyesters having reactive groups in the side chains can be obtained chemoselectively by a water phase/organic phase interfacial poly-condensation using phase transfer catalyst.   Polyesters bearing pendant carboxyl groups are prepared utilizing that the nucleophilicity of phenolate is much higher than that of carboxylate.   Polyesters bearing chlorohydrin groups are synthesized by controlling the molar ratio of bisphenolate to isophthaloyl chloride and the phase ratio of the water phase to the organic phase.   These polyesters can be utilized for the preparation of various functional polymers.

Literature Cited

1   Schnell, H. *Angel. Chem.*, **1956**, *68*, 633.
2   Ueda, M. *J. Syn. Org. Chem. Jpn.*, **1990**, *48*, 144.
3   Devour, J.; Goaded, P.; Mercer, P. *J. Polym. Sic., Polym. Phys. Ed.*, **1982**, *20*, 1895.
4   Kollodge, J. S.; Porter, R. S. *Macromolecules*, **1995**, *28*, 4089.
5   Jiang, M.; Qiu, X.; Qin, W.; Fei, L. *Macromolecules*, **1995**, *28*, 730.
6   Ladenheim, H.; Loebl, E. M.; Morawetz, H. *J. Am. Chem. Soc.*, **1959**, *81*, 20.
7   Ladenheim, H.; Morawetz, H. *J. Am. Chem. Soc.*, **1959**, *81*, 4860.
8   Kauzmann, W. *The Mechanism of Enzyme Action*, McElroy, W. D.; Glass, B., Eds., John Hopkins Press, Baltimore, MD, **1954**.
9   March, J. *Advanced Organic Chemistry*, McGraw Hill, Inc., New York, **1968**.
10 Wang, C.; Nakamura, S. *J. Polym. Sci.: Part A: Polym. Chem.*, **1995**, *33*, 2157.
11 Wang, C.; Nakamura, S. *J. Polym. Sci.: Part A: Polym. Chem.*, **1994**, *32*, 1255.
12 Wang, C.; Seko, N.; Nakamura, S. *React. Func. Polym.*, **1996**, *30*, 197.
13 Casassa, E. Z.; Chao, D. Y.; Henson, M. *J. Macromol. Sci., Chem.*, **1981**, *A15*, 799.
14 Morgan, P. W. *J. Macromol. Sci., Chem.*, **1981**, *A15*, 683.
15 Thai, H. B.; Lee, Y. D. *J. Polym.Sci., Polym. Chem. Ed.*, **1987**, *25*, 1505.
16 Iwakura, Y.; Izawa, S.; Hayano, F.; Kurita. K. *Makromol. Chem.*, **1967**, *104*, 66.
17 Iwakura, Y.; Kurita, K.; Hayano, F. *J. Polym. Sci., A - 1*, **1969**, *7*, 3075.
18 Ogata N.; Sanui, K. *J. Polym. Sci.,. A-1*, **1969**, *7*, 2847.
19 Sanui, K.; Ogata, K. *J. Polym. Sci., A-1*, **1970**, *8*, 277.
20 Nishikubo, T.; Kameyama, A.; Hayashi, N. *Polym J.*, **1993**, *25*, 1003.
21 Wang, C.; Nakamura, S. *J. Polym. Sci.: Part A: Polym. Chem.*, **1996**, *34*, 755.
22 Swain, C. G.; Swain, M. S.; Powell, A. L.; Alunni, S. *J. Am. Chem. Soc.*, **1983**, *105*, 502.

PHASE-TRANSFER CATALYSTS

Chapter 19

Polystyrene-Supported Onium Salts as Phase-Transfer Catalysts

Their State and Reactivity in Phase-Transfer Reaction Conditions

Noritaka Ohtani, Yukihiko Inoue, Jun Mukudai, and Tsuyoshi Yamashita

Department of Materials Engineering and Applied Chemistry,
Mining College, Akita University, Akita 010, Japan

Intraresin microstructure is decisive for the interpretation of the kinetics and mechanism of triphase catalysis. The linear polystyrenes containing onium salts have been used to evaluate the intraresin microstructure under liquid-solid-solid triphase conditions or liquid-solid-liquid triphase conditions. Their solution viscosity behavior under one-fluid systems and their phase separation behavior under two-fluid systems have strongly suggested that intraresin microstructure of a microporous polymer catalyst varies with the reaction conditions and governs transport phenomena of reagent ions and organic substrates. The microstructure also determines the microenvironment where polymer-bound onium salts collide to react with organic substrates. The role of ionic aggregation under liquid-solid-solid triphase conditions and the interfacial reaction mechanism under liquid-solid-liquid triphase conditions will be discussed.

A substantial effort in our laboratory has been directed toward polymer-supported reactions. The most significant examples are triphase catalysis of nucleophilic reactions (1,2), base-catalyzed reactions using a fluoride form of ammonio-attaching polystyrenes (3), hydrolysis and ester exchange of water-insoluble amino acid esters using carboxylate or ethylene diamine-attaching polystyrenes with metal ions (4), and similar reactions catalyzed by chymotrypsin that is supported on the above polymers (5). This paper deals with the recent development of the first subject promoted by our work.

Polymer-supported phase-transfer catalytic systems generally consist of three phases; a bulk organic solution phase, a bulk aqueous solution or solid reagent phase, and an insoluble polymer catalyst phase. Undoubtedly, chemical reaction takes place in the resinous phase. Besides the tremendous reports on synthetic aspects of triphase catalysis, there have been many kinetic studies on the triphase catalysis (6-12). However, most of the kinetic studies have interpreted the over-all rates without considering what is a main ionic species diffusing through the resin interior, what is a real reacting ionic species, and in what kind of microenvironment the species collide to react with the organic substrate. It is important to clarify the microstructure of the resin interior that is formed *when the resin is placed under a given reaction condition*, because the microstructure should be responsible not only for the mass-transport of the substrate and ionic species but also for the real reactant ionic species. Most

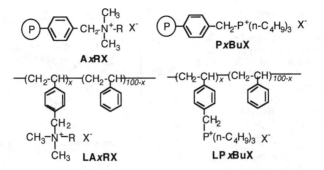

Figure 1.   Crosslinked and linear ionomers used in this study.

studies have also ignored the change in the resin microstructure according to the reaction conditions and have tried to explain the observed kinetic data based on analogous structure-activity relationships found for low-molecular soluble catalysts with the aid of the supplementary knowledge about intraresin diffusion property concerning highly-crosslinked polymers that are soaked in one-fluid. A catalyst resin imbibes liquid(s) from other phase(s) particularly when it is an active gel-type catalyst with a low crosslinking density (*13*). This paper demonstrates the occurrence of the microstructure change, which determines the apparent reactivity of onium salts and the facility of mass-transport of reagents. Some evidence to support interfacial reaction mechanism for liquid-solid-liquid triphase catalysis has been also furnished.

Liquid-Solid-Solid Triphase Catalysis

In this reaction system, the second phase that contains inorganic reagent is solid phase. Accordingly, resin imbibes organic solvent only and the resin microstructure under the triphase conditions can be assumed to be the same as that without the solid phase.

In this study, we use the classical crosslinked gel-type polystyrenes with ammonio or phosphonio groups, A*x*RX and P*x*BuX, as shown in Figure 1 where *x* represents percent ring substitution, R, alkyl group, capital letter X, the kind of anion. For comparison, we also use the corresponding linear polystyrenes, LA*x*RX and LP*x*BuX: L means linear. These polymers should be classified into cationic ionomers (*14*) because the polymers usually contain relatively low ionic groups and because their properties are completely different from those of the unfunctionalized parent polymers. After we correlate the solubility and viscosity of these linear ionomers with swelling of the corresponding crosslinked ionomers, we examine the nucleophility of the closslinked ionomers and explain the over-all catalytic activities under one-fluid conditions.

Solubility and Solution Viscosity Behavior of Linear Ionomers. The solubility of LA*x*OcCl is dependent on solvent polarity and ion loading. The ionomers with large ion loading were insoluble in nonpolar solvents that have low acceptor numbers (*AN*) such as toluene or tetrahydrofran (THF). But they were readily soluble in the solvent with a high *AN* value like chloroform (*14*). Solubility depends also on counter ion and ammonio structure. The ionomers with iodide counter ions were insoluble in methanol while the ionomers with chloride ions were readily soluble in the solvent (*15*).

In polar solvents with high dielectric constants such as nitromethane, acetonitrile, and methanol, a distinct polyelectrolyte behavior was observed (14). With a decreasing concentration of the ionomer, the reduced viscosity was increased. It is known that this behavior indicates the presence of dissociated ions. Figure 2 shows the effect of solvent on the viscosity behavior of LA10MeCl. Nitromethane and methanol showed typical polyelectrolyte effects. In contrast, in a solvent with low dielectric constant, such as chloroform or THF, we could not observe the polyelectrolyte effect. The reduced viscosity rather decreased with decreasing ionomer concentration. The ionomer behaves like non-ionic polymers.

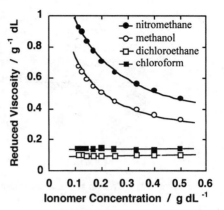

Figure 2.   Viscosity behavior of LA10MeCl in several solvents at 25 °C.

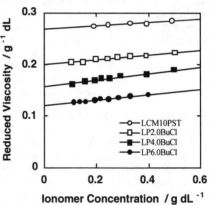

Figure 3.   Viscosity behavior of LPxBuCl in THF at 25 °C.

In toluene or THF, the reduced viscosity of an ionomer with chloride counterions was always decreased with an increasing ion content. A typical example is shown in Figure 3 and Figure 4a. This decrease in the reduced viscosity shows that an ionomer domain contracts with ion content, which indicates the formation of intrachain aggregates in such a low AN solvent. The solvation to chloride ions is very weak in such a solvent. Thus, the intrachain aggregation takes place. In chloroform, on the other hand, LAxOcCl and LPxBuCl gave a similar value of reduced viscosity irrespective of their ion content; the value was close to that of each parent chloromethylated polystyrene (Figure 4a), indicating few aggregation between ionic groups due to a strong solvation to the counter ions.

Conformations of an Ionomer Chain in Solutions. If an ionomer is dissolved in a solvent, interchain aggregation no longer plays an important role. Therefore, solvations to the ionic groups are important also for the dissolution of an ionomer. When an ionomer is solubilized in a solvent, the ionomer may be extended with the dissociation of the ionic groups. This extended conformation may be taken in methanol or acetonitrile that has a high dielectric constant. Alternatively, the ionomer may be contracted by the intrachain aggregation of ionic groups. Toluene

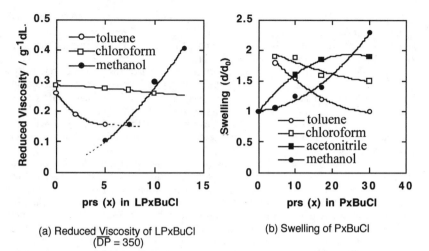

(a) Reduced Viscosity of LPxBuCl
($\overline{DP} = 350$)

(b) Swelling of PxBuCl

Figure 4. Solution viscosity and swelling of linear and crosslinked ionomers in several solvents at 25 °C.

and THF give LAxRCl or LPxBuCl this contracted conformation that resembles a reversed micelle. The volume of the ionomer domain may be very close to that of the parent polymer if a nonpolar solvent has a great affinity to solvate either small quaternary cations or small counter anions of the ionomer (*15*).

Interestingly, by adding low-molecular-weight onium salts, ionomers with high ionic loading are solubilized in such a nonpolar solvent as toluene or THF. With an increasing concentration of trioctylmethylammonium chloride (TOMAC), the reduced viscosity of LA7.5OcCl in THF increased and approached the value of its parent chloromethylated polystyrene (*16*). This shows an expansion of the ionomer chain and suggests a decrease in the intrachain aggregation among the ionomer ionic groups.

Linear ionomers are soluble even in water provided the ionic loading is not extremely low. They show polyelectrolyte behavior in water (*16*). The addition of an electrolyte to the aqueous solution diminishes the polyelectrolyte effect. As shown in Figure 5, the aqueous LA7.5OcCl solution solubilizes a water-insoluble neutral dye, sudan III. The amount of the solubilization increased with the ionomer

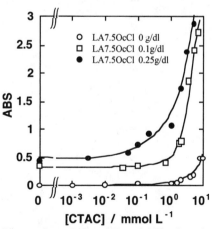

Figure 5. Sudan III solubilization by aqueous LA7.5OcCl at 25 °C.

concentration. When we add cetyltrimethylammonium chloride (CTAC) into the aqueous solution, the solubilization of sudan III increases sharply beyond the critical micelle concentration of CTAC (ca. 0.9 mmol/L). These results suggest that the ionomer takes a polysoap-like conformation in water.

Swelling of crosslinked ionomers corresponds well to their solution viscosity behavior (Figure 4b). In nonpolar solvents with low AN values such as toluene and THF, the swelling of an ionomer was relatively small. The swellability decreases with the ion loading. This is the reflection that the interchain aggregation takes place. In contrast, in a polar solvent like acetonitrile or methanol, the swelling rather increases with an increasing ion content. The appearance of dissociated ions within the resin explains this result.

In a nonpolar solvent, the swelling of a crosslinked ionomer (A12BuCl) also increases with an addition of soluble quaternary salts such as tributylhexadecylphosphonium chloride (16). The swelling in methanol presented a contrast to the behavior in toluene. The addition of a quaternary salt sharply decreased the swelling in methanol. Therefore, the ionomer swelling becomes higher in toluene rather than in methanol in the presence of the quaternary salts (16). In chloroform, however, we could not observe these salt effects for swelling behavior.

The effect of low-molecular-weight onium salts may be explained as shown in Figure 6. The ionomers with high ionic content are insoluble in nonpolar solvents due to an extensive interchain aggregation. They are only able to swell. The swelling of crosslinked ionomers is also reduced by the interchain aggregation. With an addition of a soluble onium salt, Q2X, the interchain aggregates may be replaced by the aggregates between ionomer-bound ionic groups and the added onium salt. The elimination of the inter-ionomer aggregation would induce the salting-in effect. Moreover, the decrease in the intrachain aggregation would increase the reduced viscosity. On the other hand, the behavior in polar solvents may be explained in the same way as the accepted theory about the aqueous polyelectrolyte solution (17) as well as the theory on ion-exchange resins (18).

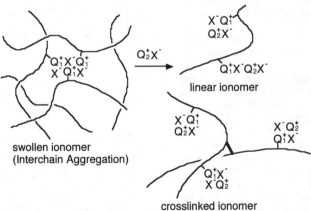

swollen ionomer
(Interchain Aggregation)

linear ionomer

crosslinked ionomer
(Aggregation with Added Salt)

Figure 6. Expansion of ionomer chain with soluble quaternary salts.

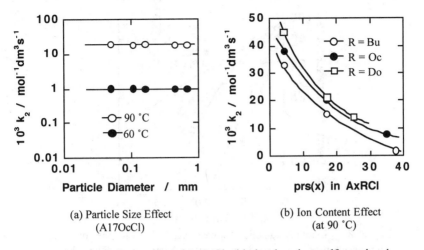

(a) Particle Size Effect
(A17OcCl)

(b) Ion Content Effect
(at 90 °C)

Figure 7.   Noncatalytic reactions of AxOcCl with decyl methanesulfonate in toluene. 4.0 mL of toluene, 0.05 mmol of quaternary salts, and 0.82 mmol of DcOMes.

Noncatalytic Reactions of Ionomers.   In nonpolar solvents such as toluene or THF, the intraresin microstructure of crosslinked ionomers may be homogeneous with the ionic aggregates scattered all within the resin interior.   The noncatalytic reaction between decyl methanesulfonate and A17OcCl in toluene gave a clean second-order kinetics.   The rate constant was barely dependent on the particle size, indicating that the intraresin diffusion of the organic substrate was very fast compared to the rate of the chemical reaction (Figure 7a).   Similar results have been obtained unless the ionomer has an extremely high ion content or a high covalent crosslinking density (2).

In toluene, however, reactivity depended on the ion content of ionomers (Figure 7b).   The apparent nucleophilicity decreased with an increasing ionic loading of AxRCl.   This indicates that the enhancement of the aggregation decreases the average reactivity of the chloride counter anions and that the life of an aggregate is much shorter than the time-scale of the reaction.   This also suggests that monomeric ion pairs are more reactive than the ionic aggregates such as triplet ions, dimers of ion pairs, and higher multiple ions.   If chloroform or acetonitrile was used as a solvent, the reactivity did not depend on the ion content irrespective of the catalyst particle size (2).   A high solvation to the small ionic groups or dissociation of the ionic groups is responsible for the disappearance of ionic aggregates.

Rates of Liquid-Solid-Solid Triphase Catalysis.   The rates of liquid-solid-solid triphase catalysis where reagent inorganic salt was in large excess over decyl methanesulfonate have been examined in detail (2).   Solvent efficiency for this triphase catalytic reaction was similar to that of non-catalytic reactions.   Most of the reactions did not follow pseudo-first-order kinetics; a large reaction rate observed at the initial stage, a slowdown after that, and a subsequent gradual autoacceleration.

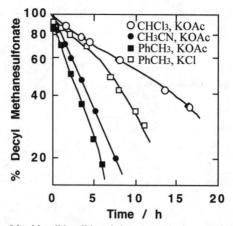

Figure 8.   Liquid-solid-solid triphase catalytic reactions of decyl methanesulfonate with solid potassium acetate or sodium chloride catalyzed by A17DoCl at 60 °C (acetonitrile and chloroform) or at 110 °C (toluene). 4.0 mL of solvent, 0.05 mmol of A17DoCl, 10 mmol of KOAc or NaCl, 0.07 mL of DcOMes.

The phenomenon was significant when toluene was used as a solvent and when the reaction temperature was high (Figure 8).   The formation of a by-product, 1-chlorodecane, was observed when the resin with chloride counter anions was used as a catalyst.   The by-product, which was caused by the residual chloride ions remaining in the catalyst resin, was formed only at the initial stage of the reaction course.   Most of the reaction rates in toluene were dependent on the particle size of ionomer catalyst.

All these features clearly indicate the presence of a strong mass-transport limitation of nucleophiles under the liquid-solid-solid conditions using toluene as a solvent.   This may be due to either low solubility of inorganic salts into the solvent, scarce contact between the ionomer particle surface and the solid reagent surface, or slow intraresin ion-transport.   However, it still remains unsolved how fast anion exchange between polymer-bound quaternary salts occurs through the transient aggregation among the ionic groups that makes the intraresin ion-transport possible.

When acetonitrile was used as an organic solvent, on the other hand, a relatively clean pseudo-first-order kinetics was observed.   A relatively high solubility of inorganic salts in the solvent may diminish the ion-transport limitation in the triphase catalysis, although noncatalytic reactions proceeds simultaneously at considerable rates in the solvent (2).

Liquid-Solid -Liquid Triphase Catalysis.

In this triphase system, crosslinked ionomers imbibe water or aqueous solution in addition to organic solvent (1, 19).   In toluene, as shown already, the ionic groups tend to aggregate, forming a reverse micelle-like structure.   In water, the polymer backbone shrinks, forming a polysoap-like or normal micelle-like structure.   These

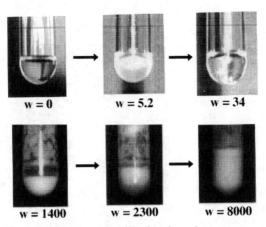

w = 0   w = 5.2   w = 34

w = 1400   w = 2300   w = 8000

Figure 9.   Ionomer solubility in toluene/water system.

conformations remind us of the property of surfactants. When two immiscible solvents are both present, however, linear ionomers are scarcely soluble in either of the two phases. The pictures in Figure 9 show the change in appearance when we add water to a 25 mmol/L solution of LA5.0OcCl in toluene at 60°C, where w represents molar ratio of water to onium groups. A minimal amount of water was enough to precipitate the ionomer. The system became turbid. Interestingly, the precipitated ionomer was then solubilized to afford an apparent homogeneous solution. With further addition of water, the ionomer gave an opaque gel. The gel further absorbed water and toluene and bulk water phase did not appear until the whole system became a gel. The phase-separation behavior depends on the salt concentration as well as the kind of salt and ionomer's ion content (Ohtani, N. unpublished data).

The results clearly show that the hydration to the ionomer ionic groups never eliminates their ionic aggregates that is formed in nonpolar solvents. The action is rather opposite. The hydration strengthens the interaction between ionic groups. The behavior strongly suggests that the interior of crosslinked ionomers is 'not' homogeneous under liquid-solid-liquid triphase conditions. The water imbibed into the crosslinked ionomer does not simply solvate the ionic groups. The water makes a certain structure with the organic solvent that was absorbed simultaneously. The presence of covalent crosslinkages should modify the structure because of the limited swellability.

Interfacial Reaction Mechanism for Triphase Catalysis. We proposed fifteen years ago a reversed micelle model for triphase catalysis (*13*) as is illustrated in Figure 10. The interior of the resin consists of two phases: one is an organic phase that contains the hydrophobic polymer backbone solvated by the organic solvent and the other is an aqueous phase that contains water and dissociated anions. Quaternary cations are present on the interface of the phases. Substrate and ionic nucleophile diffuse through the continuous organic and aqueous phases, respectively. Chemical reaction takes place at the interface: the organic substrate collides with the nucleophile that is present on the interface in the form of ion pairs. Therefore, the most important feature of this model is the continuities of organic (or oil) and aqueous (or water) phase (*1*). The continuities may be primarily determined by the volume ratio of oil to water absorbed by the crosslinked ionomer.

When an ionomer imbibes more oil than water under a given reaction condition, the microstructure of the ionomer tends to be a water-in-oil type: oil phase is continuous and water phase is discontinuous. The quaternary salts that are situated on isolated water droplets are able to react only once but the catalytic action does not continue because the quaternary salts are not recovered to their original form due to the absence of nucleophile supply from bulk aqueous phase. On the other hand, the quaternary salts that are situated on the water channel, which is continuous to bulk aqueous phase, are able to react with substrate repeatedly since

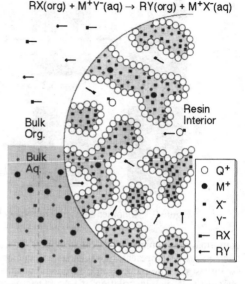

$$RX(org) + M^+Y^-(aq) \rightarrow RY(org) + M^+X^-(aq)$$

Figure 10. Reverse micelle model for liquid-solid-liquid triphase catalysis.

the counter anions may be recovered readily. Since it is known that reverse micelles exchange each other their contents after their collisions (*20*), it should be considered that the microstructure of crosslinked ionomer is not necessarily fixed and that the dispersed droplets are more or less mobile to contact each other, though the movement is extremely restricted. Such droplets behave as if they were completely isolated during the catalytic reactions. If an ionomer imbibes more water than oil, the microstructure would be reversed. In this case, organic substrate may not reach isolated oil droplets buried in deep resin interior, while nucleophile anions reach every site of the resin interior.

The importance of resin microstructure under liquid-solid-liquid triphase catalysis is summarized in Figure 11. With an increasing content of organic solvent and/or with a decreasing water content, the resin interior becomes a w/o microstructure.

In a reverse case, the resin tends to consist of a o/w microstructure that may be exemplified by lightly crosslinked ion-exchange resin. The kind of reagent salt and its concentration in bulk aqueous solution as well as the organic solvent affinity on polymer backbone directly influences the microstructure. The structural factors of catalyst ionomers influence the microstructure in consequence of the amounts of imbibed oil and water. The extents of attainable regions of substrate and nucleophile in resin interior will be varied according to the oil and water composition. The ratio of ion pair on the interface over total nucleophiles in the resin may be also varied according to the volume and anion concentration of the resinous water phase. Consequently, the over-all catalytic activity of an ionomer may follow the curve that has a maximum as shown in Figure 11(b).

Oil Content

Water Content

Microstructure

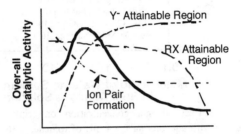

| W / O | Bicontinuos | O / W |

QX Loading   large

Q Size   large

X Size   large

Spacer Chain   long

Salt Concentration   large

Solvent Affinity   large
to Chain

(a) Factors Determining Microstructure

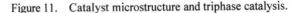

(b) Microstructure and Triphase Activity

Figure 11.   Catalyst microstructure and triphase catalysis.

    Our early study suggested the restriction of intraresin ion-transport (*21*). There have been recently reported some experimental evidence to support that simple nucleophilic substitutions *do* occur also in microemulsion systems via interfacial mechanism (*22, 23*).   When water-uptake is small compared to oil-uptake, the microstructure is assumed to be a w/o or bicontinuous structure.   The nucleophile transport may be restricted in this case.   In order to examine the influence of aqueous phase continuity in ionomer particles, the constituents in bulk aqueous phase was varied and the effect on the apparent catalyst activity was studied in this article..

Continuities of Resinous Aqueous Phase and Catalytic Activity.   The rate of triphase catalysis depends on the salt concentration of the bulk aqueous phase.   We pointed out first the phenomena (*19, 21*).   A maximum rate was usually observed at a certain concentration in each catalytic reaction.   The concentration dependence differed according to the kind of reagent salt and leaving group of the substrate.   The rate changes were relatively sharp when cyanide or acetate was a nucleophile, as shown in Figure 12.   At the saturated concentration of inorganic salts in the bulk aqueous solution, the formation of the by-product derived from the residual counter ions of the used catalysts was often observed even after prolonged preconditioning of the catalysts, indicating the presence of the intraresin region to which reagent nucleophiles could not get.

An addition of sodium hydroxide to bulk aqueous solution also affects the rate of triphase catalysis. The effect was significant when iodide ions were present in the reaction system. The rate of the reaction of 1-iododecane with sodium chloride was significantly decreased by the presence of sodium hydroxide, although the initial rate

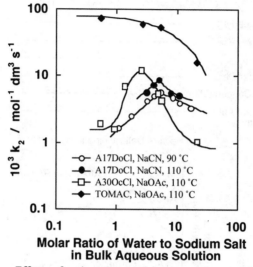

Figure 12.   Effect of salt concentration on triphase catalysis of 1-bromodecane with aqueous salt.   4 mL of toluene, 0.36 mmol of DcBr, 10 mmol of reagent salt, and 0.05 mmol of catalyst.

was quite enlarged by the presence of sodium hydroxide as shown in Figure 13.

As shown in Table I, the presence of sodium hydroxide always enhanced the reactions catalyzed by TOMAC. However, ionomer-catalyzed reactions were not always accelerated. The reaction with sodium cyanide was slightly decelerated. The reaction with sodium iodide was significantly decelerated. These effects, that is, salt concentration and sodium hydroxide addition, are explained by the change in the amount of water within the ionomer catalyst and the continuity of the resinous aqueous phase. A decrease in water content in an ionomer particle would decrease the aqueous continuity and increase the region to which reagent ions are not attainable.

Our model easily explains both pseudo-first-order kinetics and catalyst particle-size effect that are usually observed for triphase catalysis. All other models proposed for the triphase catalysis did not take into account the possible change in the resin microstructure that was induced due to the external factors of the reaction conditions. Reaction-intraparticle diffusion model adopted by reaction chemical engineers (11, 12) necessiates both continuous water-filled pores and continuous oil-filled pores in catalyst particles. It is well-known that the more active triphase catalyst has a morphological structure of gel-type resin that becomes porous only when swollen by solvent(s). Such gel-type of resin does not have large surface area that are quantitatively measured by BET method. The aqueous phase or water-filled pores are formed only because the ionomer contains hydrophilic ionic groups: unfunctionalized polystyrene hardly absorbs water, much less aqueous solution even if it has permanent micropores. Oscillation model, which assumes that polymer-bound quaternary groups go into and come back from resinous aqueous phase on the

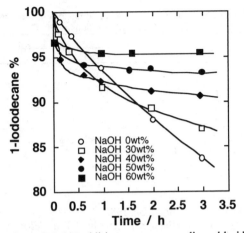

Figure 13.   Effect of NaOH addition to aqueous sodium chloride on triphase catalytic reaction of 1-iodododecane catalyzed by A17BuCl at 90 °C.   1.0 mL of toluene, 2 mL of aqueous phase (10 mmol of NaCl), 0.4 mmol of DcI, and 0.025 mmol of catalyst.

analogy of the Stark's model of phase-transfer catalysis, rules out the participation of the oscillation rates that should limit the over-all rate of the triphase reaction: the liquids imbibed by the resin should be stagnant even if the whole reaction system is vigorously stirred.   It is known that the stirring speed is one of the crucial factors for unbound phase-transfer catalysts to function well.   This model is also obliged to presume the presence of permanent pores that are fulfilled with the bulk aqueous solution.   The model that the resin interior is composed of a homogeneous microstructure without any aggregation of ionic groups seems very doubtful since the amount of water imbibed by the catalyst ionomer is much larger under the triphase conditions than the estimated hydration numbers of the corresponding low-molecular-weight quaternary salts.   This model also involves the problem as to ion transport in the catalyst particles, while there is a similar problem to be solved for liquid-solid-solid triphase catalysis.

Experimental

Materials.   Linear and crosslinked ionomers were prepared and characterized by the method reported previously (*2, 14*).   The degree of polymerization of linear ionomers is 200 unless otherwise stated.   The crosslinking density of AxRX and PxBuX was adjusted to 1% with divinylbenzene and particle size was -200+325 mesh unless otherwise stated.

General Procedures.   Viscosity and swelling measurements were done in the same way as described previously (*14*).   The solubilization of sudan III was analyzed spectrophotometrically.   The absorbance at 490 nm of the filtrate was measured after an ionomer aqueous solution and an excess amount of solid sudan III were equilibrated at 25 °C in a test tube with a Teflon-lined screw cap.   All the kinetic runs were carried out using a 40 mL culture tube equipped with a Teflon-lined screw cap and a Teflon-coated stirring bar and the reaction mixtures were analyzed by GLC in a usual way (*2*).   The reaction conditions are given in each figure or table.

260   PHASE-TRANSFER CATALYSIS

Table I Influence of Addition of Sodium Hydroxide on the Reactions of Decyl Methanesulfonate with Aqueous Sodium Salts under Phase-Transfer Conditions at 90 °C

Catalyst	Salt	$10^5 k_{obsd} / s^{-1}$	
		H2O	50 wt% NaOH
A17BuCl	NaCl	3.6	7.7
	NaCN	16	14
	NaI	13	1.6
TOMAC	NaCl	92	130
	NaI	200	260

Reaction conditions: catalyst, 57 μmol; DcOMes, 0.28 mmol; reagent salt, 10 mmol; toluene, 2 mL; aqueous phase, 2 mL of water or 50 wt% NaOH solution.

Literature Cited

1. Ohtani, N. *J. Synthetic Org. Chem. Jpn.* **1985**, *43*, 313.
2. Ohtani, N.; Nakaya, M.; Shirahata, K.; Yamashita, T. *J. Polym. Sci., Part A* **1994**, *32*, 2667.
3. Ohtani, N.; Inoue, Y.; Nomoto, A.; Ohta, S. *Reactive Polymers* **1994**, *24*, 73.
4. Ohtani, N.; Inoue, Y.; Inagaki, Y.; Fukuda, K.; Nishiyama, T. *Bull. Chem. Soc. Jpn.* **1995**, *68*, 1669.
5. Ohtani, N.; Inoue, Y.; Kobayashi, A.; Sugawara, T. *Biotechnol. Bioeng.* **1995**, *48*, 42.
6. Regen, S. L. *Angew. Chem. Int. Ed. Engl.* **1979**, *18*, 421.
7. Montanari, F.; Landini, D.; Rolla, F. *Top. Curr. Chem.* **1982**, *101*, 147.
8. Ford, W. T.; Tomoi, M. *Adv. Polym. Sci.* **1984**, *55*, 49.
9. Telford, S.; Schlunt, P.; Chau P. C. *Macromolecules* **1986**, *19*, 2435.
10. Svec, F. *Pure Appl. Chem.* **1988**, *60*, 377.
11. Wang, M. L.; Yang, H. M. *Ind. Eng. Chem. Res.* **1991**, *30*, 2384.
12. Desikan, S.; Doraiswamy, L. K. *Ind. Eng. Chem. Res.* **1995**, *34*, 3524.
13. Ohtani N.; Wilkie, C. A.; Nigam, A.; Regen, S. L. *Macromolecules* **1981**, *14*, 516.
14. Ohtani, N.; Inoue, Y.; Mizuoka, H.; Itoh, K. *J. Polym. Sci., Part A* **1994**, *32*, 2589.
15. Ohtani, N.; Inoue, Y.; Kaneko, Y.; Okumura, S. *J. Polym. Sci., Part A* **1995**, *33*, 2449.
16. Ohtani, N.; Inoue, Y.; Kaneko, Y.; Sakakida, A.; Takeishi, I.; Furutani, H. *Polymer J.* **1996**, *28*, 11.
17. Rice, S. A.; Nagasawa, M. *Polyelectrolyte Solutions*, Academic Press, New York, 1961.
18. Katchalsky, A.; Lifson, S.; Eisenberg, H. *J. Polym. Sci.* **1951**, *7*, 571.
19. Ohtani, N.; Regen, S. L. *Macromolecules* **1981**, *14*, 1594.
20. Lang, J.; Mascolo, G.; Zana, R.; Luisi, P. L. *J. Phys. Chem.* **1990**, *94*, 3069.
21. Serita, H.; Ohtani, N.; Matsunaga, T.; Kimura, C. *Kobunshi-Ronbunshu* **1979**, *36*, 527.
22. Shomäcker, R. *J. Chem. Res.* **1991**, 92.
23. Ohtani, N.; Inoue, Y.; Shinoki, N.; Nakayama, K. *Bull. Chem. Soc. Jpn.* **1995**, *68*, 2417.

Chapter 20

Rate-Limiting Step in Triphase Catalysis for the Esterification of Phenols

N. N. Dutta, A. C. Ghosh, and R. K. Mathur

Chemical Engineering Division, Regional Research Laboratory,
Jorhat 785 006, India

An analysis of the kinetic factors/rate limiting processes has been presented for the esterification of phenol, m-cresol and resorcinol in alkaline solution with benzoyl chloride dissolved in toluene using polystyrene supported tri-n-butyl phosphonium ion as the phasetransfer catalyst. Like many S_N2 displacement reactions, the aforesaid reactions under "triphase catalysis" obey pseudo-first order kinetics. In order to provide a complete description of the kinetic behaviour, the reaction engineering aspects have been addressed from experimentally observed rate determining phenomena as complimented by the theory of diffusive mass transfer in porous catalyst. The mathematical model incorporates intraparticle diffusion, intrinsic reactivity and film mass transfer. Experimental results presented could be interpreted to provide evidence of three rate processes i.e. mass transfer of benzoyl chloride (organic substrate) to the catalyst surface, diffusion of the substrate through the polymer matrix and intrinsic reactivity at the active sites. The controlling phenomenon however, actually depends on the experimental conditions and the reaction environment.

The phase transfer catalysed esterification of phenols with acid chloride gives quantitative yield (1). Phase transfer catalyst (PTC) can augment various other phenolic reactions, that can perhaps be exploited to develop method for treatment of phenolic waste-water (2). The mechanism and kinetics of the reactions of phenol, m-cresol and resorcinol with benzoyl chloride have been studied using tricaprylyl methyl ammonium chloride (Aliquat 336) and hexadecyl tributyl phosphonium bromide as the catalysts in two-phase systems (3, 4). The system that utilizes soluble PTC essentially in the organic phase suffers from the catalyst and product recovery problems. Though the catalyst can be recovered via adsorption on a suitable adsorbent (5), concommitant product adsorption can limit the applicability of such a recovery protocol. Thus for practical application, a heterogeneous PTC in the so-called phenomenon of "Triphase Catalysis" (6) would be desirable in order to simplify the catalyst recovery and reuse. We reported earlier the effect of catalyst type and operating variables on "Triphase Catalysis" for esterification of phenol with benzoyl chloride (7) and analysed the reaction kinetics in details for catalysis by a polystyrene supported tri-n-butylphosphonium ion (8). In this paper, we provide complimentary information on the kinetics for esterification of m-cresol and resorcinol along with comprehensive analysis of the rate controlling phenomenon involved in this specific reaction scheme.

The Mechanism of the Triphase Reaction

The overall reaction is

$$C_6H_5ONa + C_6H_5COCl \xrightarrow{k_{app}} C_6H_5COOC_6H_5 + NaCl \qquad \ldots \quad (1)$$

where, k_{app} is the apparent rate constant. The reactants $C_6H_5ONa_{aq}$ ($ArONa_{aq}$) and $C_6H_5COCl_{org}$ (RCl_{org}) may be assumed to undergo a substitution reaction with the active site within the catalyst particle the proposed mechanism being outlined as shown below :

Organic Phase

$$\text{(P)} \sim P^+ArO^- + C_6H_5COCl \xrightarrow{k_{org}} C_6H_5COOAr + \text{(P)} \sim P^+Cl^-$$

$$\text{(P)} \sim P^+ArO^- + Na^+Cl^- \xrightarrow{k_{aq}} Na^+ArO^- + \text{(P)} \sim P^+Cl^-$$

Aqueous Phase

ArO^- Phenolate ion in ArOH (phenols)
(P) $\sim$ Polystyrene backbone
(P) $\sim$ P^+Cl^- active site of catalyst

k_{aq} and k_{org} are the rate constant for the reactions in aqueous and organic phases, respectively. The reaction is valid when phenols are present in dissociated forms under suitable conditions of pH and reaction environment.

The majority of studies on triphase catalysis leads to the observation of pseudo-first order kinetics for the S_N2 displacement and reduction reactions. An excess of the aqueous phase ionic reagent is generally taken to ensure pseudo-first order kinetics. The apparent rate constant (k_{app}) value is generally estimated from the organic phase reagent conversion versus time profile.

$$- ln\,(1\text{-}x) = k_{app}t \qquad \ldots \quad (2)$$

In order to estimate the k_{app} value, the experimental data on conversion below 30% are considered.

The kinetics of the triphase reaction involves the following important steps (9) : (1) mass transfer of reactant from bulk solution to the catalyst surface (film diffusion), (2) diffusion of the reactant through the polymer matrix to the active sites, (3) reaction at the active site and (4) diffusion of the product to the catalyst surface and mass transfer of the product to the bulk solution. In order to understand the role of these kinetic steps, it is necessary to pursue an experimental protocol through which the effect of the pertinent variables on k_{app} can be evaluated. The pertinent variables which are important are the NaOH/ArOH mole ratio, substrate concentration, ratio of organic to aqueous phase volume, stirring speed, catalyst loading, particle size etc.

Experimental

Materials . Phenol, m-cresol, resorcinol (all designated as ArOH), benzoyl chloride (RCl) toluene and all other reagents were procured from BDH (India) and were of analytical grade. The polymer supported phase transfer catalyst was prepared from particles of microporous chloromethylstyrene (3.5 meq of Cl^-/g and 2% crosslinked divinylbenzene) and tri-n-butyl phosphine according to a procedure described in our earlier work (7). The structure of the catalyst is shown below.

$$ \text{(P)} - \langle \quad \rangle - CH_2 \sim P^+ (n\text{-}Bu)_3 Cl^- \quad (2, 1.15) $$

the values in the parenthesis indicate the percentage cross-linking and milliequivalent Cl^- per gram of catalyst respectively. The catalysts were sieved to obtain particles of 60-80 µm, 100-200µm and 180-200 µm size fractions. Arithmatic average particle sizes have been considered for interpretation of the results.

It may be noted that this catalyst was found to be more active (7) than gel-type Amberlite IRA 401(Cl) (BDH) resin containing quaternary ammonium radical in the matrix consisting of styrene-divinylbenzene co-polymer with 8% degree of cross-linking. The low activity of the resin catalyst for the reaction may be attributed to a higher percentage cross-linking and lower interionic distances imposed by the methyl chain. Even though both the catalysts are lipophilic, the catalyst with P^+ ion in a location close to the polymer backbone is more active than the catalyst with N^+ ion (9). However, the presence of n-butyl chain is likely to render the P^+ ion catalyst more lipophilic in nature, thus ensuring better solvation of the aqueous anion and high selectivity.

Method . Reactions were carried out in a 500 ml fully baffled cylindrical vessel provided with a four-bladed, pitched (45°) turbine. A schematic diagram of the reactor used is shown in Figure 1. The reactor was immersed in a constant temperature water bath, the whole assembly being housed in a fuming cup-board. Benzoyl chloride and toluene were distilled before use. Known quantities of NaOH and ArOH were dissolved in water and introduced into the reactor. Measured quantities of distilled RCl dissolved in toluene and the catalyst were then added. The stirring was immediately started maintaining the speed (rpm) at the desired level. During the reaction, an aliquot (2 ml) sample was withdrawn from the reaction mixture at an interval of 5 minutes. The organic and aqueous phases were separated after a few seconds. The organic phase was analysed by gas chromatograph (CCI, India). The aqueous phase was analysed for ArOH and chloride content spectrophotometrically (10) and by the Volhard method (11), respectively. A material balance calculation was made to cross check the ArOH and RCl converted and benzoate formed. No benzoate formed in the reaction was found to partition into the aqueous phase.

Results

In order to simplify the kinetics, triphase reactions are in general, carried out with an excess of the ionic reagents in the aqueous phase thereby ensuring pseudo-first order kinetics. In the present reaction system, an ArOH/RCl ratio of 1.5 was indeed found to satisfy the pseudo-first consumption rate of RCl (7,8). The role of the NaOH/ArOH ratio has been found to be critical as is evident from Figure 2 which is essentially a relationship of k_{app} and NaOH/ArOH mole ratio and a ratio of 1.5 appears to be optimum. In view of this, the discussion that follows is based on the experimental results pertaining to the above mentioned ratios only. The effect of various parameters on k_{app} will be discussed in relation to the possible rate determining factors.

Effect of Stirring Speed (mixing) . Plot of k_{app} vs stirring speed for different phenols are shown in Figure 3. It is apparent that the rate increases with stirring speed upto 800-1000 rpm beyond which the k_{app} values remain almost constant indicating that, at and above this speed, film diffusional limitations are eliminated. Under otherwise identical conditions, the k_{app} values for different phenols studied change in the order: phenol>m-cresol>resorcinol. The effect of catalyst particle size on k_{app} can also be assessed from Figure 3. The stirring speed effect is the same irrespective of the catalyst particle size. However, higher k_{app} values observed for smaller particles are apparently due to the lowering of intraparticle diffusional limitations under otherwise identical reaction conditions. It may however, be noted that particle size below 70 μm is not likely to enhance the rate because of mass transfer effects as will be discussed later. Indeed, particles below 45μm were found to exhibit markedly low rate (7). Such small particles suspend in the organic medium whereas particles above 65 μm settle to the interface (12). Thus low conversion with very fine particles may be attributed to insufficient contact of the particles with the aqueous phase and loss of benzoyl chloride to the aqueous phase. The mass transfer from the bulk phase to the particle surface becomes drastically low and the reaction rate is reduced, yet the rate is better than that achieved in uncatalyzed reaction (7).

Effect of Catalyst Loading . The effect of catalyst loading (M_c) on k_{app} is shown in Figure 4 for various phenols. Catalyst loading was based on molar equivalent of phosphonium ion with respect to RCl. The effect is realized differently for different

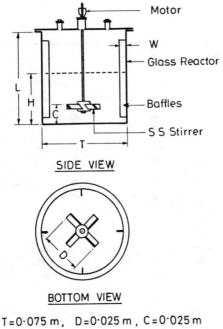

SIDE VIEW

BOTTOM VIEW

$T = 0.075\,m, \quad D = 0.025\,m, \quad C = 0.025\,m$
$W = 0.0075\,m, \quad L = 0.15 \quad H = 0.05\,m$

Figure 1 Schematic diagram of the experimental reactor

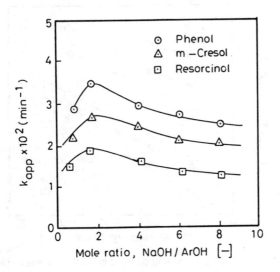

Figure 2 Effect of the NaOH/ArOH Molar Ratio on k_{app} :
$[ArOH] = 6.0 \times 10^{-2}\ mol\ l^{-1}$; $[RCl] = 4.30 \times 10^{-2}\ mol\ l^{-1}$; $d_p = 110\ \mu m$,
$V_{aq}/V_{org} = 1{:}1$; Catalyst loading = 0.55 mol eq.; agitation speed = 800 rpm.

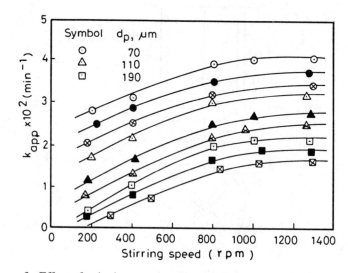

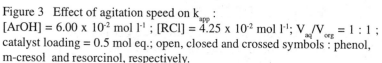

Figure 3   Effect of agitation speed on k_{app} :
[ArOH] = 6.00 x 10^{-2} mol l^{-1} ; [RCl] = 4.25 x 10^{-2} mol l^{-1}; V_{aq}/V_{org} = 1 : 1 ;
catalyst loading = 0.5 mol eq.; open, closed and crossed symbols : phenol,
m-cresol and resorcinol, respectively.

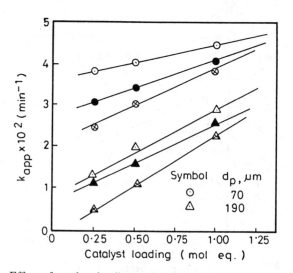

Figure 4   Effect of catalyst loading on k_{app} : conditions same as in Figure 3.

particle sizes. Increase of M_c results in increased reactive site and maximise the amount of aqueous phase anion bound to the catalyst. The influence of M_c is more pronounced for particles of larger size and for higher phenols. This effect cannot be explained easily, but it infers that though intraparticle diffusion limitations are relatively low for small particles, external mass transfer limitations are more pronounced in case of smaller particles even with high agitation speed as a result of dissimilar mass transfer behaviour exhibited by particles of different sizes in an agitated triphase system (*8, 12*).

Effect of Reactant Concentrations . An increase in aqueous phase volume to the organic phase volume ratio (V_{aq}/V_{org}) while keeping the aqueous reagent concentration constant, did not favour the reaction rate of phenol (*7*). Usually, increasing the V_{aq}/V_{org} ratio with the same initial concentration of ArOH will improve the ion-exchange rate. Decreasing the concentration of the reactant anion in the aqueous phase may lead to either a higher rate, because increased hydration of the catalyst leads to faster transport of ions to the active site, or to a lower rate, because increased hydration of the catalyst leads to decreased catalyst activity (*12, 13*). At a constant V_{aq}/V_{org} ratio of 1, increase of ArOH concentration for all the particle sizes used did not change the value of k_{app} for any of the phenols. Similarly, increase of V_{aq}/V_{org} ratio at a constant ArOH concentration did not exhibit a change in the k_{app} value. This implies that the role of ion exchange rate limitation is unimportant. In fact, a high ratio of V_{aq}/V_{org} is useful for a commercial process application. For the same aqueous phase concentration ([ArOH]), an increase in V_{aq}/V_{org} (while maintaining the level of organic substrate stoichiometrically equivalent to the phenolate ion) will increase the concentration of organic phase reactant, [RCl]. Thus, since the reaction rate is nearly first order with respect to [RCl], the rate should increase for an increase of V_{aq}/V_{org} as long as the concentration of the active catalyst ion is kept constant. But this was not actually observed under the experimental conditions used. Substantially poor rate was observed at a V_{aq}/V_{org} ratio of 4(*7*). In triphase catalysis, an environment under which the catalyst can shuttle between the two liquid phases can perhaps be maintained at a threshold level of dispersion which appears to depend on both stirring speed and V_{aq}/V_{org} ratio. It is essential that the catalyst particle should behave as dispersed phase with respect to both the liquid phases. At higher stirring speed, when V_{aq}/V_{org} is increased, there will be a subsequent increase in the liquid-liquid interfacial area, which will favour benzoyl chloride hydrolysis rather than the triphase reaction (*7*) thereby showing reduced rate due to reduction of effective benzoyl chloride concentration. Lowering of overall reaction rate at higher V_{aq}/V_{org} ratio was observed also in case of homogeneously catalysed reaction (*4*). At higher V_{aq}/V_{org} ratio, increase in stirring speed will reduce the organic droplet size (*14*) below the catalyst particle size used, rendering poor dispersion of the catalyst particle in the organic phase. While studying the reaction of 1-bromo octane with aqueous sodium cyanide at a V_{aq}/V_{org} of 1.5 : 1, Tomoi and Ford (*12*) found that a mechanically agitated reactor at 600 rpm gave a faster reaction rate than a turbulent vibromixer even though the later device generated an organic-in-water dispersion of droplet size as small as 0.05 nm. Further, the dependence of reaction rate on stirring speed decreased as the particle size decreased.

The effect of the concentration of benzoyl chloride ($[RCl]$) in the organic phase studied at V_{aq}/V_{org} ratio of 2 is shown in Figure 5. k_{app} increases with $[RCl]$ under the condition of negligible film diffusion limitation. While the inference on concommitant increase in the consumption rate of ArOH is pedastral, the more subtle inference is that the diffusive transport of RCl to the active centre is the rate limiting step in the triphase catalysis for esterification of phenols.

Effect of Temperature . From an experimental study of the effect of temperature, in terms of $\log k_{app}$ versus $1/T$ (7), the values of the apparent activation energies (E_{app}) were found to be 10.66, 11.26 and 11.86 KCal/mol for the respective phenols. E_{app} may be presented as $E_{app} = (E_{diff} + E_r)/2$, where E_{diff} is the activation energy for particle diffusion and E_r is the activation energy for intrinsic reactivity. E_{diff} value of 7.5 KCal/mol reported for identical catalyst (12) may be appropriate for pure diffusion model. So the E_r values are 12.8, 15.2 and 16.2 KCal/mol for the respective phenols. The values E_{app} obtained in the present work appear reasonable considering comparable figures reported for other triphase reactions (13). For smaller particles, the activation energy seems to be higher (7) owing to the shorter diffusional path length in the pores.

Discussion

Mass Transfer Effect . In the triphase system, both the liquid phases should be in contact with the solid catalyst to augment mass transfer on the boundary layer outside the particle. The most probable fundamental kinetic steps are (1) mass transfer of reactant from bulk solution to the catalyst surface, (2) diffusion of the reactant through the polymer matrix to the active site, (3) reaction at the active site and (4) diffusion of the product to the surface of the catalyst and mass transfer of the product to the bulk solution. In the present work, a highly active catalyst has been used and therefore, mass transfer from the bulk solution to the catalyst surface can be rate controlling under the experimental conditions studied. Mass transfer of RCl from the bulk organic phase to the surface of the catalyst particle depends on the contact between the polymer particles and organic droplets, both of which are suspended in a continuous phase of aqueous sodium phenolate. The observation of the dependence of apparent rate on agitation speed within a certain range in a well defined geometry of the reactor implies that mass transfer influences the rate in that range of stirring speed.

The conditions employed for the reaction are thought to create an environment where the factors like (1) diffusion of product out of the polymer matrix and mass transfer of product from the catalyst surface to the bulk liquid and (2) rate of ion exchange of the original Cl⁻ or the product Cl⁻ ion in the catalyst, can be neglected. The interpretation of the experimental results can be confined to three aspects, i.e. mass transfer of RCl to the catalyst surface, diffusion of RCl through the polymer matrix and intrinsic reactivity at the active sites. The one dimensional reactant concentration profile along the radius of the catalyst particle is shown in Figure 6. Intraparticle diffusion alone cannot be the rate-limiting process.

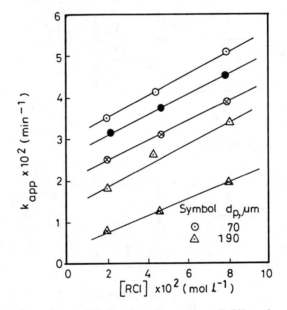

Figure 5   Effect of benzoyl chloride concentration ([RCl]) on k_{app} :
[ArOH] = 1.5 mol l⁻¹ ; catalyst loading = 0.5 mol eq. ; open, closed and crossed
symbols : phenol, m-cresol and resorcinol respectively.

A. Fast mixing

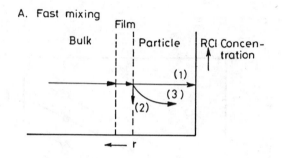

B. Slow mixing

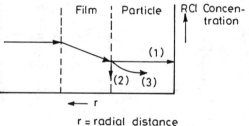

r = radial distance

(1) Reaction rate limited by intrinsic activity .
(2) Reaction rate limited by mass transfer
(3) Reaction rate limited by combination of
    intraparticle diffusion and intrinsic activity.

Figure 6   Concentration profile as a function of radial distance.

Kinetic models

From the experimentally observed rate limiting steps, kinetic models can be derived based on the theory of diffusive mass transfer in porous catalyst (*16*). Two cases may be considered.

Case I : No Film Diffusion Resistance

The molar rate of disappearance of RCl (component A) for a pseudo-first-order reaction in a triphase system may be described as

$$- \frac{dN_A}{dt} = k \, \lambda M_c \, C_{AS} \eta \qquad \qquad \ldots \qquad (3)$$

Assuming negligible reaction at the exterior surface of the catalyst and changes in the amount of reactant within the catalyst, the reaction rate for unit volume of organic phase becomes

$$- \frac{dC_A}{dt} = \frac{k \lambda \, M_c}{V_{org}} C_A \qquad \qquad \ldots \qquad (4)$$

Here, the observed pseudo-first order reaction rate constant is defined as

$$k_{app} = \frac{k \lambda M_c}{V_{org}} \qquad \qquad \ldots \qquad (5)$$

The effectiveness factor, η for spherical pellet is described by

$$\eta = \frac{3 \, \phi \, \text{Coth}(3 \, \phi) - 1}{3 \phi^2} \qquad \qquad \ldots \qquad (6)$$

Thiele modulus, ϕ is given by

$$\phi = \frac{R}{3} \sqrt{\frac{k \, \lambda \, M_c}{V_{cat} D_s}} \qquad \qquad \ldots \ldots \qquad (7)$$

From equations 5-7, one can derive

$$k_{app} = \frac{3D_s V_{cat}}{R^2 V_{org}} \left[R\sqrt{\frac{k \lambda M_c}{V_{cat} D_s}} \; Coth \left(R\sqrt{\frac{k \lambda M_c}{V_{cat} D_s}} \right) - 1 \right] \qquad \ldots \qquad (8)$$

Case II : Film Diffusion Resistance

This case is applicable when the observed rate is dependent on the stirring speed. In a triphase system, the mass transfer from a dispersed liquid phase to a solid particle in an agitated vessel is not clearly understood. In fact, the mass transfer should take place in a dispersion of two liquid phases. Catalyst particles meet dispersed organic drops periodically, and an average mass transfer coefficient is considered for component A between the organic phase and the particle surface (15). The molar rate of depletion of component A from the organic phase will be

$$- \frac{dN_A}{dt} = a_s \, k_{SL} \, (C_A - C_{As}) \, V_{cat} \qquad \ldots \qquad (9)$$

Combining equations 3 and 9 and eliminating C_{As}, the overall rate for a unit volume of organic phase is

$$- \frac{dC_A}{dt} = \left[\frac{RV_{org}}{3V_{cat} k_{SL}} + \frac{V_{org}}{k \lambda M_c \eta} \right]^{-1} C_A \qquad \ldots \qquad (10)$$

$$\frac{1}{k_{app}} = \frac{RV_{org}}{3V_{cat} k_{SL}} + \frac{V_{org}}{k \lambda M_c \eta} \qquad \ldots \qquad (11)$$

The overall catalyst effectiveness factor, η_o should account for film resistance and intraparticle diffusion. Introducing the Biot number, $B_i = k_{SL} \, R/3D_s$, η_o is given by

$$\eta_o = \frac{\eta}{1 + \phi^2 / B_i} \qquad \ldots \qquad (12)$$

where, η is given by equation 6. Equation 11 shows that the k_{app} is the result of two simultaneous phenomena, i.e. film diffusion and combination of intrinsic activity and particle diffusion. Equation 11 can be used to evaluate k_{SL} from experimental data provided that $k\lambda$ and D_s are known 'a priori'.

Estimation of $k\lambda$ and D_s can be performed by a graphical procedure described earlier (7, 12). This procedure relies on the effectiveness factor curve which for the case of phenol can be generated as shown in Figure 7. The estimated values of $k\lambda$ and D_s for the present system of esterification reaction are 2.3 x 10^{-8} cm^{2}s^{-1} and 0.35 cm^3 mol^{-1} s^{-1}, respectively. The estimated $k\lambda$ values for m-cresol and resorcinol are 0.28 and 0.22 cm^3 mol^{-1} s^{-1}, respectively. Lower values of the intrinsic reaction rate constant (k) and partition coefficient (λ) for the higher homologues may be attributed to the observed lower values of k_{app} which for the negligible film diffusion resistance is given by equation 8. The estimated values of D_s and $k\lambda$ for esterification of phenols with a polymer supported P$^-$ ion catalyst are relatively much lower than those obtained for the common nucleophilic substitution reaction between n-bromo octane and sodium cyanide reported elsewhere (15).

In order to understand the external film diffusion effect, the values of the mass transfer coefficient, k_{SL} were calculated using equation 11. Computed k_{SL} values pertaining to the reactor assembly (Figure 1) are shown in Figure 8 as a function of stirring speed for different particle sizes. The variation of k_{SL} is less sensitive to stirring speed for small particles i.e. variation of k_{SL} with agitation speed is comparatively low. Furthermore, at relatively high speed, smaller particles exhibit lower values of k_{SL}. In contrast, at low speed k_{SL} values for small particles seem to be higher than those for the largest particle used, and the mass transfer coefficient for smaller particles is relatively insensitive to increase in stirring speed. Kinetic experiments revealed that k_{app} was almost constant for stirring speed above 800-900 rpm, irrespective of the particle size used. In fact, the film diffusion resistance is reduced above 800-900 rpm. As shown in Figure 7, the average mass transfer coefficients seem to increase with particle diameter above a stirring speed of i.e. 500 rpm, and the dependence becomes stronger as the stirring speed increases.

In the triphase system considered here, k_{SL} refers to the dispersion of organic droplets in a continuous aqueous phase and particle surface. The organic droplets are dispersed in the aqueous phase owning to an impeller action, and the size of the dispersed droplets depends on the impeller speed increase of which decreases the droplet size. When the stirring speed is low, the size of the dispersed drop of the organic phase is relatively larger, and the catalyst particles behave as a dispersed phase with respect to the continuous phase of organic droplets, it may be expected that the particle hold-up in this organic droplet is larger for small particles. As a result, k_{SL} value is high for small particles below a particular stirring speed. When the stirring speed is increased, the size of the organic droplet reduces and small catalyst particles tend to remain at the organic- aqueous interface, move to the surface of the medium and cannot penetrate the organic droplet easily whereas, larger particles tend to cause drop breakage and coalescence thereby ensuring contact between particle and the bulk organic phase. This is perhaps the reason why at high agitation speed, mass transfer coefficient increases with increasing particle size. The solid density is also likely to affect this variation of mass transfer with particle size. The reactor type and geometry are likely to equally affect the hydrodynamics and mass transfer behaviour. Thus, the mass transfer effect in a triphase reaction like the one considered here cannot be predicted 'a priori' and is dependent on the system properties and reactor type and its operating condition.

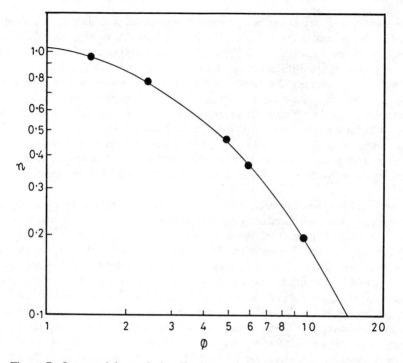

Figure 7   Intraparticle catalytic effectiveness factor versus Thiele modulus (Reproduced with permission from reference 8. Copyright 1994 Elsevier Science—NL.)

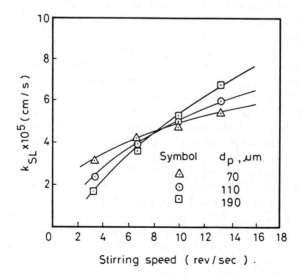

Figure 8   Effect of stirring speed on estimated average mass transfer coefficient (Reproduced with permission from reference 8. Copyright 1994 Elsevier Science—NL.)

Conclusion

The esterification of phenols with benzoyl chloride under triphase catalysis by a polymer supported phosphonium ion is controlled by diffusion of the organic substrate in the catalyst particle and the intrinsic reactivity. The ion exchange between the aqueous anion and the catalyst cation is of less importance. The external mass transfer effect is realized differently for different particle size and is dependent on the system properties and the reactor type and its operating parameters. The mass transfer effect in the triphase system can be analyzed from the well known theory of reaction-diffusion in porous catalyst. Particle-liquid mass transfer with respect to the organic phase seems to depend on the particle size and this dependence is realized differently in different ranges of stirring speed in a mechanically agitated contactor.

List of symbols

a_s External surface of particle per unit volume (cm^{-1})
B_i Biot number defined as $k_{sL} R/3D_s$ (dimensionless)
C_A Concentration of species A (mol cm^{-3})
C_{As} Concentration of species A at the surface of a catalyst particle (mol cm^{-3})
d_p Catalyst particle size (m)
D_s Effective diffusivity within catalyst particle ($cm^{-2}S^{-1}$)
E_{app} Apparent activation energy (kcal mol^{-1})
E_{diff} Activation energy for pore diffusion (kcal mol^{-1})
E_r Activation energy for intrinsic reaction (kcal mol^{-1})
k Intrinsic reaction rate constant ($cm^3 mol^{-1} S^{-1}$)
k_{app} Apparent reaction rate constant (s^{-1})
k_{sL} Mass transfer coefficient based on external surface of particle (cm s^{-1})
M_c Molar equivalent of catalyst (mol)
N_A Moles of species A (mol)
R Radius of particle (cm)
t Time (s)
T Temperature (K)
V_{aq} Volume of aqueous phase (cm^3)
V_{cat} Volume of swollen catalyst (cm^3)
V_{org} Volume of organic phase (cm^3)
x Conversion (dimensionless)
λ Partition coefficient (dimensionless)
η Intraparticle catalytic effectiveness factor (dimensionless)
η_0 Overall catalytic effectiveness factor (dimensionless)
ϕ Thiele modulus (dimensionless)

Literature Cited

1. Direktor, D.; Effenberger, R.; *J. Chem. Technol. Biotechnol.*, **1985**, *35A*, 281-284.
2. Dutta, N.N.; *Chem. Eng. World*, **1990**, *25*, 79-81.
3. Krishnakumar, V.K.; Sharma, M.M.; *Ind. Eng. Chem. Res.*, **1984**, *23*, 410-413.
4. Dutta, N.N.; Borthakur, S.; Patil, G.S.; *Sep. Sci. Technol.*, **1992**, *27*, 1435-1448.
5. Ghosh, A.C.; Srivastava, R.C.; Satyanarayana, K.; Dutta, N.N.; *Colloids and Surfaces, A.*; **1995**, *96*, 219-228.
6. Regen, S.L.; *J. Am. Chem. Soc.*, **1975**, *97*, 5956-5957.
7. Dutta, N.N.; Borthakur, S.; Patil, G.S.; *Ind. Eng. Chem. Res.*, **1992**, *31*, 2727-2731.
8. Dutta, N.N.; Pangarkar, V.G.; *Reactive Polymers*, **1994**, *22*, 9-17.
9. Ford, W.T.; Tomoi, M.; *Adv. Polym. Sci.*, Spinger Verlag, Berlin,**1984**, *55*, 49-75.
10. Whitlock, L.R.; Siggia, S.; Smolla, J.E.; *Anal. Chem.*, **1992**, *44*, 532.
11. Vogel, A.I.A.; *Text Book of Quantitative Inorganic Analysis Including Elementary Instrumental Analysis*, Longman, London, **1978**, pp 340-342.
12. Tomoi, M.; Ford, W.T.; *J. Am. Chem. Soc.*, **1981**, *103*, 3821-3828.
13. Wang, M.L.; Tang, H.M.; *Ind. Eng. Chem. Res.*, **1991**, *30*, 2384-2388.
14. Tavlarides, L.L.; *Chem. Eng. Commun*, **1981**, *8*, 133-164.
15. Marconi, P.F.; Ford, W.T.; *J. Catal.*, **1983**, *83*, 160-167.
16. Carberry, J.J.; Varma, A.S.; *Chemical Reactions and Reaction Engineering,* Marcel Dekker, New York, **1987**

Chapter 21

Multisite Phase-Transfer Catalyst for Organic Transformations

T. Balakrishnan and J. Paul Jayachandran

Department of Physical Chemistry, Guindy Campus,
University of Madras, Madras 600 025, India

A three step synthesis of a new, water soluble, "multi-site" phase transfer catalyst viz., 2-benzylidine-N,N,N,N',N',N'-hexaethyl propane-1,3 diammonium dichloride is described. The potentiality of "multi-site" phase transfer catalyst under study is demonstrated by studying the kinetic aspects of reactions viz., Dichlorocarbene addition to styrene double bond and C-alkylation of phenylacetone with ethyl iodide.

Phase transfer catalysis (PTC) has been a fascinating area of research interest to the synthetic chemists. It is well recognised as a general and more versatile technic applicable to a number of organic biphase reactions (1,2). The applications of PTC include the preparation of compounds from an unreactive starting material, dramatic enhancement in yields and product selectivity. The soluble 'single-site' phase transfer catalysts are immensely popular due to their availability and easy reaction work up. The important considerations in the selection of the catalyst are economy of scale and efficiency of the phase transfer catalyst specifically on the explicit industrial scale preparations of organic compounds. In order to satisfy these needs, novel "multi-site" phase transfer catalysts have been developed which consists more than one catalytic active site.

Critical survey of literature reveals the studies of Idoux et al (3) dealing with "multi-site" PTCs based on phosphonium type. The authors have also studied the efficacy of the catalysts for simple Sn2 reactions and for weak nucleophile electrophile SnAr reactions. However, studies dealing with "multi-site" PTC of ammonium type are not known in literature. This article focuses on the synthesis of "multi-site" PTC based on ammonium salt and covers the detailed kinetic study of reactions viz., Dichlorocarbene addition to styrene double bond and C-alkylation of phenylacetone with ethyl iodide in the presence of the synthesised "multi-site" phase transfer catalyst.

Synthesis of "Multi-site" Phase Transfer Catalyst (MPTC)

A detailed procedure for the synthesis of 2-benzylidine-N,N,N,N',N',N'-hexaethyl propane-1,3-diammonium dichloride (MPTC) are discussed elsewhere (4,5). Scheme 1. shows the steps involved in the synthesis of the same.

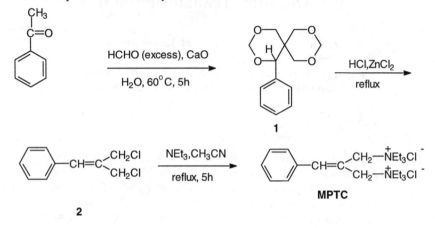

Scheme 1

Reaction of acetophenone with excess aqueous formaldehyde in the presence of CaO at 60°C afforded 1-phenyl-2,4,8,10-tetra oxaspiro [5.5] undecane (1). The spiro compound 1 was chlorinated by refluxing with aqueous HCl (35% solution) and catalytic amount of $ZnCl_2$ resulting in the formation of a dichlorocompound (2). The compound 2 was quaternised with excess triethylamine using dry acetonitrile as a solvent under reflux to yield the desired **MPTC**. It was found to be highly hygroscopic. The structures were confirmed using elemental analysis and by spectroscopic means.

Preliminary studies (4) on the catalytic abilities of the new MPTC revealed its synthetic utility under PTC/OH⁻ conditions. Control experiments of dichlorocarbene addition to olefins in the absence of the catalyst resulted in 1% conversion in three hours. Control experiments for alkylation reactions in the absence of the catalyst under specified conditions proceed only with low conversions. For majority of the reactions, improved yields of products were observed using the new "multi-site" PTC as compared to those catalysed by 'single-site' PTC (namely triethylbenzylammonium chloride) under identical conditions.

Kinetics of dichlorocarbene addition to styrene double bond using MPTC

The kinetic experiments (followed by GC) of the dichlorocarbene addition to styrene were carried out under pseudo-first order conditions, taking chloroform and 30% aqueous sodium hydroxide in excess at 40°C (Scheme.2).

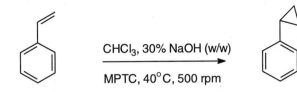

Scheme 2. Dichlorocarbene addition to styrene under PTC conditions.

Effect of stirring speed. The effect of varying stirring speed on the rate of dichlorocarbene addition to styrene reaction using MPTC was studied in the range 0-700 rpm. An increase in the rate of reaction was noticed as the stirring speed enhanced. The effect of varying stirring speed is well documented (*1,6,7*) for interfacial mechanisms which are transfer rate limited (the rate constants increases with stirring) below a given stirring speed (600 - 800 rpm) and intrinsic reaction rate limited (the rate constant is nearly a constant) above this stirring speed. Similar behaviour is displayed by reactions with a real 'phase transfer' (Starks' extraction mechanism) but with a smaller limit of stirring speed between physical and chemical control (100 - 300 rpm).

In the present study, the rate constants of the reaction increases as stirring speed increases and levels off to a constant value above the optimum stirring speed (500 rpm). The interfacial area per unit volume of dispersion increased linearly with increasing speed till a stage is reached where there is no significant increase in the interfacial area per unit volume of dispersion with the corresponding increase in the speed (*1*). Thus, on increasing the stirring speed changes the particle size of the dispersed phase. Above certain stirring speed (500 rpm), the particle size does not change. The constancy of the reaction rate constants is observed not because the process is necessarily reaction rate limited but because the mass transfer rate has reached constant value. Therefore Figure 1 is indicative of an interfacial mechanism and not of a real 'phase transfer'.

Effect of catalyst amount. The effect of catalyst amount on the rate of dichlorocarbene addition reaction to styrene was studied in the range 0.6 to 1.4 mol% of the catalyst (based on the substrate amount). The rate constant of the reaction is proportional to the amount of catalyst added (Figure 2). No reaction was observed even after three hours of stirring in control experiments. The linear dependence of reaction rate constants on catalyst concentration shows that the reaction is believed to proceed through the extraction mechanism. A bilogarithmic plot of the reaction rate versus the concentration of the catalyst gives a straight line over a wide range of concentration as shown in Figure 3. The slope of 0.5 for dichlorocarbene addition to styrene was found to be identical with the slope of the same reaction reported by Balakrishnan et al (*8*) in the presence of a 'single-site' PTC (triethylbenzylammonium chloride). This suggests that the chemical reaction between the ion pair and the organic substrate is not the sole rate determining step. The foregiven observations enable us to predict that the carbanions formed cannot leave the phase boundary to go into the organic phase since their counter ions Na^+ are strongly solvated in the aqueous phase and poorly in the organic phase. In this state, the carbanions are very unreactive being able to react with strong electrophiles. The "multi-site" quaternary ammonium cations

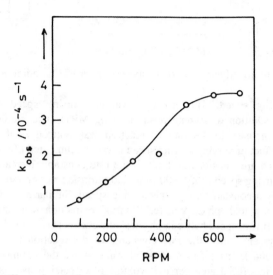

Figure 1. Effect of variation of stirring speed (Reproduced with permission
from ref. 5 copyright 1995 Royal Society of Chemistry)

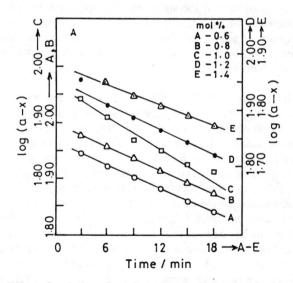

Figure 2. Effect of variation of catalyst amount (Reproduced with permission
from ref. 5 copyright 1995 Royal Society of Chemistry)

serve as a source of organic cations, thus transferring them into the organic phase for further transformation. The remarkable increase in yield of the dichlorocarbene adduct reflects the ability of the quaternary salt to cause :CCl_2 to be generated in or transferred to the organic phase, where its reaction rate with substrate is much greater than with water as reported by Starks in the study of dichlorocarbene addition to cyclohexene using tridecylmethylammonium chloride (9). In the kinetic study of ethylation of phenylacetonitrile by Chiellini et al (16), a bilogarithmic plot of initial rate versus catalyst concentration, gave a slope equal to 0.6 for which an interfacial mechanism was proposed.

Effect of substrate amount. Kinetic experiments were performed by varying the substrate amount ranging from 6.52 to 21.75 mmol, maintaining other reactants such as chloroform and 30% w/w NaOH in excess. The observed rate constants decrease as the [substrate] increase. It is clear from Table I. that the molar ratio of substrate to catalyst increases considerably for small increments in substrate amount. The decrease in rate constants may be attributed to the decrease in the ratio of the number of active sites of the catalyst to the corresponding amount of substrate present.

Table I. Effect of variation of substrate amount

Entry	Substrate amount mmol	Substrate : Catalyst	$k_{obs}/10^{-4}$, s^{-1}
A	6.5	74.97	11.13
B	8.7	100.03	3.49
C	13.1	150.05	3.08
D	17.4	200.07	2.91
E	21.8	250.66	2.73

Source: Reproduced with permission from reference 5. Copyright 1995 Royal Society of Chemistry.)

Effect of sodium hydroxide concentration. The reaction rates were measured in the range 7.89 to 14.06 mol dm^{-3}. The rate constants of dichlorocarbene addition strongly depend on the concentration of NaOH (10). The rate constants were found to increase with an increase in sodium hydroxide concentration (Figure 4). A bilogarithmic plot of the reaction rate against sodium hydroxide concentrations gives a straight line having a slope 5.5 (Figure 3). This may be attributed to the fact that hydroxide ions are less solvated by water molecules and thereby the activity of hydroxide ion increases. In the reaction dichlorocarbene addition to styrene using triethylbenzylammonium chloride, 40% w/w NaOH was employed whereas 30% w/w NaOH is the optimum concentration used in the present study.

Influence of temperature variation. The effect of varying temperature on the rate of dichlorocarbene addition reaction to styrene was studied in the temperature range 20 to 40°C. The kinetic profile of the reaction is obtained by plotting log(a-x) versus time. The rate constants increase with the increase in temperature. The energy of activation is calculated from an Arrhenius plot, E_a=15.07 kcal mol^{-1}.

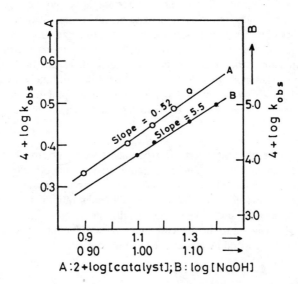

Figure 3. Effect of catalyst amount and [NaOH] on the observed rate constant (Reproduced with permission from ref. 5 copyright 1995 Royal Society of Chemistry)

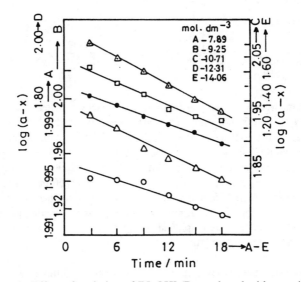

Figure 4. Effect of variation of [NaOH] (Reproduced with permission from ref. 5 copyright 1995 Royal Society of Chemistry)

The other thermodynamic parameters $\Delta S^{\ddagger}$, $\Delta G^{\ddagger}$ and $\Delta H^{\ddagger}$ were evaluated from Eyring plot as 25.11 cal K^{-1} mol^{-1}, 21.99 kcal mol^{-1} and 14.47 kcal mol^{-1} respectively.

Comparison of reaction rate constants. The reaction, dichlorocarbene addition to styrene, has been chosen to investigate the comparative reactivities of three different catalysts, viz., "multi-site" PTC (MPTC), soluble 'single-site' PTC [triethylbenzylammonium chloride (TEBA)] and polymer-bound 'single-site' [triethylbenzylammonium chloride (2% cross-linked, 200 mesh, 18-20% ring substitution)] (*11*). Based on the observed rate constants, it is evident that the MPTC is 37% and 95% more active than the 'single-site' TEBA and polymer-bound TEBA respectively (Table.II). The much lower rate constants obtained on using polymer-bound catalyst were attributed to the diffusional limitations of the rates under triphase conditions (*12*).

Table II. Comparison of k_{obs} using different phase transfer catalysts

Entry	Catalyst	Amount (mol%)	$k_{obs}/10^{-4}, s^{-1}$
A	None	None	Nil[*]
B	MPTC	1.0	4.01
C	TEBA	1.0	2.50
D	Polymer-bound TEBA	1.0	0.20

* No conversion after 3 h.
Source: Reproduced with permission from reference 5. Copyright 1995 Royal Society of Chemistry.)

Mechanism. Haloform treated with concentrated aqueous sodium hydroxide and a quaternary ammonium salt, $Q^+ X^-$ (as a PTC) generates trihalomethyl anion which further splits into dichlorocarbene (*13*). The selectivities of dichloro and dibromo carbenes (generated under PTC conditions) towards alkenes are independent of the structure of catalysts. This indicates that free $:CX_2$ is involved in all cases inspite of the fact that there is a strong catalyst influence on the reaction path starting from $CX_3 \rightleftharpoons :CX_2$ (*14*). Scheme 3 shows the mechanism for the dichlorocarbene addition to styrene under PTC conditions. According to Starks' extraction mechanism, it was thought that the hydroxide-ion may be extracted from an aqueous reservoir into an organic phase with the help of quaternary onium cations. Makosza and Bialecka proposed (*15*) an alternative mechanism for dichlorocarbene addition reactions in which deprotonation of the organic substrate by the hydroxide-ion occurs at the interface into bulk organic phase for subsequent reaction. Several studies (*16,17*) have provided support for various aspects of Makosza's mechanism. It has been established by Makosza and Fedorynski (*18*) that the slowest reaction is the addition of $:CCl_2$ to alkenes, considering the other steps as fast equilibrium processes. In our study, a fractional order with respect to the catalyst concentration suggests that step (2) is not the sole rate-determining one and that of the chemical reaction in the organic phase is also rate- determining. The effects of other experimental results such as stirring speed, sodium hydroxide concentration and temperature over the observed rate of the reaction support the interfacial mechanism proposed by Makosza for PTC/OH⁻ systems. The generation and reaction of carbene with styrene may be represented as :

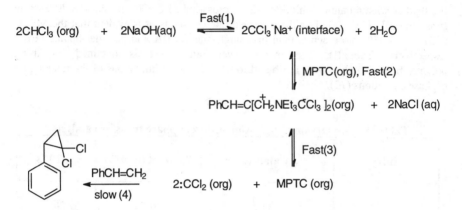

Scheme 3. Mechanism for the Dichlorocarbene addition to styrene under PTC conditions.

Kinetics of C-Alkylation of Phenylacetone with Ethyl iodide

In our study the C-alkylation of phenylacetone with ethyl iodide was chosen to investigate the kinetic aspects using the new, water soluble, "multi-site" Phase Transfer Catalyst (MPTC).

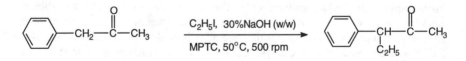

Scheme 4. C-Alkylation of phenylacetone with C_2H_5I under PTC Conditions

The kinetic experiments for the C-alkylation of phenylacetone (Scheme 4) were conducted under biphase conditions with excess of aqueous sodium hydroxide and ethyl iodide under pseudo-first order conditions and the data reported are for the formation of monoalkylated product. The reaction was studied at a stirring speed of 500 rpm in the temperature range 30 to 50°C. Before the kinetic run was started, the catalyst was conditioned with aqueous sodium hydroxide, ethyl iodide for 10 minutes. The substrate phenylacetone preheated at the appropriate temperature was added to the reaction mixture. The samples were collected from the organic layer at regular intervals of time. The kinetics of ethylation was followed by the disappearance of phenylacetone using gas chromatograph. The effect of various experimental parameters such as stirring speed, amount of catalyst, substrate concentration, sodium hydroxide concentration and temperature on the reaction rate constants were studied. The kinetics was measured upto 30% conversion of product .

Effect of varying stirring speed. The effect of varying stirring speed on the rate of C-alkylation of phenylacetone with ethyl iodide was studied in the range 200-900 rpm. From the plots of log(a-x) versus time, the pseudo-first order rate constants were evaluated. A plot of k_{obs} against stirring speed is shown in Figure 5. Balakrishnan et al (*19*) has documented the kinetic studies of C-alkylation of phenylacetone using n-butyl bromide catalysed by aqueous sodium hydroxide and triethylbenzylammonium chloride (TEBA). This reaction was also carried out by Jonczyk et al (*20*) using 50% aqueous sodium hydroxide and catalytic amount of TEBA and obtained a monoalkylated product, 3-phenyl-2-heptanone. The rate constant is dependent on the stirring speed upto 500 rpm and independent beyond this limit. The anion exchange equilibrium between the anions in the aqueous phase and those associated with quaternary salt in the organic phase was very fast relative to the organic phase displacement reaction. The mass transfer across the interface is retarded as evident from Figure 5 at low stirring speed. At agitation level 500 rpm, anion exchange equilibrium is very fast relative to the organic displacement reaction and the substrate consumption rate becomes independent of the stirring rate as observed by Landini et al (*21*) in the study of n-octyl methane sulfonate catalysed by quaternary salts $Q^+ X^-$ under PTC conditions in a chlorobenzene-water two-phase system. A similar observation had been reported independently by Starks (*22*), Herriott (*23*) and Freedman (*24*), reflect kinetic control by chemical reaction in which [$Q^+ X^-$] is at a steady state concentration. Below 500 rpm the requirement for sufficiently rapid mass transfer of the reacting anion is not met and diffusion controlled kinetics is observed. Hence, the independence of the reaction rate constants on the stirring speed above 500 rpm in the present study is indicative of extraction mechanism. This behaviour is in sharp contrast to reactions operative through interfacial mechanism, where the reaction rate is directly proportional to the stirring speed. Chiellini et al (*16*) reported the continuous increase in the rate of ethylation even upto stirring speeds of 1950 rpm. It has also been reported that the rate of an interfacial reaction is proportional to the speed of stirring in the range of 600 upto 1700 rpm.

Effect of varying catalyst amount. The amount of catalyst was varied from 0.3 to 1.1 mol% (based on the substrate amount) and the experiments were conducted using 30 % w/w aqueous NaOH solution. The rate constants are calculated from the plots of log(a-x) versus time (Figure 6). The rate constants are dependent on the amount of catalyst used in each reaction (Table III). The increased rate constants are attributed to the increase in the number of active sites. Control experiments were performed and no product was detected even after 2 hours of the reaction. Only a catalytic amount (0.3 mol% based on the substrate amount) is required in order to obtain good yields of the product emphasising the indispensability of the catalyst. Molinari et al (*25*) observed a similar dependence of pseudo-first order rate constants on the amount of heterogenised phosphonium groups for Br-I exchange reaction of 1-bromooctane.

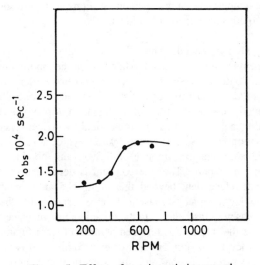

Figure 5. Effect of varying stirring speed

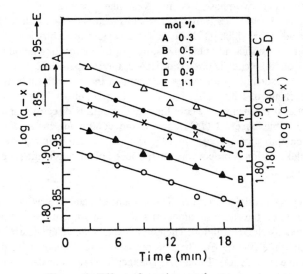

Figure 6. Effect of varying catalyst amount

A bilogarithmic plot of the reaction rate constants versus the concentrations of catalyst gave a straight line having slope 0.16. In the study (26) of dehydrobromination of phenethyl bromide in the presence of tetraoctylammonium bromide, zero order kinetics with respect to the catalyst was observed.

Table III. Effect of varying catalyst amount

Entry	Catalyst amount mol%	$k_{obs}/10^{-4}$, s^{-1}
A	0.3	1.79
B	0.5	1.92
C	0.7	1.97
D	0.9	2.20
E	1.1	2.19

Catalyst Poisoning. Many PTC/OH⁻ reactions are totally or partially suppressed by 'catalyst poisoning' (27). This occurs in cases in which the deprotonated organic substrate (24) (when the interfacial mechanism is operative) or an extracted OH⁻ (6, 28) cannot compete successfully with a highly lipophilic anion (for example; iodide) for association with the quat in the organic phase. In the present study, the lipophilic anion, I^- originates as a leaving group eliminated from the alkylating agent, ethyl iodide. As the reaction progresses the concentration of the lipophilic anion increases, as a result, we observed faster rate of reaction initially in individual runs. It is clear from Table III, the rate constants are initially dependent on using MPTC from 0.3 to 0.9 mol% and attains constancy thereafter. This behaviour is also an indication of extraction mechanism operative in the present study. This claim was further supported by comparing the reactivities of MPTC and 'single-site' PTC, triethylbenzylammonium chloride, based on the observed rate constants values. Contrary to our expectation, small increase in k_{obs} value on using TEBA (2.41 x 10^{-4}, s^{-1}) is noticed than on using MPTC (2.29 x 10^{-4}, s^{-1}) which is an indication of marginal catalyst poisoning prevailing in the present study. This further confirms the extraction mechanism operative in the PTC/OH⁻ system of current interest. It is also to be mentioned that the amount of catalyst used is small quantity (1 mol% based on substrate amount) and the percentage conversion of the product was about 30%. Under these circumstances, catalyst poisoning by I^- is negligible. In a systematic study (6) of isomerisation of allyl benzene, in the presence of 100 fold excess of hydroxide over the catalyst counter ion, a change from HSO_4^- (effective ion : SO_4^{2-}) to Br^-, resulted in a 50 fold decrease in the rate of the reaction. The order of lipophilicity is reported to be $SO_4^{2-} < Cl^- < Br^-$. In the case of deuteration (29, 30), it was shown that initial rate of reaction was faster in the presence of bromide relative to chloride. Dehmlow et al (31), overcame the catalyst poisoning problem by employing a very organophilic quat.

Effect of varying substrate amount. Kinetic experiments were performed by varying the amount of phenylacetone from 5.60 mmol to 18.66 mmol and keeping other reagents such as ethyl iodide and NaOH in excess. Pseudo-first order rate constants are evaluated from the linear plots of log(a-x) versus time. The observed reaction rate constants decrease as the amount of substrate increases. The decrease in rate may be attributed to the proportionate decrease in the number of catalytic active sites available. The results suggest that the concentrations of the substrate in the organic phase is not important and the concentration of the substrate at the interface may be vital. The molar ratio of the substrate to catalyst and corresponding rate constants are given in Table IV.

Table. IV Effect of variation of substrate amount

Entry	Substrate amount mmol	Substrate : Catalyst	$k_{obs}/10^{-4}$, s^{-1}
A	5.60	98.24	3.44
B	7.46	130.87	2.17
C	11.19	196.30	1.92
D	14.93	261.91	1.45
E	18.66	327.35	1.33

Effect of varying sodium hydroxide concentration. The rate of C-alkylation of phenylacetone with ethyl iodide strongly depends on the strength of sodium hydroxide. Kinetic experiments were carried out employing 7.89 to 14.06 mol dm^{-3} aqueous NaOH. Pseudo-first order rate constants are evaluated from the plots of log(a-x) versus time (Figure 7). The reaction rate constants are strongly influenced by the concentration of aqueous NaOH. The observed rate constants tremendously increased with increase in basicity of hydroxide-ion. A bilogarithmic plot of the reaction rate against sodium hydroxide concentrations gives a straight line having a slope of 2.0. In the case of the isomerisation of allylbenzene, the effective kinetic order with respect to aqueous NaOH concentration was reported to be 5.0, where extraction mechanism is operative. Makosza (33) reported the successful application of two phase catalytic and ion-pair extractive methods in the reactions of carbanions and halocarbenes. The instability of some C-H acids in the presence of aqueous alkali (such as hydrolysis of esters of other functional groups) was pointed out.The remedy suggested was the use of less concentrated solutions of the base. This requirement is met by employing MPTC as the optimum NaOH concentration is 30% as against 40% used in the case of TEBA. Generally, quaternary salts are stable upto ~150°C but less stable at temperatures 70 to 80°C when strong alkalis are used. While employing MPTC the need to use low concentrations of alkali inspite of the fact that the reaction is in a great measure dependent on it reveals the higher activity and applicability of the MPTC.

Influence of temperature variation. The effect of varying temperature on the rate of C-alkylation of phenylacetone with ethyl iodide was studied in the temperature range 30 to 50°C. The kinetic profile of the reaction is obtained by plotting log (a-x) versus time (Figure 8). The rate constants increase with the increase in temperature. The energy of activation is calculated from Arrhenius plot, E_a=7.09 kcal mol^{-1}.

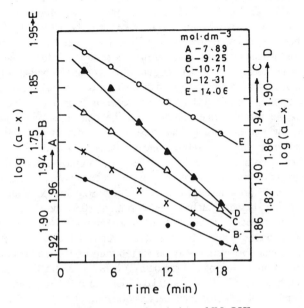

Figure 7. Effect of variation of [NaOH]

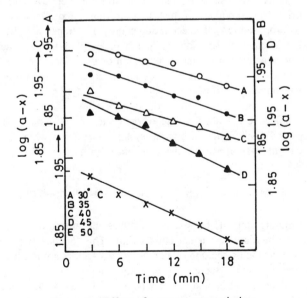

Figure 8. Effect of temperature variation

The other thermodynamic parameters, $\Delta S^{\ddagger}$, $\Delta G^{\ddagger}$ and $\Delta H^{\ddagger}$, were evaluated and was found to be 53.03 cal K^{-1} mol^{-1}, 22.95 kcal mol^{-1} and 6.48 kcal mol^{-1} respectively.

The very low activation energy indicate that the step (2) (i.e., chemical reaction) is not the rate-determining step. On the other hand, these results are characteristic of diffusion control. At a given temperature, stirring speed and shape of the reaction vessel, there exists a definite rate of mass transfer between two phases in contact (the aqueous-organic phase boundary), which is also dependent on the catalyst concentration upon a certain value. It can be seen from the Table III that the dependence of the reaction rate constants on the concentration of the catalyst is stronger at the low concentration (in the present case, between 0.3 - 0.9 mol%) and is weakly dependent at higher catalyst concentrations (0.9 and 1.1 mol%). In the concentration range examined, the dependence of k_{obs} on the [catalyst] does not attain linearity. Therefore, it may be concluded that a process involving maximum rate of diffusion occurs in the present reaction and that the maximum rate of diffusion is being approached at the higher catalyst concentration (0.9 mol%). At still higher concentration, the capacity of the system to accommodate diffusion of additional catalyst is smaller or negligible. Thus we observe constant k_{obs} value after 0.9 mol% [catalyst] run. Similar observation was earlier reported by Rabinovitz et al (26) in the study of dehydrobromination of (2-bromoethyl) benzene where E_a is reported to be 8 kcal mol^{-1}. In general (32), $E_a = 0 - 10$ kcal mol^{-1} for diffusion-controlled processes and usually over 15 kcal mol^{-1} for chemical reaction control.

A higher E_a value has been reported (34) for the polystyrene-bound trimethyl ammonium ion catalysed reaction which was controlled by strict intrinsic reactivity under triphase reactions. The activation energy for the heterogenous ethylation of phenylacetonitrile was reported to be 20 kcal mol^{-1} and for this an interfacial mechanism was proposed (16). The activation energy of intraparticle diffusion of anion exchange resins in aqueous solutions is of the order of 5-10 kcal mol^{-1} (35). The observed energy of activation for the C-alkylation of phenylacetone is 7.09 kcal mol^{-1} and hence we propose a hydroxide ion extraction mechanism for the reaction under study which is governed by diffusion control.

Mechanism. Selective solvation of the oxygen site of the enolate anion by water molecules favoured C-alkylation in biphase system. The anion which is derived from phenylacetone in this system exists in two cannonical forms :

$$Ph\text{-}CH_2\text{-}\overset{\overset{\displaystyle O}{\|}}{C}\text{-}CH_3 \ + \ OH^- \ \longleftrightarrow \ Ph\text{-}\underset{}{C}H\text{-}\overset{\overset{\displaystyle O}{\|}}{C}\text{-}CH_3 \ \longleftrightarrow \ Ph\text{-}CH=\overset{\overset{\displaystyle O^-}{|}}{C}\text{-}CH_3$$

The rate determining step of the reaction involves the attack of an ambident O^-/C^- enolate anion on the alkylating agent ethyl iodide. This kinetically controlled step is greatly influenced by the orientation of all the relevant factors around the enolate anion during the attack. From GC studies, it is clear that there is no O-alkylated product, at all hydroxide ion concentration and this may be due to the association of the active site and the weak electrophilic nature of ethyl iodide.

Two major mechanisms proposed for phase transfer catalytic reactions are the Starks' extraction mechanism (9) and Makosza's interfacial mechanism (15). Reactions believed to proceed through the extraction mechanism are characterised by 1. increased reaction rate with increased organophilicity or with larger symmetrical tetraalkyl ammonium ions (*21,23,36*), 2. independence of reaction rate on stirring speed above a certain value (*9,23,36*) and 3. linear dependence of reaction rate on catalyst concentration (*9,23,36*). On the contrary, reactions believed to proceed via the Makosza's interfacial mechanism are characterised by 1. maximum reactivity with relatively hydrophilic quats, usually alkyltriethyl ammonium quats (*17,37,38*), 2. increased reaction rate with increased stirring speed even up to 1950 rpm and 3. fractional order with respect to catalyst.

From the observed experimental results, we conclude that the dependency of the kinetic data on the stirring speed up to 500 rpm, concentrations of catalyst and aqueous hydroxide ions and temperature are consistent with a hydroxide ion extraction mechanism.

The mechanism for the C-alkylation of phenylacetone with C_2H_5I may be written as :

Aqueous phase :

$$PhCH=C[CH_2\overset{+}{N}Et_3Cl^-]_2 \ + \ 2NaOH \ \overset{(1)}{\rightleftharpoons} \ PhCH=C[CH_2\overset{+}{N}Et_3OH^-]_2 \ + \ 2NaCl$$

(aq/org)   (aq)   (aq/org)   (aq)

Organic phase :

$$PhCH=C[CH_2\overset{+}{N}Et_3OH^-]_2 \ + \ 2Ph\text{-}CH_2\text{-}CO\text{-}CH_3 \ \xrightarrow{\text{slow}(2)} \ PhCH=C[CH_2\overset{+}{N}Et_3 \ \overset{-}{C}H\text{-}CO\text{-}CH_3]_2 \ + 2H_2O$$
$$\underset{Ph}{|}$$

$$\downarrow 2C_2H_5I \text{ (org), (3)}$$

$$2Ph\text{-}\underset{\underset{(org)}{C_2H_5}}{\overset{|}{C}}H\text{-}CO\text{-}CH_3$$

Summary

The MPTC has been employed in various hydroxide ion initiated reactions and was found to have high catalytic activity. Studies pertaining to the kinetics and mechanistic aspects of dichlorocarbene addition to styrene and C-alkylation of phenylacetone with ethyl iodide have also been systematically carried out. Besides the above reactions, the MPTC also catalyses N-alkylation of pyrrole with n-butyl bromide and Darzen's condensation of cyclohexanone with chloroacetonitrile reactions for which a detailed kinetics have been studied (Balakrishnan, T., Paul Jayachandran, J., University of Madras, Unpublished data). The MPTC under study is found to be consistent with Halpern's pKa guide lines and could be applicable for the substrates having pKa values in the range 16-23. The preparation of the catalyst needs inexpensive chemicals and the time required is less.

Acknowledgments

The authors are grateful to the Council of Scientific and Industrial Research, New Delhi for the award of senior research fellowship to J.P.J.

Literature Cited

1. Starks, C. M.; Liotta, C. L.; Halpern, M. In *Phase Transfer Catalysis;* Chapman and Hall : New York, 1994.
2. Dehmlow, E. V.; Dehmlow, S. S. In *Phase Transfer Catalysis;* Verlag; Weinheim: 1993.
3. Idoux, J. P.; Wysocki, R.; Young, S.; Turcot, J.; Ohlman, C.; Leonard, R. *Synth. Commun.* **1983**, *13*, 139.
4. Balakrishnan, T.; Paul Jayachandran, J. *Synth.Commun.* **1995**, *25* (23), 3821.
5. Balakrishnan, T.; Paul Jayachandran, J. *J.Chem.Soc.Perkin Trans.2*, **1995**, *11*, 2081.
6. Rabinovitz, M.; Sasson, Y.; Halpern, M. *J.Org. Chem.* **1983**, *48*, 1022.
7. Reeves, W. P.; Hilbrich, R. G. *Tetrahedron*, **1976**, *32*, 2235.
8. Balakrishnan, T.; Shabeer, T.K.; Nellie, K. *Proc. Indian Acad. Sci.(Chem.Sci.).* **1991**, *103*, 785.
9. Starks, C.M. *J.Am.Chem.Soc.* **1971**, *93*, 195.
10. Halpern, M.; Sasson, Y.; Willner, I.; Rabinovitz, M. *Tetrahedron Lett.* **1981**, 1719.
11. Balakrishnan, T.; Haribabu, S.; Shabeer, T. K. *J.Polym.Sci. Chem.Ed.* **1993**, *31*, 317.
12. Ford, W. T.; Tomoi, M. *Adv. Polym.Sci.* **1984**, *51*.
13. Fedorynski, M.; Ziolkowska, W.; Jonczyk, A. *J.Org.Chem.* **1993**, *58*, 6120.
14. Dehmlow, E. V.; Fastabend, U. *J.Chem.Soc.Chem.Commun.* **1993**, *16*, 1241.
15. Makosza, M.; Bialecka, E. *Tetrahedron Lett.* **1977**, 183.
16. Chiellini, E.; Solaro, R.; Antone, S.D. *J.Org.Chem.* **1980**, *45*, 4179.
17. Halpern, M.; Sasson, Y.; Rabinovitz, M. *Tetrahedron*, **1982**, *38*, 3183.
18. Makosza, M.; Fedorynski, M. *Adv.Catal.* **1987**, *35*, 375.
19. Balakrishnan, T.; Arivalagan, K.; Vadukut, R. *Indian.J.Chem.* **1992**, *31B*, 338.
20. Jonczyk, A.; Serafinova, B.; Makosza, M. *Tetrahedron Lett.* **1971**, 1351.
21. Landini, D.; Maia, A.; Montanari, F. *J.Am.Chem. Soc.* **1978**, *100*, 2796.
22. Starks, C. M.; Owens, R. M. *J.Am.Chem.Soc.* **1973**, *95*, 3613.
23. Herriott, A. W.; Picker, D. *J.Am.Chem.Soc.* **1975**, *97*, 2345.
24. Freedman, H. H.; Dubois, R. A. *Tetrahedron Lett.* **1975**, 3251.
25. Molinari, H.; Montanari, F.; Quici, S.; Tundo, P. *J.Am.Chem.Soc.* **1979**, *101*, 3920.
26. Halpern, M.; Sasson, Y.; Rabinovitz, M. *J.Org.Chem.* **1984**, *49*, 2011.
27. Herriott, A.; Picker, D. *Tetrahedron Lett.* **1972**, 4521.
28. Gorgues, A.; Lecoq, A. *Tetrahedron Lett.* **1976**, 4723.
29. Halpern, M.; Cohen, Y.; Sasson, Y.; Rabinovitz, M. *Nouv.J.Chim.* **1984**, 8, 443.
30. Dehmlow, E. V.; Barahona-Naranjo, S. *J.Chem.Res.Synop.* **1982**, 186.
31. Dehmlow, E. V.; Lissel, M. *Tetrahedron.* **1981**, 37, 1653.
32. Levennspiel, O. In *Chemical Reaction Engineering*; Wiley: New York, **1972**.
33. Makosza, M. *Pure Appl. Chem.* **1975**, *43*, 439.
34. Tomoi, M.; Ford, W.T. *J.Am.Chem.Soc.* **1981**, *103*, 3821.
35. Helfferich, F. In *Ion Exchange*, McGraw Hill: New York, **1962**.
36. Landini, D.; Maia, A.; Montanari, F. *J.Chem.Soc.Chem.Commun.* **1975**, 95.
37. Makosza, M. and Serafinowa, B. Rocz. Chem. **1965**, *39*, 1223.
38. Dehmlow, E. V.; Lissel, M. *Tetrahedron Lett.* **1976**, 1783.

Chapter 22

Cation-Binding Properties of Sodium-Selective 16-Crown-5 Derivatives

Mikio Ouchi[1], Kenji Mishima[2], Reizo Dohno[1], and Tadao Hakushi[1]

[1]Department of Applied Chemistry, Faculty of Engineering, Himeji Institute of Technology, 2167 Shosha, Himeji, Hyogo 671–22, Japan
[2]Department of Chemical Engineering, Fukuoka University, Nanakuma, Fukuoka 814–80, Japan

New 16-crown-5 ether derivatives possessing a variety of side arm(s) are synthesized, and their cation-binding abilities are evaluated by solvent extraction technique. They show high sodium selectivity over potassium compared with typical crown ethers. The effect of side-arm, complex formation, and the role of donor atom on cation-binding ability/selectivity especially for sodium ion are discussed.

Crown Ethers have been considered as the phase transfer catalyst (PTC) for industrial applications, however, money wise, quartenary ammonium salts are used in most phase-transfer reactions. The advantage of using a crown ether as a PTC compared with ammonium salts is their thermal stability, and the following factors are suggested when considering a crown ether as PTC (*1*); catalytic ability, stability, separability, toxicity, and cost. Of the crown ethers reported, dibenzo-18-crown-6 has been typically used as the PTC, and further investigation of crown ethers has not been done yet. When dibenzo-18-crown-6 is used in the reaction, potassium salt is needed in view of the hole-size concept (*2*), while sodium salts are generally less expensive than the analogous potassium salts. It has also been suggested that PTC possessing high sodium ion binding ability/selectivity have the advantage of being polymer-bound and readily recoverable (*3*). In this sense, when screening new PTC applications, a crown ether possessing high sodium ion binding ability should be considered. This chapter deals with the cation-binding ability/selectivity for sodium ion against alkali metal ions evaluated from the solvent extraction of metal picrates.

Crown Ethers of Low Symmetry

So called "Host Guest Chemistry" has been developed since the discovery of dibenzo-18-crown-6 ether (*2*), and now a number of papers in this field have been published to investigate the attractive filed of "Supramolecular Chemistry". Despite the extensive work in this field, the effect of methylene chain length between two adjacent oxygen atoms of a crown ether (a crown ether of low symmetry) on its complexation ability has not been investigated systematically until very recently. We have reviewed complexations by crown ethers of low symmetry, and demonstrated the specific alkali metal ion selectivity of unsymmetrical crown ethers (*4*). Compared with symmetrical crown ethers, crown ethers of low symmetry, possessing (3m+n)-crown-m skeletons showed lower cation-binding ability in general, however, exhibit drastically different and , in some cases, higher selectivities for specific cations. In particular, 16-crown-5 (1,4,7,10,13-pentaoxacyclohexadecane) shows much higher cation selectivity for sodium than symmetrical 15-crown-5 (5).

Cation Binding by 16-crown-5

The specific interaction of 16-crown-5 with sodium may be explained by cavity size and orientation of the donor oxygen atoms. The examination of a CPK molecular model shows that Na^+ (cation diameter, 2.04 Å (6)) is better accommodated in the cavity (1.8 Å) of 16-crown-5 than in the cavity (1.7 Å) of 15-crown-5. The five oxygen atoms are directed inside because of the larger ring structure compared with that of 15-crown-5, which makes 1:1 complex formation with Na^+ favorable but with larger cations (K^+, Rb^+, and Cs^+) unfavorable. This specific behavior prompted our group to use the 16-crown-5 framework for the design of sodium selective crown ethers. Their cation-binding abilities are evaluated from solvent extraction technique. When we discuss the cation-binding ability of crown ethers, at the beginning, the complexation phenomena are often considered by "Hole-size " concept (2). It was reported that this relationship in homogeneous solution fails to explain the cation selectivity in flexible macrocycles and lariat ethers (7). On the other hand, the cation-selectivity in the solvent extraction of metal picrates can be rationalized in terms of this concept. This indicates that the concept is still valid at least in the solvent extraction technique (8).

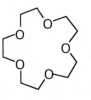

15-Crown-5   16-Crown-5

Solvent Extraction of Metal Picrates

A variety of measurement techniques for recording the cation-binding ability of crown ethers have been reported (9). Notable among them are ion-selective electrode techniques, conductance methods, NMR methods, solvent extraction tequniques, and calorimetry. In this paper, data obtained by solvent extraction of metal picrates are presented for general discussion.

The solvent extraction of metal picrates has been employed as a convenient method for evaluating cation-binding ability of crown ethers, and , as far as alkali metal ions are concerned, it affords quantitative binding constants compatible with those obtained in homogeneous phase complexation (10). The solvent extraction of aqueous metal picrates were performed with crown ethers in dichloromethane. The percentage extractabilities (% Ex), defined as percentage picrate extracted into the organic phase, are measured. In order to discuss the interaction of crown ethers with cations from a quantitative point of view, extraction studies are carried out at different ligand concentrations to determine the extraction equilibrium constant (Kex) and the complex stoichiometry (8).

This extraction experiment must be used with caution since the data obtained depend on several variables (9): solvent, temperature, salt, and ionic strength. When the conditions of extraction are fixed, the data obtained give invaluable information (11).

The data presented in this paper are obtained under fixed and controlled conditions. Metal picrates are used as salts, and the dichloromethane-water system at 25 °C is adopted.

16-Crown-5 Lariats

Lariat ethers have been designed to enhance the cation-binding ability of crown ethers by introducing a side arm carrying extra donor group (*12*). It has been demonstrated that some carbon- and nitrogen-pivot lariat ethers exhibit higher cation-binding abilities than the parent crown ethers. From the point of synthetic feasibility, lariat ethers reported are based on symmetrical crown ethers, e.g. 15-crown-5 lariats, or their aza crown ethers. Based on the sodium selectivity, 16-crown-5 with a variety of side arm(s) should be of interest. We have reported the syntheses and the cation-binding ability of double-armed 16-crown-5 ethers (*8*). The discussion on the number of donor atoms in a side arm, the difference between the single- and double arms, and the lipophilicity of crown ethers was made. The results suggest that the single arm with two oxygens of 16-crown-5 should be the factor of designing sodium selective 16-crown-5 ethers.

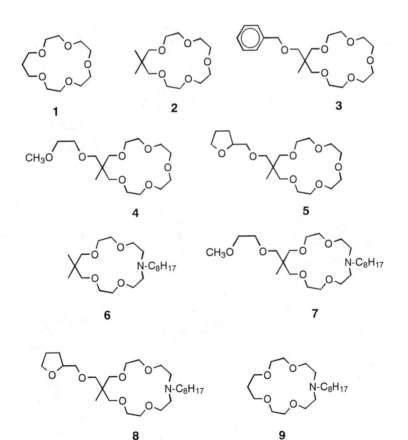

In this paper, we discuss the cation-binding ability of 16-crown-5 derivatives (**1-9**) from the data of solvent extraction experiments (*13*). All crown ethers in this study are synthesized by reaction of the corresponding diols and the tosylates in the presence of NaOH in THF as shown in scheme. The solvent extraction of aqueous metal picrates are performed with crown ethers in dichloromethane. The extraction equilibrium constants (Kex) are determined. (Table I, and II)

As seen from Table I, lariat 16-crown-5 ethers, **4-5** with two oxygen atoms in a side arm show higher sodium ion binding ability/selectivity compared with **1-3**. Especially, the crown ether, **5**, with tetrahydrofurfuryl group in a side arm, showed the highest selectivity among the crown ethers used. On the other hand, as seen from Table II, aza-16-crown-5 derivatives, **6-9**, do not show the drastic change of Na^+/K^+ selectivity with the change of the substituent on the side arm. The reason why aza-16-crown-5 derivatives do not show the enhancement of the sodium selectivity from the results of extraction experiment is not clear from the data.

It is natural that complexation in homogeneous solution should be compared with those in solvent extraction system. The effect of a side arm of 16-crown-5 derivatives (**1,2**, and **4**) in homogeneous phase (methanol) is examined by conductance method (*14*).(Table III)   Contrary to the pronounced lariat effect reported for several lariat

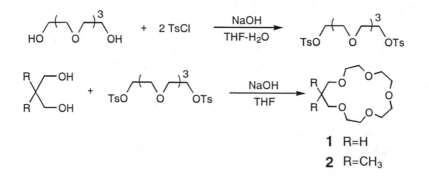

1 R=H
2 R=CH₃

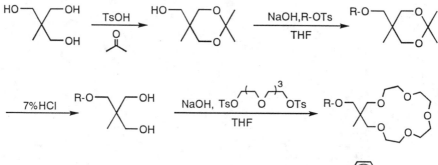

3 = ◯–CH₂

4 = CH₃OCH₂CH₂

5 = ◯–CH₂ (tetrahydrofurfuryl)

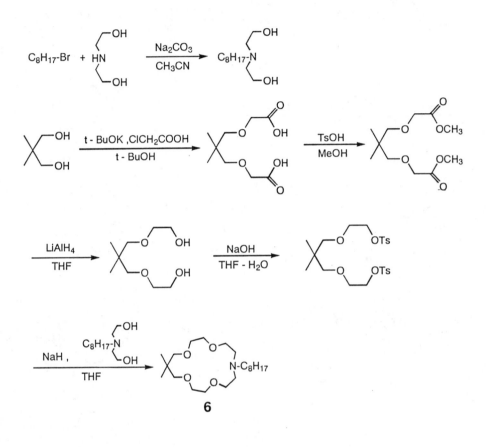

6

Table I. Extraction Equilibrium Constants (K_{ex})[a]

Ligand	log K_{ex}[b]			selectivity[c]
	Li+	Na+	K+	Na+/K+
1 [d]	e)	4.55	3.65	7.9
2	2.20	4.12	3.09	10.7
3	2.49	4.22	3.21	10.2
4	2.80	4.57[d]	3.39[d]	15.1
5	2.76	4.65	3.18	29.5

a) Temperature 25°C; aqueous phase (10ml):[Picrate]=3mM;
 organic phase (CH2Cl2,10ml):[Crown ether]=1~15mM.
b) Valus for 1:1 cation-crown ether complexes.
c) Relative cation selectivity determined by K_{ex}.
d) Reference 8.
e) Not measured.

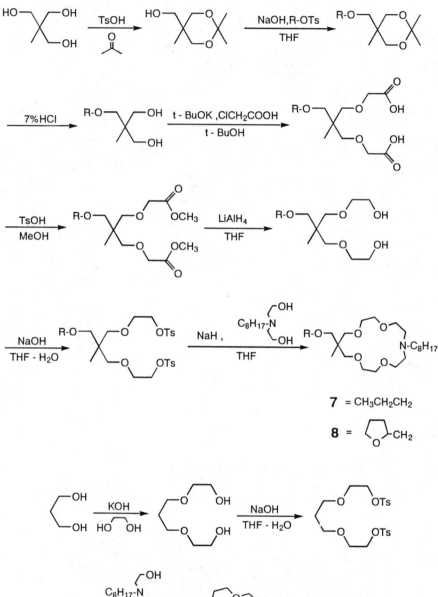

7 = CH$_3$CH$_2$CH$_2$

8 =

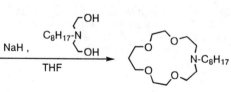

9

Table II. Extraction Equilibrium Constants $(K_{ex})^{a)}$

	log $K_{ex}^{b)}$		selectivity$^{c)}$
Ligand	Na$^+$	K$^+$	Na$^+$/K$^+$
6	4.26	3.00	18.2
7	4.39	3.15	17.4
8	4.56	3.32	17.4
9	4.78	3.69	12.3

a)Temperature 25°C;aqueous phase (10ml):
[Hpic]=3mM,[MOH]=10mM;organic phase
(CH$_2$Cl$_2$,10ml):[Crown ether]=0.8~30mM.
b)Valus for 1:1 cation-crown ether complexes.
c)Relative cation selectivity determined by K_{ex}.

Table III. Complex Stability
Constants (Ks) in methanol
at 25°C$^{a)}$

	log Ks	
ligand	Na$^+$	K$^+$
1	4.10	2.92
2	3.73	2.54
4	3.53	2.43

a) Reference 14.

ethers with symmetrical crown ethers (*15*, *16*), the homogeneous phase complexation by lariat 16-crown-5, **4** , does not display any positive lariat effect, while sodium selectivity is hold. This obviously conflicts with the basic concept of the lariat effect and may require further verifications of its applicability to solvent systems.

Conclusion

The authors intended to provide an evident criterion, which is helpful for designing sodium selective crown ethers through the investigation of the cation-binding ability of 16-crown-5 derivatives. One should not automatically presume a symmetrical crown ether structure. e.g. 15-crown-5 and 18-crown-6, but rather utilize the low symmetry, like 16-crown-5 skeleton. A crown ether possessing sodium ion selectivity is a promising PTC catalyst since sodium is a favorable cation for salts used in phase transfer process. We expect sodium selective 16-crown-5 derivatives would be applied as phase transfer catalysts in the future.

Acknowledgments

We thank the Kawanishi-Shinmeiwa Foundation for the partial support of this work.

Literature Cited

1. Halpern, M.: *Phase-Transfer Catalysis Commun.*, **1995**, 1, 17.
2. Pedersen, C. J.: J. Am. Chem. Soc., **1967**, 89, 7017.
3. Gokel, G. W.: Arnold, K.: Cleary, T.: Friese, R.: Gatto, V.: Goli, D.: Hanlon, C.: Kim, M.: Miller, S.: Ouchi, M.: Posey I.: Sandler, A.: Viscariello, A.: White, B.: Wolfe, J.: Yoo, H.: *ACS Symp. Ser.*, **1987**, 326, 24.
4. Ouchi, M.: Hakushi, T.: Inoue, Y.: in Inoue, Y.: Gokel, G. W. (Eds), *Cation Binding by Macrocycles*, Marcel Dekker, New York, **1990**, Chapter 14.

5. Ouchi, M.: Inoue, Y.: Kanzaki, T.: Hakushi, T.: *J. Org. Chem.,* **1984,** 49, 1408.
6. Shannon, R. D.: *Acta Crystallogr.,* **1976,** A32, 751.
7. Schultz, R. A.: White, B. D.: Dishong, D. M.: Arnold, K. A.: Gokel, G. W.: *J. Am. Chem. Soc.,* **1985,** 107, 6659.
8. Ouchi, M.: Inoue, Y.: Wada, K.: Iketani, S.: Hakushi, T.: Weber, E.: *J. Org. Chem.,* **1987,** 52, 2420.
9. Gokel, G. W.: Trafton, J. E.: in Inoue, Y.: Gokel, G. W. (Eds), *Cation Binding by Macrocycles,* Marcel Dekker, New York, **1990,** Chapter 6.
10. Inoue, Y.: Ouchi, M.: Hakushi, T.: *Bull. Chem. Soc. Jpn.,* **1985,** 58, 525.
11. Ouchi, M.: Fujiwara, H.: Zheng, D.: Shintani, H.: Fujitani, S.: Morita, Y.: Hakushi, T.: *Chem. Express,* **1993,** 8, 153.
12. Dishong, D. M.: Diamond, C. J.: Cinaman, M. I.: Gokel, G. W.: *J. Am. Chem. Soc.,* **1983,** 105, 586.
13. Preliminary reports: Mishima, K.: Ouchi, M.: Dohno, R.: Hakushi, T.: Tanaka, M: Abstract II of the 1995 International Chemical Congress of Pacific Basin Societies, ORGN-9, No. 565.
14. Inoue, Y.: Ouchi, M.: Hakushi, T.: Liu, Y.: Takeda, Y.: J. Chem.Soc. Dalton Trans., **1991,** 1291.
15. Davidson, R. B.: Izatt, R. M.: Christensen, J. J.: Schultz, E. A.: Dishong, D. M.: Gokel, G. W.: *J. Org. Chem.,* **1984,** 49, 5080.
16. Nakatsuji, Y.: Nakamura, T.: Yonetani, M.: Yuya, H.: Okahara, M.: *J. Am. Chem.Soc.,* **1988,** 110, 531.

INDEXES

Author Index

Affiliation Index

Subject Index

Highlights from ACS Books

Chemical Research Faculties, An International Directory
1,300 pp; clothbound ISBN 0–8412–3301–2

College Chemistry Faculties 1996, Tenth Edition
300 pp; paperback ISBN 0–8412–3300–4

Visualizing Chemistry: Investigations for Teachers
By Julie B. Ealy and James L. Ealy
456 pp; paperback ISBN 0–8412–2919–8

Principles of Environmental Sampling, Second Edition
Edited by Lawrence H. Keith
700 pp; clothbound ISBN 0–8412–3152–4

Enough for One Lifetime: Wallace Carothers, Inventor of Nylon
By Matthew E. Hermes
364 pp; clothbound ISBN 0–8412–3331–4

Peptide-Based Drug Design
Edited by Michael D. Taylor and Gordon Amidon
650 pp; clothbound ISBN 0–8412–3058–7

Attenuated Total Reflectance Spectroscopy of Polymers: Theory and Practice
By Marek W. Urban
232 pp; clothbound ISBN 0–8412–3348–9

Teaching General Chemistry: A Materials Science Companion
By Arthur B. Ellis, Margaret J. Geselbracht, Brian J. Johnson, George C. Lisensky,
and William R. Robinson
576 pp; paperback ISBN 0–8412–2725–X

Understanding Medications: What the Label Doesn't Tell You
By Alfred Burger
220 pp; clothbound ISBN 0–8412–3210–5; paperback ISBN 0–8412–3246–6

For further information contact:

American Chemical Society
Customer Service and Sales
1155 Sixteenth Street, NW
Washington, DC 20036

Telephone 800–227–9919
202–776–8100 (outside U.S.)

The ACS Publications Catalog is available on the Internet at
http://pubs.acs.org/books

Bestsellers from ACS Books

The ACS Style Guide: A Manual for Authors and Editors
Edited by Janet S. Dodd
264 pp; clothbound ISBN 0–8412–0917–0; paperback ISBN 0–8412–0943–X

Writing the Laboratory Notebook
By Howard M. Kanare
145 pp; clothbound ISBN 0–8412–0906–5; paperback ISBN 0–8412–0933–2

Career Transitions for Chemists
By Dorothy P. Rodmann, Donald D. Bly, Frederick H. Owens, and Anne-Claire Anderson
240 pp; clothbound ISBN 0–8412–3052–8; paperback ISBN 0–8412–3038–2

Chemical Activities (student and teacher editions)
By Christie L. Borgford and Lee R. Summerlin
330 pp; spiralbound ISBN 0–8412–1417–4; teacher edition, ISBN 0–8412–1416–6

Chemical Demonstrations: A Sourcebook for Teachers, Volumes 1 and 2, Second Edition
Volume 1 by Lee R. Summerlin and James L. Ealy, Jr.
198 pp; spiralbound ISBN 0–8412–1481–6
Volume 2 by Lee R. Summerlin, Christie L. Borgford, and Julie B. Ealy
234 pp; spiralbound ISBN 0–8412–1535–9

From Caveman to Chemist
By Hugh W. Salzberg
300 pp; clothbound ISBN 0–8412–1786–6; paperback ISBN 0–8412–1787–4

The Internet: A Guide for Chemists
Edited by Steven M. Bachrach
360 pp; clothbound ISBN 0–8412–3223–7; paperback ISBN 0–8412–3224–5

Laboratory Waste Management: A Guidebook
ACS Task Force on Laboratory Waste Management
250 pp; clothbound ISBN 0–8412–2735–7; paperback ISBN 0–8412–2849–3

Reagent Chemicals, Eighth Edition
700 pp; clothbound ISBN 0–8412–2502–8

Good Laboratory Practice Standards: Applications for Field and Laboratory Studies
Edited by Willa Y. Garner, Maureen S. Barge, and James P. Ussary
571 pp; clothbound ISBN 0–8412–2192–8

For further information contact:

American Chemical Society
1155 Sixteenth Street, NW ◆ Washington, DC 20036
Telephone 800–227–9919 ◆ 202–776–8100 (outside U.S.)

The ACS Publications Catalog is available on the Internet at
http://pubs.acs.org/books